Instruments of Knowledge

Finding Meaning in Objects, Habits, and Museums

By

Jean-François Gauvin

BRILL

LEIDEN | BOSTON

Cover illustration: painting known as *A Philosopher Giving that Lecture on the Orrery, in which a Lamp is put in place of the Sun or The Orrery*. By Joseph Wright of Derby, ca 1766. Derby Museum and Art Gallery, Courtesy of Wikimedia Commons: https://commons.wikimedia.org/wiki/File:Wright_of_Derby,_The_Orrery.jpg

The Library of Congress Cataloging-in-Publication Data is available online at https://catalog.loc.gov
LC record available at https://lccn.loc.gov/2023018225

Typeface for the Latin, Greek, and Cyrillic scripts: "Brill". See and download: brill.com/brill-typeface.

ISSN 2405-5077
ISBN 978-90-04-50460-8 (hardback)
ISBN 978-90-04-50461-5 (e-book)

This book is printed on acid-free paper and produced in a sustainable manner.

Contents

Acknowledgments

I have anticipated this moment with pleasure and dread for a couple of years now, torn between the joy of thanking those who helped me make it through and the fear of forgetting someone or being incapable of finding the appropriate words. That said, one thing is clear: my deepest gratitude goes to Mario Biagioli. During my all-to unorthodox graduate school journey and beyond, Mario has been there for moral support (sometimes badly needed), for opening doors and providing academic opportunities of all sorts. Most importantly, he helped me understand what it really means to study history of science. His intellectual acumen, persistent questioning, and incisive remarks have forced me to go beyond what I thought I could accomplish. Involved but not always looking over my shoulder, Mario encouraged me to explore and tackle issues about the early modern material culture of natural philosophy that by myself I would have never contemplated nor written about. (Ending, for him, in having to read too many half-baked historical and theoretical assumptions.) He had faith in the project from the onset, which gave me the necessary confidence to pull it through. *Mille grazie Mario.*

Ann Blair and Peter Galison were likewise of seminal importance during my formative years and since then. Ann has given me the training in early modern France that was lacking to embark on this project. Thanks to her teaching, I have acquired a *goût prononcé* and a basic expertise in early modern French thought and culture. Though I tried my best, I am afraid I have not always been able to find *le mot juste*, as she pointed out to me so many times. I feel fortunate to have been guided by such an accomplished and generous scholar. Peter is the reason why I went to Harvard. In fact, he was the one I contacted when I decided to apply to Harvard—after I finished reading his monumental *Image & Logic*. Over the past twenty years now, Peter's scholarship has been a constant source of inspiration, somewhat reflected in my views of early modern material culture and museums. His assistance, availability, and encouragement as Faculty Director of the Collection of Historical Scientific Instruments (CHSI), when I was a curator (2007–2009) and later his 'number one' as the Director of Administration (2011–2018), provided me with everything I could ever expect or hoped for in launching my career.

Finally, over the past several years, Matthew L. Jones has been a source of inspiration and a wonderful reader. He provided thoughtful insights on various technical fine points and offered warnings when I went too far astray in intellectual analyses of early modern material culture (I'm sure he still doesn't agree with some assertions I make regarding Pascal and the arithmetical machine). Matt has still a lot to teach me, and I remain one of his most eager students.

Several other people (now begins my fear) have helped tremendously by reading, criticizing, and discussing with me various drafts of the following chapters. I benefited over the years from the expertise and intellectual generosity of Jim Bennett, Tim Boon, Paolo Brenni, Lorraine Daston, Nick Dew, Emily Dolan, Sven Dupré, H. Floris Cohen, Jean-Baptiste Fressoz, Daniel Garber, Justin Grosslight, Rob Iliffe, Myles Jackson, David Kaiser, Alex Marr, David Pantalony, Katy Park, Simon Schaffer, Alison Simmons, Will Thomas, and Elly Truitt. Special thanks go to Dániel Margócsy, Paola Bertucci, and two anonymous reviewers, whose more recent comments have strengthened the overall argument of the book. Leib Celnik carefully reviewed the manuscript to remove as many 'Frenglish' sentence structures as possible; I take full responsibility on the remaining ones.

Combining my intellectual journey in the history of science and my professional one in museums, this book also benefited from two unique experiences. Primo, several years ago, the Max Planck Institute for the History of Science in Berlin provided me with a unique opportunity to travel across Europe during the summer of 2006 to visit and study museum collections and practices. This project, dubbed the Wandering Seminar, opened my eyes to a lot of issues regarding the relationships between museum work, academic traditions, and the significance of the material culture of science in studying history of science. Thanks to Lorraine Daston, then head of this pioneering project, I connected with fourteen other 'wanderers' to share and discuss about the theory and formal practices of historical objects—dating from the Renaissance to the present day. The intellectual rewards and lessons learned during this experience are in many ways at the core of this book. Secondo, the nine years I worked at the CHSI under the leadership of Peter Galison, forged the hybrid scholar I eventually became. From documenting scientific instruments to teaching with them and preparing exhibitions, the CHSI focused and honed my skills, leading me to the Chair in museum studies I was offered at Université Laval in 2018. For that, besides Peter's unflinching support, I have to thank sincerely Sara Schechner, Marty Richardson, Sara Frankel, Samantha van Gerbig, Michael Kelley, Juan Andrés León, Judith Lajoie, Ken Launie, and Richard Wright.

Several institutions over the years have made this book possible. Besides Harvard, I want to recognize the Dibner Institute (then at MIT), McGill University through the Andrew W. Mellon Foundation Postdoctoral Fellowship, the Bibliothèque nationale de France, the Institut de France, and my current home, Université Laval. I presented my work at various venues and conferences throughout the years and I would like to thank all the participants who were present: Yale University, Harvard University, History of Science Society (several meetings), Musée de l'air et de l'espace (Paris), Acfas, Science History Institute (Philadelphia), Centro Fermi (Rome), Science Museum

(London), Kings' College (London), University of Saskatchewan, University of Manchester, Centre Jacques Cartier (Saint-Malo), McGill University, Université de Montréal, and the Max Planck Institute for the History of Science (Berlin).

Although the book in its structure and argument is an original contribution to scholarship, some parts are based on articles going back fifteen years. I would like to acknowledge here the ones I borrowed excerpts from:

JFG. "Réparation, restauration, exposition: valoriser les instruments disloqués des musées de sciences et de technologie." In *Les Réparations dans l'histoire. Cultures techniques et savoir-faire dans la longue durée*, eds. Gianenrico Bernasconi, Guillaume Carnino, Liliane Hilaire-Pérez, and Olivier Raveux, 259–69 (Paris: Presses des mines, 2022).

JFG. "The Next Level of Play: How Cartesian Devils, Volcanos, and Quantum Toys Taught General Science at Harvard, 1730–1970." *Journal of the History of Collections* 31 (2019): 531–46.

JFG. (with Maxime Sainte-Marie and Vincent Larivière). "Les collections muséales au service de la science: Une analyse de la Collection of Historical Scientific Instruments de l'université Harvard." *Document numérique* 20 (2–3) (2017): 137–58.

JFG. "Functionless: Science Museums and the Display of 'Pure Objects'." *Science Museum Group Journal*, vol. 5 (Spring 2016). http://journal.sciencemuseum.ac.uk/browse/issue-05/functionless/.

JFG. "The Instrument That Never Was: Inventing, Manufacturing, and Branding Réaumur's Thermometer During the Enlightenment." *Annals of Science* 69 (2012): 515–49.

JFG. "Instruments of Knowledge." In *The Oxford Handbook of Philosophy in Early Modern Europe*, eds. Desmond Clarke and Catherine Wilson, 315–37 (Oxford: Oxford University Press, 2011).

JFG. "Artisans, Machines, and Descartes's *Organon*." *History of Science* 44 (2006): 187–216.

Lastly, on a more personal note, my mom and dad were pillars of support, psychologically and materially throughout the years. Nathalie Martimbeau took care of me and the family in two countries, *contre vents et marées*, for twenty years. Most of all, I owe much to my adorable Camille and Simon, for their love, patience, attention, support, tolerance, and fortitude (did I mention patience?) in putting up for so many years with my absences, mood swings, anxieties, and writing dry spells (did I mention absences?) so I could finish my *gros livre*, as Camille said a long time ago. This book is dedicated to you both.

Québec City, December 2022

Illustrations

Introduction

Thousands of scientific instruments and machines are safeguarded as curiosities, artworks, and 'relics' privately held or found in academies, societies, universities, and museums around the world. What can we learn from them? Do they have any use to us beyond the 'scientific' representation of a glorious past? What are the role and value of instrument collections to our understanding of scientific knowledge production?[1] In *Do Museums Still Need Objects?*, Steven Conn examines the evolving role of objects in museums since the end of the nineteenth century. While artifacts appeared then at the center of the museum's discourse and the focus of the visitor's visual landscape, today we can easily foresee (especially after a global pandemic) a 'museum' almost entirely devoid of objects—though equipped with a café, a virtual room and dematerialized exhibitions, a gift shop, a high-end dining venue, and an IMAX cinema. Conn also notices a direct correlation between the rise of educational programs and the sharp decrease of exhibited artifacts over the last few decades. He observes, though most curators would categorically refute such an assertion, that objects "are no longer as central to the conception and function of the museum." And, when made the focus of an exhibition, those chosen for display "are selected for aesthetic, rather than cultural or scientific reasons."[2] Scientific instruments, one might suggest, suffer a similar treatment in science museums. Most of the time, as we discuss in chapter 5, they are restored—sometimes buffed until they shine (how dare we see dirt and grime!)—before being carefully put inside modern, sparkling glass cases. Can we learn anything from these 'beautiful' objects? Not if they are gazed at fleetingly. As Jean Baudrillard rightfully declared decades ago, "any object immediately becomes the foundation of a network of habits, the focus of a set of behavioural routines. Conversely, there is probably no habit that does not center on an object. In everyday existence the two are inextricably bound up with each other."[3] Habit demands careful attention. If no means of access is provided to an object's specific 'set of behavioural routines,' especially those

1 Rebekah Higgitt, "Instruments and Relics: The History and Use of the Royal Society's Object Collections c. 1850–1950," *Journal of the History of Collections* 31 (2019): 469–85.

2 Steven Conn, *Do Museums Still Need Objects?* (Philadelphia: The University of Pennsylvania Press, 2010), 58 and 35, respectively. I examine this idea in detail elsewhere, Jean-François Gauvin, "Functionless: Science Museums and the Display of 'Pure Objects'," *Science Museum Group Journal* 5 (2016), online http://journal.sciencemuseum.ac.uk/browse/issue-05/functionless/.

3 Jean Baudrillard, *The System of Objects* (London: Verso, 2005), 100, n. 23.

 | DOI:10.1163/9789004504615_002

related to the practice of science, can anything be learned from it? The object has been converted—distorted—into a pure and functionless thing.

This is capital for this book because, I claim, 'scientific objects' require a significant amount of conceptual labor; they necessitate a good dose of what has been called by some applied metaphysics, which cannot be fully appreciated and processed when material objects are locked behind physical exhibit cases and described by mere textual presentations. One needs to move orthogonally from the plane of debate about what should count as a 'natural' (real) in opposition to a 'constructed' (historical) object of scientific inquiry.[4] Such a process of knowing produces work. It is "linked with ways of *production*—to ways of *making* things." For John Pickstone, there exist three ideal types of making: craft, rationalized production, and systematic invention. These are akin to all the various ways of knowing, which are divided into three fundamental components: (extended) natural history (the describing and classification of things), analyzing (the reduction into assorted elements), and experimenting (the command of phenomena to create novelties). Ways of making and ways of knowing follow parallel paths and are nested similarly into the history of knowledge production. What is important to understand, however, is "that all our knowings and all our creations are underpinned by systems of meaning and values." This course of action is what Pickstone calls 'world-readings,' i.e., a hermeneutic designed to help us decode the world under specific systems of meaning.[5] For anthropologist Tim Ingold, ways of making and ways of knowing work together as an intertwined operation—performance—he calls an 'art of inquiry.' Ingold, however, believes that material culture studies have for too long emphasized the end product of objects, their life biography recounting their interactions with people through social systems of meaning.[6] The result is one of erosion: the "processes of making appear swallowed up in objects made." *Contra* the conventional hylomorphic model, Ingold proposes instead that an artisan's whole life, his/her "own movements and gestures," be coupled with the "becoming of his materials, joining with and following the forces and flows that bring his work to fruition." Such movements and gestures suggest that making should not be considered an *assembly* but rather a *procession*, not a 'building *up*' from tiny parts into a 'hierarchically organised totality,' but

4 Lorraine Daston, ed., *Biographies of Scientific Objects* (Chicago: The University of Chicago Press, 2000).

5 John Pickstone, *Ways of Knowing: A New History of Science, Technology and Medicine* (Chicago: The University of Chicago Press, 2000), 3–20. For an important discussion of 'values,' see Nathalie Heinich, *Des valeurs. Une approche sociologique* (Paris: Gallimard, 2017).

6 See, for instance, the influential edited volume from Arjun Appadurai, ed., *The Social Life of Things. Commodities in Cultural Perspective* (Cambridge: Cambridge University Press, 1986).

rather a 'carrying *on*,' a pathway filled with twists and turns but where every step brings one closer to its destination.[7] Yet an artisan's life is not entirely orthogonal to his local systems of meaning. Richard Sennett argues that making things is not disconnected from the social. On the contrary, craftsmanship "shows the continuum between the organic and the social put in action." Sennett claims that social consequences "are built into the structure and the functioning of the human body, as in the working of the human hand." Echoing the pragmatist movement, Sennett contends that "the capacities our bodies have to shape physical things are the same capacities we draw on in social relations."[8] Our body is a conduit between the social, the material and the intellect. It is malleable. It is molded by as much as it molds the production of knowledge.

1 A Short Case Study: the Early Académie royale des sciences in Paris

The early history of the Académie royale des sciences in Paris, though well known and carefully studied by scholars, exemplify in many ways the overall argument of this book—and what was briefly broached above.[9] My focus here primarily relates to how integrated and contributive instruments and artisans were to the theoretical and experimental practices of the *académiciens* toward the production of scientific knowledge. Christiaan Huygens's air pump is an obvious and documented example,[10] but it was by no means the only machine

---

7 Tim Ingold, *Making. Anthropology, Archaeology, Art and Architecture* (New York: Routledge, 2013), 6–7, 31, and 45 for the quotes. On the role of hand and head toward the production of knowledge, see Christian Jacob, ed., *Lieux de savoir, 2: Les mains de l'intellect* (Paris: Albin Michel, 2011) and Lissa Roberts, Simon Schaffer, and Peter Dear, eds., *The Mindful Hand. Inquiry and Invention from the Late Renaissance to Early Industrialisation* (Amsterdam: KNAW, 2007).

8 Richard Sennett, *The Craftstman* (New Haven: Yale University Press, 2008), 290.

9 The classic studies remain Roger Hahn, *The Anatomy of a Scientific Institution: The Paris Academy of Sciences, 1666–1803* (Berkeley: University of California Press, 1971). David J. Sturdy, *Science and Social Status. The Members of the Académie des sciences, 1666–1750* (Woodbridge, UK: The Boydell Press, 1995). Alice Stroup, *A Company of Scientists. Botany, Patronage, and Community at the Seventeenth-Century Parisian Academy of Sciences* (Los Angeles: University of California Press, 1990). Anita Guerrini, *The Courtiers' Anatomists. Animals and Humans in Louis XIV's Paris* (Chicago: The University of Chicago Press, 2015), esp. chapters 2 and 3. The two-volume institutional history written by Fontenelle is *Histoire de l'Académie royale des sciences*, 2 vols. (Paris, 1733) [hereafter cited as *HARS*].

10 Alice Stroup, "Christiaan Huygens and the Development of the Air-Pump," *Janus* 68 (1981): 129–58. Stroup, *A Company of Scientists*, 159–66. Steven Shapin and Simon Schaffer, *Leviathan and the Air-Pump. Hobbes, Boyle, and the Experimental Life* (Princeton: Princeton University Press, 1985), chapter 6. For a description of the machine and the first

at the academicians' disposal.[11] It played, in fact, a somewhat marginal role in comparison to the vast array of apparatuses dedicated to astronomy and found in the newly built observatory.[12] Instruments rapidly became an integral part of the academicians' daily business. They oversaw their construction and judged their merit—without necessarily focusing on utility or industrial applications.[13] Instruments and machines were refereed by ad hoc committees, which reported to a general assembly in order to reach a collective agreement.[14] Although the social and epistemic boundaries between theory and practice did not fade within the confines of the Academy, it was blurred considerably owing to instruments and the individuals who invented, made, and repaired them.[15]

Artisans and academicians were in frequent contact, whether directly (for special projects, like Huygens's air pump [with Denis Papin] and balance spring watch [with Thuret], and Cassini's silver planisphere [with Michael Butterfield]) or via a third party, through one of the *élèves* or *commis*, when dealing with a common project investigated by the Academy. In the social body of the Academy, these *élèves*, like Claude-Antoine Couplet, were paid "to execute the decisions of the company, and especially conduct any observations

---

experiments performed at the Academy, see Académie des sciences, *Procès-verbaux de l'Académie royale des sciences*, Registre de physique, vol. 4 (1668), over several meetings, 1r–27v.

11 After an initial month of experimentation with Huygens's air pump, and hearing reports about Robert Boyle and the older experiments done at the Accademia del Cimento, "la Compagnie a jugé que la matiere du vuide auoit esté suffisamment examinée et qu'il falloit passer a quelque autre matiere." Académie des sciences, *Procès-verbaux de l'Académie royale des sciences*, Registre de physique, vol. 4 (1668), séance du samedi 19 mai 1668, 27r. Some twenty years later, Wilhelm (Guillaume) Homberg, a chemist appointed academician in 1691, proposed a new air pump deemed "différente de l'ordinaire" by the assembly. Académie des sciences, *Procès-verbaux de l'Académie royale des sciences*, Registre de physique, vol. 12 (1686–1689), séance du mardi 7 mai 1687, 60r–v. On Homberg, Alice Stroup, "Wilhelm Homberg and the Search for the Constituents of Plants at the 17th-Century Académie royale des sciences," *Ambix* 26 (1979): 184–201. Mi Gyung Kim, "Chemical Analysis and the Domains of Reality: Wilhelm Homberg's *Essays de chimie*, 1702–1709," *Studies in the History and Philosophy of Science* 31 (2000): 37–69.

12 Stroup, *A Company of Scientists*, tables 3, 5, and 7 on pp. 246–7, 250–51, and 254.

13 Robin Briggs, "The Académie des sciences and the Pursuit of Utility," *Past and Present* 131 (1991): 38–88.

14 Hahn, *The Anatomy of a Scientific Institution*, 24. Hahn ascribes this similarity in procedure to a bureaucratic purpose, "where all that counted was the rendering of a considered collective judgment."

15 Richard Westfall, "Science and Technology During the Scientific Revolution: An Empirical Approach," in *Renaissance and Revolution: Humanists, Scholars, Craftsmen and Natural Philosophers in Early Modern Europe*, eds. J. V. Field and Frank A. J. L. James (Cambridge: Cambridge University Press, 1993), 63–72.

which required it." *Élèves* more often than academicians received the mandate to have an instrument or machine made by an artisan for the Academy.[16] Examples abound from the *Procès-verbaux*, usually following this sequence: academician, *élève*, artisan.[17] Couplet and the other *élèves* were in charge of setting up and carrying out experiments, whether at the Observatory, the king's library, or out in the field (and onto the sea, in the case of La Voye-Mignot to test Huygens's pendulum clocks for longitude determination; Jean Richer later received such assignments.)[18] Couplet, who resided at the observatory with his family, was also accountable for providing domestic and wild animals to the Company so its members may conduct their numerous dissections.[19]

The artisans working for the Academy were numerous. For mathematical and astronomical devices, one finds the armorers (*arquebusiers*) Sevin, Gosselin, Guerne, Tanguy, and Lagny. To these, we can add the mathematical instrument makers Ph. Le Bas, Migon, Butterfield, and Chapotot. For clocks and clock mechanisms in general, Thuret was working alone (though Gosselin was responsible, with Thuret, for keeping the clocks and instruments of the Academy and Observatory in good order).[20] Although lenses were often manufactured by the academicians themselves (Huygens, Auzout, Borel), instrument makers such as Pasquin, Le Bas, and Hartsoëker in France, Campani and Divini in Italy, were regularly employed.[21] Regarding the construction and repair of mechanical models, found in the *chambre des machines*, specialized artisans and engineers were also hired: Danglebert, Gayon, Potel, Buirette,

---

16 Sturdy, *Science and Social Status*, 127–37, quote on p. 127. The other *élèves* were Antoine de Niquet, La Voye-Mignot, Pivert, and Jean Richer. On the term *commis*, see Christiaan Huygens to his brother Constantyn, 30 November 1668, in *Oeuvres complètes*, 22 vols. (The Hague: M. Nijhoff, 1888–1950), vol. 6, no. 1677, 300. [Hereafter cited as *OC*(6), no. 1677, 300].

17 See, for instance, Académie des sciences, *Procès-verbaux de l'Académie royale des sciences*, Registre de mathématique, vol. 3 (1668–1669), séance du mercredi 5 septembre 1668, 132r, where it is written: "La Compagnie a resolu qu'on traiteroit de cette matiere dans la 2eme assemblée, et a donné ordre a Mr Couplet de faire preparer ce qui sera necessaire pour cette experience, suivant les ordres que Mr de Roberval luy en donnera."

18 Sturdy, *Science and Social Status*, 128. On Richer, see Nicholas Dew, "'Vers la ligne': Circulating Measurements Around the French Atlantic," in *Science and Empire in the Atlantic World*, eds. James Delbourgo and Nicholas Dew (New York: Routledge, 2008), 53–72.

19 Guerrini, *The Courtiers' Anatomists*, 117.

20 Regarding Thuret, the *Procès-verbaux* say: "Monsr. Thuret oultre les Pendules qu'il fait, reglera les autres qui sont desja faites." Académie des sciences, *Procès-verbaux de l'Académie royale des sciences*, Registre de mathématique, vol. 2 (1667–1668), séance du 30 mars 1667, 161.

21 C. Wolf, *Histoire de l'Observatoire de Paris, de sa fondation à 1793* (Paris: Gauthier-Villars, 1902), 137. See also Maurice Daumas, *Les Instruments scientifiques aux XVIIe et XVIIIe siècles* (Paris: PUF, 1953).

Cosson, Langenach, and Colson—as well as the *élèves* Antoine de Niquet and Couplet in a few instances.[22] The Academy's dealings with expert instrument makers and skilled *ouvriers* happened regularly—if not daily.[23] Auzout suggested early that someone—Couplet was appointed—should drop in the artisans' workshops to examine their instruments and learn their needs, secrets, and deceptions (*sophisteries*). Pierre de Carcavy, the king's librarian, thought it could be done easily by visiting the king's artisans.[24] Above them all, the Academy positioned itself as the foremost arbiter of inventions. In fact, a favorable endorsement was rapidly turning into a *sine qua non* condition for a royal privilege.[25]

The number of machine projects presented to the Academy was sizable. Gentlemen, engineers and artisans alike were all looking for the Academy's seal of approval, increasingly using informal networks—bypassing government channels—to place their projects under the scrutiny of academicians.[26] The Academy was thus significantly changing the function of patronage in matters

22 Stroup, *A Company of Scientists*, table 5, 250–51. For an interesting historical aperçu regarding instrument collecting, see Anthony J. Turner, "From Mathematical Practice to the History of Science," *Journal of the History of Collections* 7 (1995): 135–50.

23 For instance, in Auzout's "Memoire des Instrumens & autres choses necessaires dont il faudra fournir ceux qui iront à Magadascar [sic]," a long list of instruments mentions: "Deux grands quarts de Cercle," "Vn ou deux sextans de la mesme grandeur," "Deux ou trois boussoles de 6 ou 8 pouces," "De grande Lunettes de 30, 40 ou 60 pieds &c, et d'autres de moindres longueur," "Plusieurs oculaires de touttes grandeurs & foyers," etc. However, there is no indication as to who should be hired to make these. Some could just be bought from trustworthy makers. Others, like Huygens's clock, Auzout's micrometer, and Thévenot's level would have been built by other artisans working specifically for these academicians. Académie des sciences, *Procès-verbaux de l'Académie royale des sciences*, Registre de mathématique, vol. 2 (1667–1668), [January 1667], 37–40. Auzout followed this *mémoire* with another one on how observations needed to be made in Madagascar.

24 Académie des sciences, *Procès-verbaux de l'Académie royale des sciences*, Registre de mathématique, vol. 2 (1667–1668), séance du 30 mars 1667, 161. Going to the king's artisans, living in the Galeries du Louvre, was indeed the obvious choice since they were not regulated by the Parisian guilds. Trade secrets, in other words, had a different meaning for them (if at all). And since they were working for and paid by the king, as were the academicians, they most likely considered themselves equal to the latter. This idea of visiting the artisans' workshop was already in the anonymous memoir Huygens received in circa 1663 about the project of an Academy: "On taschera aussi d'apprendre toutes les tromperies des Artisans et des Marchands et leurs Sophistiqueries avec les Moyens pour les decouurir, que l'on publiera pour empescher le public d'y estre trompé, et pour obliger les ouuriers a trauailler plus fidelement." [?] to Huygens, [1663?], in OC(4), no. 1105, 326.

25 Only in the regulations of 1699 was this advisory function to the king defined in writing. On the evolution and role of the Academy as arbiter of inventions, see Liliane Hilaire-Pérez, *L'Invention technique au siècle des Lumières* (Paris: Albin Michel, 2000).

26 Hahn, *The Anatomy of a Scientific Institution*, 22.

of invention, technology, and their value to the kingdom. An expert—and purportedly neutral—third party had been crucially positioned by Colbert between the king (the ultimate patron) and his potential clients. In this most important role, the Academy acted henceforth as an authoritative power-broker only in the case of bona fide inventions. And these were relatively few (and most came from academicians themselves) over the first three decades of the Academy's existence.[27] According to Charles Perrault, Colbert was in fact extremely pleased that individuals were rarely given certificates of approval after having presented their invention to the Academy.[28]

One example of the various mandates tackled by the Academy and requiring novel thinking about instrumentation comes from the numerous hydraulic projects around Paris, especially the planning and completion of an adequate water supply for the fountains at Versailles.[29] The latter project led to major leveling efforts in order to accurately measure the relative slope (*dénivellation*) between distant water supplies and Versailles's gardens. Jean Picard was the chief instigator of this project and his level, first described in the *Traité de la mesure de la terre* (1671) and later improved—a description of which was published posthumously in his 1684 *Traité du nivellement*—was a gold standard in surveying instruments until the end of the century.[30] Picard's chief development was to replace the level's sighting vanes with a telescope (*lunette*), increasing the instrument's precision. Although Auzout may have been the original inventor, there is no concrete evidence he ever developed it.

27 The most important of these were gathered in the eighteenth century in the well-known *Machines et inventions approuvées par l'Académie royale des sciences, depuis son établissement jusqu'à present; avec leur Description. Dessinées & publiées du consentement de l'Académie, par M. Gallon*, 7 vols (Paris, 1735–1777), vol. 1 for the period between 1666 to 1699.

28 Académie des sciences, *Procès-verbaux de l'Académie royale des sciences*, Registre de physique, vol. 7 (1675–1679), séance du 8 mai 1677, 103r–v.

29 On the significance of this project, Chandra Mukerji, *Territorial Ambitions and the Gardens of Versailles* (Cambridge: Cambridge University Press, 1997).

30 Picard, *Traité du nivellement*, in *Mémoires de l'Académie royale des sciences*, vol. 6, 631–713 where the use of his level for the measurements he took are also described. See also *HARS*, vol. 2, 198–200 explaining that de la Hire was responsible for the posthumous editing and publishing of this treatise in 1684. Picard gave a general talk at the Academy on levels. See Académie des sciences, *Procès-verbaux de l'Académie royale des sciences*, Registre de physique, vol. 7 (1675–1679), séance du samedi 21 mai 1678, 158v–159r. On Picard's level and his involvement with Versailles, see Michel Kasser, "Picard et le renouvellement de l'art du nivellement à la fin du XVIIe siècle," in *Jean Picard et les débuts de l'astronomie de précision au XVIIe siècle*, ed. Guy Picolet (Paris: Editions du CNRS, 1987), 255–63 and Hubert Loriferne, "L'influence de Picard dans les travaux d'alimentation en eau du château de Versailles sous Louis XIV," in *ibid.*, 275–311.

Following Picard's improvement, similar levels were invented by artisans, *honnêtes hommes* (gentlemen) and academicians alike between 1677 and 1686.[31] All incorporated a telescopic sight. Michael Butterfield and Louis Chapotot, well-known Parisian instrument makers, each designed two during that period.[32] (Fig. 1L–R) Jean de Hautefeuille and the Sieur de Puyrichard, *honnêtes hommes*, invented one each, the former with what seems to have been two air thermometers à la Hubin connected together.[33] At the Academy, Cassini, Rømer, de la Hire, and Huygens also claimed to have improved the level.[34] Of all these, Huygens's appeared to have enjoyed the most success.[35]

31 Other inventions, such as Thévenot's bubble level in 1661 and Mariotte's 1772 level, described in his own *Traité du nivellement*, did not use telescopic sights. They remained marginal discoveries—though the bubble level became a common commodity in the eighteenth century. Mariotte's treatise was reviewed in the *Journal des sçavans*, 12 December 1672, 147–8.

32 On Butterfield, see the *Journal des sçavans*, 6 September 1677, 227–8; ibid., 19 September 1678, 440 and ibid., 16 December 1678, 441–3. For Chapotot, see the *Journal des sçavans*, 17 June 1680, 174–6 and ibid., 20 May 1686, 141–4. Chapotot also demonstrated his level at the Academy. See Académie des sciences, *Procès-verbaux de l'Académie royale des sciences*, Registre de physique, vol. 10 (1679–1683), séance du samedi 18 mai 1680, 60r. A short general overview on the development of levels is found in Daumas, *Les Instruments scientifiques aux XVIIe et XVIIIe siècles*, 76–8. On Butterfield and the representation of his instruments, Samuel Gessner, "The Journals and the Instrument Maker: Visuality of Butterfield's Instruments in the *Journal des Sçavans* and the *Philosophical Transactions* around 1680," *Nuncius* 30, no. 3 (2015): 610–36.

33 Académie des sciences, *Procès-verbaux de l'Académie royale des sciences*, Registre de mathématique, vol. 8 (1675–1679), séance du mercredi 17 mai 1679, 202v. Also described in the *Journal des sçavans*, 7 August 1679, 215–6. There is only a brief mention, with engraving, of Puyrichard's level in the *Journal des sçavans*, 16 September 1680, 276.

34 For Rømer, see Académie des sciences, *Procès-verbaux de l'Académie royale des sciences*, Registre de physique, vol. 7 (1675–1679), séance du samedi 23 mai 1676, 46r–50v; ibid., séance du samedi 28 novembre 1676, 64v. For Cassini, ibid., séance du samedi 31 juillet 1677, 117v; ibid., Registre de physique, vol. 10 (1679–1683), séance du samedi 25 novembre 1679, 2r. For Huygens, see ibid., séance du samedi 13 janvier 1680, 9v and séance du samedi 30 mars 1680, 35v–44r. Also published in the *Journal des sçavans*, 29 January 1680, 21–4 and ibid., 26 February 1680, 57–60. For de la Hire, see Picard's *Traité du nivellement*, in which Rømer's and Huygens's levels are also described.

35 In 1684, the instrument maker Butterfield reduced the size of Huygens's level to a pocket level. It was advertised in the *Journal des sçavans* with the following preface: "La bonté du Niveau de M. Hugens a fait rechercher avec soin le moyen de le pouvoir porter commodément." *Journal des sçavans*, 5 June 1684, 192. In 1686, several years after Huygens left Paris definitely, Philippe de la Hire wrote to Huygens that "Vostre niueau est celuy de tous les nieueaux qui est le plus en uogue et ie croy que ce que uous uoudrez bien nous donner la dessus sera toujours tres bien receu." de la Hire to Huygens, 5 December 1686, in *OC*(9), no. 2447, 114. There are several other correspondences mentioning Huygens's level as well

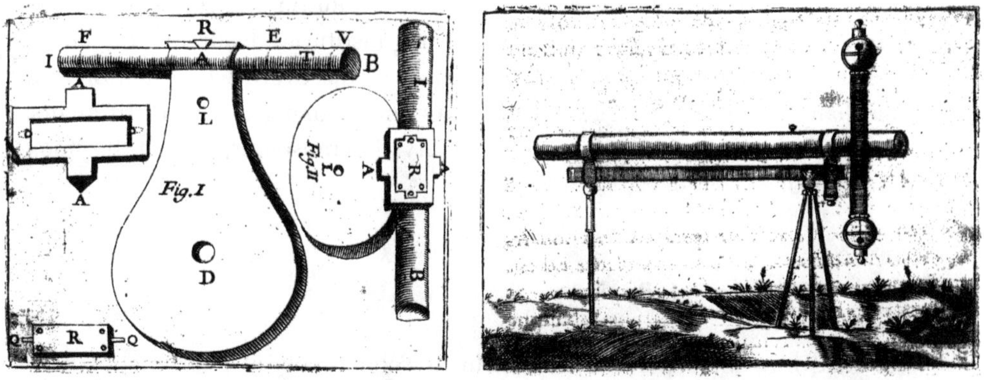

FIGURE 1.L–R Two types of seventeenth-century French levels. Left, Butterfield's level with a telescopic sight, three years before Huygens's. Right, Chapotot's level published the same year as Huygens's in the *Journal des sçavans*.
BUTTERFIELD, *JOURNAL DES SÇAVANS*, 6 SEPTEMBER 1677, 227–8
CHAPOTOT, *JOURNAL DES SÇAVANS*, 17 JUNE 1680, 174–6

One sees how a specific problem (exact leveling) raised by a special need (royal gardens) put to work not only the Academy but the entire learned and artisanal community of Paris. The Academy did not work isolated from technical expertise. Colbert positioned the Company at the center of a growing scientific networking encompassing all kinds of people and specialized branches of craftsmanship and learning.

From an *honnête homme* to the Academy, to an artisan close to the Academy, and ultimately to the public: such an instrument's intellectual and material trajectory increasingly characterized the life of early modern French machines. As long as authorship was well established and respected, this technological itinerary—with its unexpected and necessary adaptations—ensured the superiority and value of mechanical devices. We see that the manufacturing of knowledge and machines involved several authors (or instigators) working in separate and sometimes clearly delineated spheres of knowledge production. Yet what brought them together repeatedly again were the tools, contraptions, mechanisms, and complex pieces of equipment and instruments that

as a chapter in *OC(21)*, 81–108. In 1686, before describing a new level by Chapotot, the author of the *Journal des sçavans* wrote the following: "Il semble qu'après l'exacte précision des Niveaux que nous ont donnéz feu M. Picard & M. Hugens, il soit assez difficile de porter le nivellement à un plus haut degré de perfection qu'il l'est aujourd'huy." *Journal des sçavans*, 20 May 1686, 141.

came to epitomize the so-called Scientific Revolution and modern science. The Academy's role as an authority-granting institution facilitated the emergence of this pattern of natural philosophical discovery, where reason and mechanical skills coalesced around instruments. Looking at these instruments, some of which (or similar ones) still preserved in museums, is what this book is all about.

2 Reading This Book

Divided into three parts, the book looks sequentially at instruments, habits, and museums. The goal is to uncover how, together, these material and immaterial activities, rules, and commitments form one meaningful and credible blueprint revealing the building blocks of knowledge production. Part 1, *Organum*, offers a novel reading and understanding of the concept of instrument in early modern Europe. Chapter 1 takes a notional and theoretical approach, looking at multiple definitions and examples taken from a variety of sources in order to better comprehend and appreciate what an instrument—*instrumentum/organum*—was to early modern practitioners. Whether a material (Galileo's telescope) or immaterial (Bacon's *Novum organum*) instrument, an *organum* was primarily a means to an end. It couldn't do anything itself; it needed a guiding hand or mind, hence the prescribed—and often binding—method (logic) or practice (repetition, routine = habits) invariably accompanying an instrument. Such a 'ratiocination of action' (*rationalité de l'action*)[36] is what Christian Jacob calls the 'hands of intellect.' Jacob, a historian of cartography and knowledge transmission in Antiquity, sees a radical and epistemic rapprochement between hands-on practices and rational interpretation. To him, "the work of the hand and mind, the handling of words and tools in the workshop, the laboratory, the library or sitting in front of a computer can be construed as making practices [*arts de faire*], where jolts and operations go one after the other, where gestures and body positioning take turns according to the numerous rules of the game of knowledge."[37] An instrument alone is inert. It requires a specific or tacit course of action to play the game of knowledge. Chapter 2 provides an original case study apropos Marin Mersenne's description of musical instrument making in the 1630s, especially the organ. The latter was generally celebrated as the most perfect of musical

36 Liliane Hilaire-Pérez, *La pièce et le geste. Artisans, marchands et savoir technique à Londres au XVIIIᵉ siècle* (Paris: Albin Michel, 2013).

37 Jacob, *Lieux de savoir*, 19.

instruments; mechanically speaking, it was undoubtedly the most complex of them all. Looking at Mersenne's writings on the organ, one discovers it was more than a Church instrument for delivering and conveying the Word of God. To Mersenne, it encompassed the ideal means of studying sound. He performed experiments with pipes, visited organ makers' workshops, reported on artisanal practices, and applied pure mathematical concepts to improve the material components of the organ—thereby, the overall tonality of the instrument. It was an archetypal natural philosophical *organum*, exhibiting to those willing to move beyond its apparent musical and devotional purpose a comprehensive interposition of theoretical, experimental, and artisanal activities.

In Part 2 of the book, we explore how the game of knowledge is apprehended through the early modern conception of *habitus*, both the *habitus in corpore* and *habitus in anima* (habit of the body and habit of the mind). Chapter 3 takes once again a more notional and theoretical approach, identifying what *habitus* was to early modern scholars and finding meaning in René Descartes's method, or logic of practice. We examine as well, especially with Blaise Pascal's celebrated arithmetical machine, how *coutume* is linked to *habitude*, its benefit for mastering such a mechanical device and, more broadly, its influence (desirable or not) toward natural philosophical knowledge production. We continue focusing our attention on *habitus* in chapter 4, one more time with Descartes as our heuristic guide, to examine how this notion can advance our understanding of the meaning and epistemology of the early modern use and manufacturing of an *organum*. To do so, we look at artisans and their interactions with savants. As we saw briefly above with the example given by the Academy of Sciences in Paris, and from the scholarship over the past couple of decades, the relationship between artisans and natural philosophers is no longer in doubt. What interests me here is looking at some taxonomies regarding artisans, from an *idiota* in the sixteenth century to an *artiste* in the eighteenth century. Though not a comprehensive analysis of these shifts and redefinitions, such a focus provides conceptual keys for understanding how makers were perceived and how they adapted to changes in early modern Europe.

In the final part, Part 3, on *Museum*, I do not offer an early modern notional and theoretical account, as I did for *organum* and *habitus*, since it has authoritatively been done by Paula Findlen three decades ago.[38] Studies on the meanings and purposes of 'museum' abound. Recently, for instance, M. Norton Wise explains in a thought-provoking book how in Germany "the pursuit of experimental science emerged from within the museum" in the first half of the

38 Paula Findlen, "The Museum: Its Classical Etymology and Renaissance Genealogy," *Journal of the History of Collections* 1 (1989): 59–78.

nineteenth century. The collections assembled around the founding of the University of Berlin in 1810, he claims, were regarded as the core elements of a developing experimental ethos useful to fields like agriculture, industry, and medicine. Whether for art, science or industry, museums in this specific context became, "so to speak, the culture medium for experimentation."[39] Although collections and museums were being challenged by the 1870s "as crucial site of scientific inquiry," Sophie Forgan reminds us that "laboratories grew up as it were in the shadow of the museum."[40] Rather than outlining what a museum is, I decided in this last part to prioritize my experience and current research in the field of museum studies. Chapter 5 explores the activity and function of conservation/restoration to demonstrate the epistemic problems occurring when the materiality of objects is confronted to museum practices and instrument *habitus*. Through examples taken from the eighteenth to the twentieth centuries, especially a case study considering the restoration of abbé Jean-Antoine Nollet's instruments, I argue that the conservation/restoration of scientific instruments ought to evaluate better the place given to performances, operations, craftsmanship, and practices of these same instruments. Chapter 6, finally, examines three specific activities linked to the study of instruments in a museum. First, I look at 'collecting,' focusing my intellectual lens on a turning point in the nineteenth century for the current Collection of Historical Scientific Instruments (CHSI) at Harvard University, where I worked for over nine years as a curator and an administrative director. Next, I consider 'playing' as a valuable, and often essential, way of investigating how instruments in collections were wielded and deployed in the context of teaching and demonstrating scientific principles. Lastly, I briefly look at 'digitizing' collections, i.e., at the power of data visualization and the level of understanding one gains by considering the bulk of data extracted from whole collections instead of singling out individual objects—ditching the proverbial 'can't see

---

39 M. Norton Wise, *Aesthetics, Industry & Science. Hermann von Helmholtz and the Berlin Physical Society* (Chicago: The University of Chicago Press, 2018), quotes on pp. xv and 35 respectively.

40 Sophie Forgan, "The Architecture of Display: Museums, Universities and Objects in Nineteenth-Century Britain," *History of Science* 32 (1994): 139–62, quote on p. 155. For a complementary view, Forgan, "Building the Museum: Knowledge, Conflict, and the Power of Place," *Isis* 96 (2005): 572–85. On the role of history and memory in the lab through collections of instruments, Boris Jardine, "The Museum in the Lab: Historical Practice in the Experimental Sciences at Cambridge, 1874–1936," *BJHS: Themes* 4 (2019): 245–71, where he writes that these collections bind the laboratory to society in 'integral networks,' "networks of affect, commerce and authority: the museum in the lab is the space where a form of sociable materiality can coexist with, augment and contrast with a form of (supposedly) asocial materiality." (p. 271)

the forest for the trees.' Data visualization provides a formidable new method of exploring museum collections, greatly facilitating inquiries about objects and collections heretofore unimagined.

Organum, *habitus*, and *museum* form a coherent whole allowing us to conceptualize and to examine the entire life of an instrument: from its ideation and craft to its use, reuse, circulation, recycling, and (if not obliterated) its final accession into a museum. My goal is not to generalize what I describe in this book to all periods: the early modern notions of *organum* and *habitus* are significantly different than what could be inferred from the nineteenth or twentieth century. Although the book focuses principally on the early modern era, *organum*, *habitus*, and *museum* compose an analytical triptych adaptable to a variety of circumstances, modern or not. Together, they have guided my scholarship for the last twenty years.

PART 1

Organum

Introduction to Part 1

Chapters 1 and 2 work as a tight unit and ought to be considered the cornerstone of the book. Together, they set up the themes developed in Part 2 (*Habitus*) and Part 3 (*Museum*) by focusing respectively on the materiality of scientific and museum practices. First, chapter 1 provides an original definition and description of what an instrument was in the early modern period. Such a concept, I claim, is best depicted in terms of an *organum*, an abstract or material tool designed for the sake of an end. In the case of René Descartes's *méthode* or Francis Bacon's *novum organum*, these abstract (rational) instruments were truth-seeking methods formulated with one specific end: acquiring systematic truth. A material contraption such as Galileo's telescope, in contrast, was instrumentalized to look at the stars and to see what had been hitherto invisible to the naked eye. Although both distinct ends from these familiar examples are simply stated and well defined, the knowledge claims they asserted were completely open-ended, not predetermined. Truth-seeking methods, by definition, promised one would discover something true—whatever it might be. Likewise, Galileo's telescope was meant to observe something never seen before, though precisely what remained undetermined. Hence, an early modern *organum* was no longer an instrument restricted to a *terminus ad quem*, in the strictest scholastic interpretation (the hammer in a blacksmith's hand being the most common example used). Instead, an *organum*, in the hands of a growing number of natural philosophers, grew into an instrument guaranteeing a *terminus a quo*—a starting point, if used (and manufactured) properly. Though *organum* and *instrumentum* were often employed interchangeably during the early modern period, I suggest that the former appellation articulates a broader epistemic framework as regards the core image and conception of what an instrument was and should be. The best example I found was through a musical instrument: the organ.

Focusing on Marin Mersenne's organ, I show in chapter 2 that the celebrated natural philosopher wanted to convey in his magisterial opus *Harmonie universelle* one of the chief principles of seventeenth-century natural philosophy: i.e., instruments were a fundamental part of knowledge production. Through the organ, Mersenne displayed how much artisan, experiment and theory were tightly meshed into the nascent mechanical philosophy. The organ, owing to its complexity, socio-cultural, and religious status in the early modern period, became for Mersenne the paradigmatic material emblem of the new science of sound—a special instance of the more general natural philosophy. Not only did the organ epitomize the nature of sound, it proved his *harmonie universelle*

 | DOI:10.1163/9789004504615_003

wasn't purely conjectural and metaphysical, confirming he had *concretely* surpassed ancient knowledge. The organ, and most musical instruments, once they had received the lights of theory, would embody the mathematical knowledge at the center of the *harmonie universelle*. Yet these same instruments, employed as natural philosophical tools, generated experimental data from which a theory of sound could be 'soundly' established. And to ensure these musical instruments were adequately manufactured, a thorough knowledge of artisanal craftsmanship was necessary. I demonstrate, in short, that the organ in Mersenne's view became the instrument, or *organum*, of universal harmony, the material extension of music theory and mechanical philosophy. It is quite clear, moreover, that such an *organum* requires special skills in making it, playing it, or experimenting with the nature of sound. This is what Part 2 on *habitus* is mostly about.

In my historical and epistemological description of an early modern *organum*, I didn't try to move orthogonally from the plane of the debate about what should count as a 'natural' (real) in opposition to a 'constructed' (historical) object of scientific inquiry. The applied metaphysics of scientific objects, as described and understood by Lorraine Daston *et al.*, refers to objects, whether in substance or not, that display "the dynamic world of what emerges and disappears from the horizon of working scientists." The material and immaterial things forming the core of my study around the concept of instrument are more 'quotidian' and certainly less unusual than monsters, dreams, cytoplasmic particles or even culture itself. Applied metaphysics is a valuable epistemic concept, one in fact that helps focus attention to a multifold of 'things' that otherwise would never be perceived as scientific. Yet applied metaphysics bears the risk of losing contact with what it really means to live in a material world.[1]

That said, material and immaterial things are no longer perceived as inert and passive objects of study solely defined by thinking subjects.[2] Historians, sociologists and especially anthropologists recognized several decades before

1 Lorraine Daston, ed., *Biographies of Scientific Objects* (Chicago: The University of Chicago Press, 2000), 1. Ian Hacking has come up with a somewhat similar notion in coining the expression 'historical ontology.' To him, "Historical ontology is about the ways in which the possibilities for choice, and for being, arise in history." It is not "about the coming into being of objects of study" but rather about "the coming into being of objects, period." The essence of these objects is historical and found somewhere in the grid formed by the three axes of knowledge, power, and ethics. Hacking, *Historical Ontology* (Cambridge, MA.: Harvard University Press, 2002), 23 and 11 respectively for the quotes. Latour's "actor-network theory" and Pickering's "mangle of practice" are other manifestations of applied metaphysics.

2 On the rapid expansion of the field of immaterial culture, see Chiara Bortolotto, ed., *Le Patrimoine culturel immatériel: enjeux d'une nouvelle catégorie* (Paris: Maison des Sciences de l'Homme, 2011).

historians of science that things had a life of their own: they shape the subject and its social context as much as they are shaped by the latter. Without losing any of the things' distinctive essences, gaining rather a better understanding of it, scholars have worked out 'thing theories' that have become significant in explaining and confronting their interaction with the human world.[3] Historians and philosophers of science have barely begun sketching such materialist theories of scientific knowledge, where instruments are considered more than 'inscription devices,' exclusively providing visual displays in scientific texts. Instruments and machines have matured in some cases into 'thing knowledge,' not only raised up to the level of theory and experiment but literally on a par with the spoken word.[4] They do not need, that is, to be associated with a written text to signify something. As a thing knowledge, an instrument bears insight into the craft, skill, experimental practice, and theory involved in its coming into being: it can (and does) speaks volume about the practice of science if one avoids letting it drown in words.[5] Instruments have acquired multiple folds, layer upon layer of meanings in recent decades. They cannot be reduced anymore to simple matrices of functional performances. They now have to be considered 'thick things,' objects representing multiple (and divergent) points of view.[6] Instruments are 'thick' because they are never created, crafted, devised, apprehended, and conceived without a purpose; they do not exist isolated from a specific socio-cultural, experimental, and rational context.

---

3 Bill Brown, "Thing Theory," *Critical Inquiry* 28 (2001): 1–22. The whole issue is dedicated to this topic. The literature is rich and vast here. For a historical perspective, see Daniel Roche, *Histoire des choses banales. Naissance de la consommation dans les sociétés traditionnelles* (*XVII*[e]*–XIX*[e] *siècle*) (Paris: Fayard, 1997) and Curtis Perry, ed., *Material Culture and Cultural Materialisms in the Middle Ages and Renaissance* (Turnhout, Belgium: Brepols, 2001). A good summary on the theoretical concepts of things is found in Christopher Tilley, ed., *Reading Material Culture: Structuralism, Hermeneutics, and Post-Structuralism* (Cambridge, MA.: B. Blackwell, 1990).

4 Davis Baird, *Thing Knowledge: A Philosophy of Scientific Instruments* (Berkeley: University of California Press, 2004). See also Robert J. Ackermann, *Data, Instruments, and Theory: A Dialectical Approach to Understanding* (Princeton: Princeton University Press, 1985) and Don Ihde, *Instrumental Realism: The Interface Between Philosophy of Science and Philosophy of Technology* (Bloomington: Indiana University Press, 1991). On inscription devices, Bruno Latour, *Science in Action: How to Follow Scientists and Engineers Through Society* (Cambridge, MA: Harvard University Press, 1987), 64–70, where he writes: "What is behind a scientific text? Inscriptions. How are these inscriptions obtained? By setting up instruments." (p. 69)

5 Lorraine Daston, ed., *Things that Talk: Object Lessons from Art and Science* (New York: Zone Books, 2004). In the introduction, Daston pointedly remarks that "If things are 'speechless,' perhaps it is because they are drowned out by all the talk about them." (p. 9)

6 Ken Alder, ed., "Focus: Thick Things," *Isis* 98 (2007): 80–142. See also Alder, "Making Things the Same: Representation, Tolerance, and the End of the Old Regime in France," *Social Studies of Science* 28 (1998): 499–545.

Part 1's goal is to work (sometimes struggle) with these sorts of ideas in order to better understand the definition, role, and value of instruments in early modern Europe. As it will become clearer in Part 2 of the book, an *organum* cannot be fully decoded without an appreciation of the environment and skills involved in handling it—its *habitus*. And without these two, how can such an instrument, and the material culture of science in general, make any sense in the context of a *museum*?

CHAPTER 1

Quid organum erat? The Idea of Instrument in Early Modern Europe

In *Thinking with Objects*, Domenico Bertoloni Meli gives a fascinating account of the tremendous variety of commonplace objects put to work by late Renaissance and seventeenth-century savants, engineers, and natural philosophers. Their study of mechanics and the laws of motion using pendulums, inclined planes, beams, projected balls, vibrating strings, and rolling spheres are "just some notable examples of the engagement of scholars with the material world around them." By focusing too heavily on the notion of 'laws,' historians before Bertoloni Meli were unable, he says, to "capture the complexity of the field." According to him this "[u]nbalanced attention to laws fails to convey the scholars' *practices* and their engagement with the features and peculiarities of the objects of mechanics."[1] These abundant objects, however, shouldn't be regarded as 'instruments' because, as Bertoloni Meli claims reluctantly, the term conveys the idea of something more elaborate than the mundane tools and things of everyday life. Yet, could have they been acknowledged as 'instruments' in the sixteenth century?[2] Galileo himself seems to believe so, writing in the introduction of his *Delle macchine* of 1592–1593:

> we will start speculating from the first and simplest instruments [*più semplici instromenti*], to which all the others reduce, or of which they are compounded, and they are called first instruments [*primi instromenti*].

1 Domenico Bertoloni Meli, *Thinking with Objects: The Transformation of Mechanics in the Seventeenth Century* (Baltimore: The Johns Hopkins University Press, 2006), 2. (my emphasis.) See more recently Bertoloni Meli, *Mechanism: A Visual, Lexical, and Conceptual History* (Pittsburg: University of Pittsburg Press, 2019).

2 On the 'real' instruments Guidobaldo del Monte designed and used, Enrico Gamba and Roberto Mantovani, "Gli strumenti scientifici di Guidobaldo del Monte," in *Guidobaldo del Monte (1545–1607): Theory and Practice of the Mathematical Disciplines from Urbino to Europe*, eds. Antonio Becchi, Domenico Bertoloni Meli, and Enrico Gamba (Berlin: Max Planck Research Library for the History and Development of Knowledge, Proceedings 4, 2013), online http://edition-open-access.de/proceedings/4/index.html (accessed 9 August 2022).

 | DOI:10.1163/9789004504615_004

> First of all, there are the lever, the winch, the pulley, the screw and the wedge. All of which also reduce themselves, in some way, to only one, that is, the scales, or, the balance.[3]

Perhaps it is not a surprise to notice that this way of thinking about *instrument* has a direct link to Leonardo's own consideration of mechanical devices. Though the savant from Vinci, according to Paolo Galluzzi, pursues a "foundational analysis of the general principles of mechanics" in what he labels '*elementi macchinali*,' mirroring Euclid's *Elements of Geometry*, Leonardo is also clearly after the 'organs of mechanical devices,' conducting what Galluzzi calls an exhaustive 'anatomy of machines.' These machines, or *strumenti* as Leonardo calls them, can be intricate or as elementary as a single rope. Whatever their configuration, the efficacy of instruments rests on the organization of their internal parts, which can be of infinite variation.[4]

So, what are Galileo's *instromenti* (or *engins*, as Descartes called them in his *Traité de mécanique*—and 'instruments' in Jean-Daniel Mayor's 1672 Latin translation of the same book, *Inventio omnium horum instrumentorum*) if not the simple objects described by Bertoloni Meli? It is their elemental operations and universal meaning that conferred them the notion of instrument.[5] Take another such ordinary object as the prism, for instance, which was chiefly used before the mid-1650s as a common decorative and entertaining trinket. In the hands of experimental practitioners (this is the key argument here)—and especially Isaac Newton—it was converted into a genuine instrument of

3 Galilei, *Delle macchine* (Hamburg manuscript, 1592–93a), quoted in Matteo Valleriani, *Galileo Engineer* (Dordrecht: Springer [Boston Studies in the Philosophy of Science, vol. 269], 2010), 97.

4 Paolo Galluzzi, *The Italian Renaissance of Machines* (Cambridge, MA: Harvard University Press, 2020), 101–16. Leonardo writes in the Madrid manuscript I (written in the 1490s), p. 96v: "Creato che ffia lo sstrumento, la neciessità della sua operatione dimanda la forma delle sue menbra. Le quale son d'infinite figure, ma pure tutte fien sottoposste a queste regole de' 4 volumi." About the instrument consisting of a single rope, p. 152v: "E cquesto accade perché nello strumento era prima una sola linia di corda e poi se l'agiugnie 99 linie d'essa corda, che a ogni linia toca una libra." The quotes from the original manuscripts are taken from https://www.leonardodigitale.com/ (accessed 28 October 2020).

5 Though Galileo calls them explicitly *instromenti*, Bertoloni Meli still refers to them as 'tools' and 'objects.' Bertoloni Meli, *Thinking with Objects*, 67. Mersenne published a French translation in 1634 of Galileo's *Delle macchine*, where he talks about "les instrumens de la Mechanique." *Machine* is also considered a synonym at the time (abundantly used by Mersenne and Descartes).

knowledge.[6] Robert Boyle emphasized early the 'instrumental' status of the prism when he wrote in his 1664 treatise on color that it was "the usefullest Instrument Men have yet imploy'd about the Contemplation of Colours."[7]

What attributes make an 'instrument' is not unlike asking the question: what characterizes a work of art? What are the human agency and intentions at play in both cases? How do recycled and discarded objects become, on the one hand, instruments for the study of Renaissance mechanics, and, on the other hand, authentic and respectable displays in a museum of contemporary art? One has to reach beyond the (intricately) inherent scientific and aesthetic values, meanings, and prescriptions as these new objects emerge and evolve into instruments and artworks. As Peter Lamarque explains in *Work & Object*, "It is only through a kind of social compact between people with like-minded interests that the recognition of a work as a work is possible." It is essential to acknowledge the 'cultural embeddedness of works,' to see them as 'institutional entities.'[8] It is true of instruments as well. With the rise of institutional and cultural powerhouses such as the Royal Society in London and the Académie royale des sciences in Paris, technicians and engineers who previously studied the laws of motion were increasingly replaced by bona fide philosophers—mechanical natural philosophers, who came to see the world as machinelike. Not only did it impact the social and intellectual status of both scholars and engineers, it resulted in a "shift in the nature of the objects themselves, from technical devices such as the lever, the wedge, the beam, and the hydraulic press to moving objects such as the pendulum and colliding, falling, projected and orbiting bodies," which embraced better the cosmological Weltanschauung of this new breed of scholars.[9] Such 'scientific objects,' to borrow the (much broader) ontological category outlined by Lorraine Daston,

6 Jed Buchwald has reproduced Descartes's earlier experiments with prisms, showing how important they were as well in working out the Cartesian model of colors. In fact, according to Buchwald, Descartes's optical explanatory novelties were similar to Newton's own. Buchwald, "Descartes's Experimental Journey Past the Prism and Through the Invisible World to the Rainbow," *Annals of Science* 65 (2008): 1–46.

7 Simon Schaffer, "Glass Works: Newton's Prisms and the Uses of Experiment," in *The Uses of Experiment. Studies in the Natural Sciences*, eds. D. Gooding, T. Pinch and S. Schaffer (Cambridge: Cambridge University Press, 1989), 67–104, quote on p. 76. On the importance of ordinary things in early modern Europe, see Daniel Roche, *Histoire des choses banales. Naissance de la consommation* (*XVII*^e^–*XIX*^e^ *siècle*) (Paris: Fayard, 1997). For a *longue durée* analysis of all the things scientists work with, Françoise Waquet, *L'Ordre matériel du savoir: Comment les savants travaillent, XVI*^e^*–XXI*^e^ *siècles* (Paris: CNRS Éditions, 2015).

8 Peter Lamarque, *Work & Object: Explorations in the Metaphysics of Art* (Oxford: Oxford University Press, 2010), 53–54.

9 Bertoloni Meli, *Thinking with Objects*, 164.

come and go, produce results, surprises, connections, and explanations, and "grow more richly real as they become entangled in webs of cultural significance, material practices, and theoretical derivations."[10] Ursula Klein and Emma Spary likewise suggest going outside the common study of mechanical devices because they form "only one type of material object involved in artisanal labor and learned inquiry into nature." Raw materials, stuff like ink, milk, blood, metals, and mineral water put to quotidian use in craftsmanship and artisanal probing can "extend and enrich our understanding of the early modern sciences and material culture beyond a narrow focus on mathematics, mechanics, and experimental philosophy."[11] As Thomas Sprat, the biographer of the Royal Society in London explained, a history of trades would "bring innumerable benefits to all practical *Arts*: When all the secrets of *Manufactures* shall be discover'd, their *Materials* describ'd, their *Instruments* figur'd, their *Products* represented: It will sonn be determin'd, how far they themselves may be promoted, and what new consequences may thence be dedu'd."[12] For instruments and works of art alike, I would argue, "[b]oth the context of viewing and appropriate background knowledge contribute to the experience demanded by, and realizable in, a work when perceived correctly."[13] As Simon Werrett explains in looking at the thrifty practices of savants, from the use of kitchen utensils to cleaning devices, "[s]cientific instruments might perhaps be better thought of as *processes* rather than as fixed, static objects. … 'Instrument' should be thought of as a practice, rather than a thing."[14] Said otherwise, instruments are situated and contextualized—not typified, epitomized, and presupposed.

An early modern instrument is not as explicitly defined as it is often believed to be. Blaise Pascal's arithmetical machine, for instance, is described in the *Lettre dédicatoire* to the chancelier Séguier as an '*instrument*' and

10 Lorraine Daston, ed., *Biographies of Scientific Objects* (Chicago: The University of Chicago Press, 2000), quote on p. 13. Daston identifies four ontological categories: salience, emergence, productivity, and embeddedness. On objects and things in early modern Europe, Paula Findlen, ed., *Early Modern Things: Objects and Their Histories, 1500–1800* (New York: Routledge, 2013). In the introduction Findlen writes: "Virtually every society vacillates between the euphoria of its materialism and episodes of introspective anxiety about the passion for stuff … Every object takes its place in a system of use and meaning in which value is constantly being renegotiated." (p. 7)

11 Ursula Klein and Emma C. Spary, eds., *Materials and Expertise in Early Modern Europe: Between Market and Laboratory* (Chicago: The University of Chicago Press, 2010), 16.

12 Thomas Sprat, *History of the Royal Society* (London, 1667), 310 (emphasis original).

13 Lamarque, *Work & Object*, 17. He writes elsewhere: "Without appropriate cultural and institutional background conditions there would be no works of art even if there were by chance objects physically indistinguishable from those we call works of art. Works of art are not natural kinds, nor physical objects *tout court*." (p. 105)

14 Simon Werrett, *Thrifty Science. Making the Most of Materials in the History of Experiment* (Chicago: The University of Chicago Press, 2019), 82 (emphasis original).

'*machine*' interchangeably, with few apparent distinctions between the two appellations.[15] Pascal wasn't alone in conflating these terms. For William Harvey, instruments were embedded within the teleology of their ultimate user, God. Though he conceived of a mechanical account of the human body, comparing human physiology to hydraulic engineering, he did not acknowledge the open-endedness of instruments. Harvey's concept of instrumentalism was based on a world where all machines came from the Divine Being—*machina ex Deo*.[16] Later in the century, Gottfried Wilhelm Leibniz drew upon the concept of 'organism'—very close in meaning to 'organization'—in describing organic bodies, or machines of nature. He appropriated the notion from the Greek *organon*, i.e., 'instrument' or 'tool,' suggesting that 'organic' was synonymous to, and not antithetical to, mechanical. In Leibniz's corpus, according to philosopher of science Justin Smith, "organism is to natural machines what mechanism is to artificial machines, and this organism is not contrasted with mechanism, but rather is conceived as a variety of it ... It is the general condition of organic bodies."[17] From the tiny to the large, instruments manifest and articulate—Gaston Bachelard would say "reify"—the theoretical and experimental practices of natural philosophers. When the architect Claude Perrault was asked to design the Observatoire de Paris in the mid 1660s, he made it clear that the building itself could substitute for key astronomical instruments: its alignment and bulk supplied a *meridian line* of seventeen *toises long*; two octagonal towers were set with windows facing sunrises and sunsets to determine the equinoxes and solstices; a hole pierced *de fond en comble* let light in to ascertain the zenith, thus avoiding the use of a *quadrant*.[18] Other such solar observatories were devised in majestic Italian and French cathedrals from the

15 Blaise Pascal, *Oeuvres complètes*, ed. by Jean Mesnard, 4 vols. (Paris: Desclée de Brouwer, 1964–1992).

16 Don Bates, "*Machina ex Deo*. William Harvey and the Meaning of Instrument," *Journal of the History of Ideas* 61 (2000): 577–93. Marjorie O'Rourke Boyle, "Harvey in the Sluice: From Hydraulic Engineering to Human Physiology," *History and Technology* 24 (2008): 1–22.

17 Justin E. H. Smith, *Divine Machines: Leibniz and the Sciences of Life* (Princeton: Princeton University Press, 2011), 97–110, quote on p. 106. For Leibniz, "All organism is in truth mechanism, but more exquisite." (p. 97) In Johann Hübners, *Curieuses und Eales Natur- Kunst- Berg- Gewerck- und Handlungs-Lexicon ...*, 7th ed. (Leipzig, 1736) one finds the following definition: "Organismus, f. Mechanismus." (column 1468)

18 C. Wolf, *Histoire de l'Observatoire de Paris, de sa fondation à 1793* (Paris: Gauthier-Villars, 1902), 22–3. "Le bastiment de l'Observatoire est construit de telle sorte qu'il peut suppléer tout seul à tous les principaux instrumens d'Astronomie dont on se sert pour les observations." Perrault's original plans were unsuitable for the astronomers' work, and Jean-Dominique Cassini, the first Royal Astronomer, was vocal about it. Oded Rabinovitch, *The Perraults: A Family of Letters in Early Modern France* (Ithaca: Cornell University Press, 2018).

sixteenth to the early eighteenth century; equipped with carefully measured meridian lines, these church observatories became a type of *organum Dei*, an instrument of God providing true astronomical time—especially for the determination of Easter.[19] On a treacherous mountainside in Scotland in 1774, lastly, the Astronomer Royal Nevil Maskelyne used the gravitational attraction of the Schiehallion Mountain not as much to prove the universality of Newtonian gravitation as to demonstrate the 'exactness' of a new instrument, a zenith sector, and the 'exactness' of his observational techniques. The mountain was turned into a calibration 'instrument' in ascertaining the precision of the zenith sector, which Maskelyne then employed to correct errors in the large instruments at the Greenwich Observatory.[20] (As was well known in the early modern period, Plato understood Nature as "*Dei artem, vel artificiosum Dei Organum; The art or artificiall Organ of God.*"[21])

So, in early modern Europe, *quid organum erat*? To begin with, it wasn't 'scientific'—as the common terminology proposes since the mid-nineteenth century. Secondly, it was more diverse and wide-ranging than the telescope, microscope, thermometer, barometer, air pump and pendulum clock—sometimes identified as 'philosophical' instruments.[22] Peter Dear historicized not too long ago the transformation of nature's study, from a natural philosophical enterprise to a modern scientific one. This gradual shift started in the early seventeenth century when an emphasis on the 'instrumentality' of knowledge

---

19 John Heilbron, *The Sun in the Church. Cathedrals as Solar Observatories* (Cambridge, MA: Harvard University Press, 1999). Heilbron uses this theme to put forward a 'Catholic Merton thesis,' disclosing an early modern significant support of the Catholic Church toward the sciences.

20 Nicky Reeves, "'To Demonstrate the Exactness of the Instrument': Mountainside Trials of Precision in Scotland, 1774," *Science in Context* 22 (2009): 323–40, quote on p. 325.

21 Cited by Sir Walter Raleigh, *The History of the World* (London, 1617), 20, where he adds that Cusanus called Nature "*Diuini praecepti instrumentum; The instrument of the diuine precept.*"

22 I leave aside the ambivalence of this conventional expression—scientific instrument—which needs some explanation, especially when dealing with early modern Europe. On this topic see Deborah J. Warner, "What is a Scientific Instrument, When Did it Become One, and Why?," *British Journal for the History of Science* 23 (1990): 83–93, Judith V. Field, "What is Scientific About a Scientific Instrument?," *Nuncius* 3 (1988): 3–26, and Liba Taub, "Introduction: Reengaging with Scientific Instruments," *Isis* 102 (2011): 689–96. Taub has recently revisited this notion in "What is a Scientific Instrument, Now?," *Journal of the History of Collections* 31 (2019): 453–67. On the conventional notion of an early modern 'scientific' instrument, Albert van Helden, "The Birth of the Modern Scientific Instrument, 1550–1700," in *The Uses of Science in the Age of Newton*, ed. J. G. Burke (Berkeley: University of California Press, 1983), 49–84. Jérôme Fatet, ed., *A quoi servent les instruments scientifiques? Réflexions et études de cas sur les rôles et les fonctions des instruments dans la pratique scientifique* (Limoges: Presses universitaires de Limoges, 2016).

production, advocated initially by Francis Bacon, changed the fundamental character of natural philosophy. The latter field was no longer solely identified with *theorica*, or the speculative understanding and explaining of the natural world. Useful matter, *practica*, was slowly integrated into the practice of natural philosophy for the benefit of mankind. According to Dear, this shift "resulted in the establishment of a new enterprise that took the old 'natural philosophy' and articulated it in the quite alien terms of instrumentality—science was born a hybrid of two formerly distinct endeavors."[23] *Theorica* and *practica* became two essential components of natural philosophy during the so-called Scientific Revolution, with knowledge-producing instruments often bridging—unifying—divergent branches of learning into *scientia*.[24] The chapter elaborates on this notion of instrumentality, not so much by providing a long list of instruments along with their typical functions, users, and makers, but rather by examining the concept of instrument—*organum*—in early modern Europe. Exploring the etymology and usage of the word instrument, I suggest that an early modern *instrument* embraced not only mundane and elaborate objects of craftsmanship, but a variety of simple objects and things as well as abstract instruments of knowledge, such as Bacon's *novum organum* and Descartes's *méthode*.[25]

1 *Organum scientiae*: Definitions and Examples

In what is commonly considered the first European museum treatise, Samuel Quiccheberg's *Inscriptiones vel tituli theatri amplissimi* (1565) outlined how to build a princely *Kunst-* or *Wunderkammer*. Presented as a vast organizational

23 Peter Dear, *The Intelligibility of Nature. How Science Makes Sense of the World* (Chicago: The University of Chicago Press, 2006), 11.

24 Ian Hacking argued that during the twentieth century, instruments were often better unifiers than grand unified theories. So much so in fact that an 'instrumentalist' thesis could easily be put forward. It would argue, for instance, that "it is not high-level theory that has stopped the innumerable branches of science from flying off in all directions, but the pervasiveness of a widely shared family of experimental practices and instrumentation." Hacking, "The Disunities of Science," in *The Disunity of Science: Boundaries, Contexts, and Power*, eds. P. Galison & D. J. Stump (Stanford: Stanford University Press, 1996), 37–74, on pp. 68–9.

25 Mario Biagioli's concept of 'instruments of credit' ascribed to Galileo is very interesting. These instruments were not only the geometrical compasses or telescopes he designed and built, but were also all the "techniques he used to maximize the credit he could receive from readers, students, employers, and patrons." Biagioli, *Galileo's Instruments of Credit: Telescopes, Images, Secrecy* (Chicago: The University of Chicago Press, 2006), 1–3.

and archetypical system of things, the short book, though important today, remained mostly ignored by Quiccheberg's contemporaries. As the librarian to Duke Albrecht V of Bavaria, Quiccheberg was put in charge of the growing ducal collections in Munich—the core of which was donated by the powerful German banker Hans Jakob Fugger. For the Duke's librarian and collection curator, the sum total of objects accumulated in the Munich *Kunstwunderkammer* (as idiosyncratically termed by his brother Leo in the dedicatory preface to Emperor Maximilian II) ought to be deemed a *theatrum sapientiae*, a theater of wisdom of universal utility: "Indeed, I also judge that it cannot be expressed by any person's eloquence how much wisdom and utility in administering the state—as much in the civil and military spheres as in the ecclesiastical and cultural—can be gained from the examination and study of images and objects [*imaginum et rerum*] that we are prescribing."[26] These images and objects were actually nothing less than instruments:

> For there is no discipline under the sun, no field of study, no practice, that might not most properly seek its instruments [*instrumenta*] from these prescribed furnishings. Yet, in fact, it would require divine genius to assemble from all places and to order all these things that, once concisely and comprehensively collected, might be capable of instructing the mind of any reasonable refined person in countless matters. Nonetheless, we shall not refuse to at least give our advice about this elsewhere.[27]

A system of classification and organization based on a series of five interrelated classes, forming a linked cycle, ensured that the end product of the *Kunstwunderkammer* was a comprehensive and meaningful representation, godly-given theater of the world.[28]

26 Samuel Quiccheberg, *Inscriptions vel tituli theatri amplissimi* (Munich, 1565), Dij. The best study of this book, with an English translation, is from Mark A. Meadow and Bruce Robertson, *The First Museum Treatise on Museums: Samuel Quiccheberg's* Inscriptiones, *1565* (Los Angeles: The Getty Research Institute Publications Program, 2013), 74.

27 Ibid.

28 Ibid., especially pp. 23–5, where the authors schematize a few of the 'cycle' possibilities. The literature on *Kunstkammer* is vast. Useful here is Sven Dupré and Michael Korey, "Inside the *Kunstkammer*: The Circulation of Optical Knowledge and Instruments at the Dresden Court," *Studies in History and Philosophy of Science* 40 (2009): 405–20. The *Kunstkammer* in Dresden was founded roughly at the same time Quiccheberg published his treatise, i.e. 1560. It housed the richest collection of tools and scientific instruments in its day. The Munich cabinet held a total of 3407 objects according to an inventory made in 1598, only a fraction of which were mathematical instruments. (p. 407)

In the first section of the book, Quiccheberg briefly describes these five classes, each divided into ten or eleven subclasses called *inscriptiones*—inscriptions or titles. The latter provide a list of objects essential to building one's own universal cabinet. The fourth class comprised a series of various instrument *inscriptiones*, including *instrumenta musica* (flutes, organs, war trumpets, etc.), *instrumenta mathematica* (astrolabes, clocks, surveying instruments, etc.), *instrumenta & supelle scriptoria, pictoriaque* (vellum, paper, writing tablets, styluses, print type, etc.), *instrumenta violenta* (instruments for lifting weights and breaking gates, for pulling and dragging, etc.), *instrumenta officianarum, & tabernarum laboratoriarum* (used by the more skilled artisans [*artificiosioribus artificibus*] such as tools for sculptors, turners, goldsmiths, and indeed all artisans whom this world supports in our age), *instrumenta chirurgica, & anatomica* (scissors, lancets, bone-saws, syringes, ear probes, etc.), *instrumenta venatoria & quaecunque ruri, ac in hortorum cultura necessaria sunt* (whether specialized hunting instruments such as bird whistles, hooks, and tridents or other devices for planting, pruning, weeding, and cultivating the earth), and *instrumenta lusoria* (chess, javelin, dice, darts, ball games, etc.).[29] Quiccheberg's expansive notion of 'instrument' is not at all unusual in early modern Europe. Rather, it reflected, as we will see below, a general understanding of what an *instrumentum* was at the time and how it was used as a transformative agent.

Instrument, in English and French vernaculars, comes from the fourteenth-century *estrument* and before that from the Latin *instrumentum* (provision, apparatus, furniture, tool)—itself derived from *instruere*, i.e., to fit out, equip, or instruct. The primary definition of an instrument is, "That which is used by an agent in or for the performance of an action." It is "a thing with or through which something is done or effected."[30] An instrument, at its core, is thus "anything that serves or contributes to the accomplishment of a purpose or end." The 'thing' here mentioned is not only material, like a craftsman's hammer, a natural philosopher's air pump, or a musician's viola da gamba but also *immaterial* such as faith, 'Fides enim Organum est,' and geometry, 'il più potente strumento d'ogni altro per acuir l'ingegno,' according to Galileo,[31] a *person*—the instrument of God, or more specifically, as Pope Innocent IV was then known: *Doctor Pater et Organum Veritatis*—, an *organ* or body part—the instruments

29 Quiccheberg, *Inscriptions vel tituli theatri amplissimi*, B4r–C1r. Meadow and Robertson, *The First Museum Treatise on Museums*, 65–6.

30 In Johannes Micraelius, *Lexicon philosophicum* (Jena, 1653), *s.v. instrumentum: "Instrumentum* est, quod assumitur â causa principali, & in virtute ejus producit effectum."

31 Galileo Galilei, *Discorsi e dimostrazioni matematiche, intorno à due nuove scienze* (Leiden, 1638), 135.

of breathing—, or a *formal legal document* in law or an authenticated record of a transaction drawn up by a notary public.[32] It could also mean a religious or sacred text, such as Erasmus's *Novum instrumentum* (1516), an authoritative Latin translation of the Greek New Testament. (The Latin translation, alongside the Greek original, was not meant to replace the Latin Vulgate but to be used alongside it, as a kind of study Bible.[33] Samuel Hartlib, to take but one famous gentleman, offers in his papers and correspondence a good survey of the many meanings of 'instrument' described above in the first half of the seventeenth century.[34]) Etymologically speaking, *instrumentum* is derived from the Greek ὄργανον—organon—or *organum* in Latin. Though the meanings of these two words were alike, as one can ascertain from contemporary dictionaries—for instance, in Johannes Micraelius's *Lexicon philosophicum*, *organum* "est idem,

---

32 Quotes taken from the *Oxford English Dictionary Online* (2007), *s.v.* instrument (accessed via Harvard College Libraries, 7 July 2007). A common definition is found in *A New and Complete Dictionary of Arts and Sciences; Comprehending all Branches of Useful Knowledge* ... (London, 1763), vol. 2, p. 1789: "Instrument, in general, whatever is subservient to a cause in producing any effect. See CAUSE and EFFECT." The French definition for instrument is analogous, see *Trésor de la langue française informatisé* (accessed online, 9 May 2018). Robert Estienne's *Dictionarium latinogallicum* (1552) also offers a very similar definition when one looks under the headword *instrumentum*. For similar contemporary definitions, see the first edition of the *Dictionnaire de l'Académie française* (1694) and Antoine Furetière, *Dictionnaire universel* (1690). Micraelius gives similar examples in his *Lexicon philosophicum*, such as "estque *aliud* instrumentum supernaturale, ut verbum & Sacramenta in Ecc[l]esiâ, que sunt instrumenta gratiam offerentia; & fides, quae est instrumentum gratiam recipiens." (columns 549) On the body as "both an *object* of research and an *instrument* of experience" see the several interesting contributions in Charles T. Wolfe and Ofer Gal, eds., *The Body as Object and Instrument of Knowledge: Embodied Empiricism in Early Modern Science* (Dordrecht: Springer, 2010).

33 The second edition of Erasmus's text was published in 1519 and was renamed *Novum testamentum*, its conventional title. It appears that the title *Novum instrumentum* surprised many at the time. In a letter to Robert Aldrige in 1527, Erasmus explained that people who complained about his choice of term were overdramatizing, since he wrote it only twice in all of his writings. Yet the choice of *instrumentum* was certainly not wrong, as it was used by Jerome and Augustine themselves when they mentioned in *Contra duas epistolas Pelagionorum* that it was better to say *Instrumentun* than *Testamentum*. Erasmus probably knew he would draw some outrage from this title. One could explain the choice of *instrumentum* by Erasmus's desire to renew with the *prisca theologia* and also to signify the philological intention of the work. Erasmus, *Les Préfaces au* Novum testamentum (*1516*), ed. Yves Delègue, with the collaboration of J. P. Gillet (Geneva: Labor et Fides, 1990), 21. Henk Jan de Jonge, "Erasmus's Translation of the New Testament: Aim and Method," *The Bible Translator* 67 (2016): 29–41. On the role of such hermeneutics for the interpretation of nature, Stephen Gaukroger, *The Emergence of a Scientific Culture: Science and the Shaping of Modernity, 1210–1685* (Oxford: Oxford University Press, 2006), 139–48.

34 See the Hartlib Papers, online here: https://www.dhi.ac.uk/hartlib/context (accessed 21 July 2022).

quod instrumentum de quo vide supra" (column 772)[35]—distinctions existed.[36] *Organum*, however, is more interesting as a word concept than *instrumentum* owing to its clear association with Aristotle's *Organon*—or logic—and especially with Francis Bacon's criticism of it, the *Novum Organum* (1620), as we will see shortly.

Organum was sometimes endowed with what might be called a more universal overtone, lacking the affordance of the word *instrumentum*. A book such as *Instrumentum pacis Constantiæ* (Venice, 1497) is about a specific event, the 1183 peace treaty of Constance. Leonhard Zubler's *Novum instrumentum geometricum* (Basel, 1607) applies to a specific instrument, i.e., a triangulation instrument used for military purposes (range-finding) and surveying. Even Erasmus's *Novum instrumentum*, as mentioned above, deals with a specific sacred text, the New Testament. *Organum*, on the other hand, can mean all styles of vocal polyphonic music. Michael Praetorius's *De organographia* (Wolfenbüttel, 1618), book two of his *Syntagma musicum*, describes all known musical instruments. Similarly, Johannes Hevelius's *Organographiam, sive,*

35 Jean Nicot's *Thresor de la langue françoyse* (1606) puts under the headword 'instrument' the Latin *instrumentum* and *organum*. The latter two words are indeed closely related to one another. Actually, 'instrument' in Nicot's *Thresor de la langue françoyse* and 'organum' in Estienne's *Dictionarium latinogallicum* (1552) both have the exact same definition: "Instrument à faire quelque chose que ce soit." The very same definition is found in a French children's dictionary, *Les Mots francoys selon l'ordre des lettres, ainsi qu'il les faut escrire, tournez en latin pour les enfans* (Lyon, 1567). The 1600 edition has the same definition. The Italian *Vocabolario degli Accademici della Crusca* (1612) points toward the same definition for *strumento*: "Quello, col quale, o per mezzo del quale, noi operiamo." The *Lexicon philosophicum* d'Étienne Chauvin provides a similar take: "ORGANUM generaliter significat rem omnem actioni certae & usui vel opeationi dicatam." *Lexicon philosophicum*, 2nd ed. (Leeuwarden, 1713), 463. Hübners, *Curieuses und Eales Natur- Kunst- Berg-Gewerck- und Handlungs-Lexicon …*, defines it thusly: "Instrumentum, Organon, ein Werkzeug, vermittelst welches die menschliche hand ihre vorgesesste Operation verrichten muss." (column 1051)

36 See the following very useful online resource: Ludwig Schütz, *Thomas-Lexikon: Sammlung, Übersetzung und Erklärung der in sämtlichen Werken des hl. Thomas von Aquin vorkommenden Kunstausdrücke und wissenschaftlichen Aussprüche*. Available at: http://www.corpusthomisticum.org/tl.html (accessed 9 May 2018). For instance, in circa 400, Augustine favored the term *organum* to specifically designate the pipe organ (its true Latin meaning according to him), whereas *instrumentum* was to be used for all other musical instruments. In music, especially in the medieval period, *organum* had several meanings besides a musical instrument and an organ: the simple and elaborate (polyphonic) vocal ensembles (to sing *in hymnis et organis*), the canonical liturgy, its chant and even the book containing the authorized texts. On all these different possibilities, see Peter Williams, "The Meaning of *Organum:* Some Case Studies," *Plainsong and Medieval Music* 10 (2001): 103–20. See also the very complete definition of organ in the *Oxford English Dictionary Online*.

Instrumentorum astronomicorum omnium ... accuratam delineationem ... exhibens (Gdansk, 1673) is meant as a description and representation of all astronomical instruments. And Oronce Fine's *organum universale*, part of his *De mundi sphaera* (Paris, 1542), was intended as much for geometry as for all the canons of astronomy (*tum omnes astronomici canones*), endowed with a remarkable facility of use (*mira facilitate practicantur*).[37] That said, more often than not, *organum* and *instrumentum* were used interchangeably in the core of a text, such as in Tycho Brahe's *Astronomiae instauratae mechanica* (1598) and Peter Apianus's *Cosmographia, sive descriptio universi orbis* (1584), to take two known examples.

A nice illustration of *organum*'s universal use is Athanasius Kircher's *Organum mathematicum*, a 'mathematical instrument' or box—also called *cista*—within which the basic principles of arithmetic, geometry, fortification, chronology, horography, astronomy, astrology, steganography (encryption), and music were explained.[38] (Fig. 2) Like the other mathematical disciplines embodied in the *organum*, these wooden rods were meant to assist its owner in making calculations without the exhaustion of one's head (*capitis defatigatione*). In a letter acknowledging the reception of Kircher's *organum mathematicum*, the young Austrian Archduke Joseph Karl wrote how pleased he was to receive a gift that would help him better understand the rudiments of the mathematical disciplines, which were truly worthy of a king: "my *organum* pleases me more and more."[39] There was thus the idea of universality behind some uses of the word *organum*, such that Aristotle's *Organon*, as dubbed by

37 On Fine, Alexander Marr, ed., *The Worlds of Oronce Fine: Mathematics, Instruments and Print in Renaissance France* (Donington: Shaun Tyas, 2009).

38 Gaspar Schott wrote the description of Kircher's *organum mathematicum*. Schott, *Organum mathematicum libris IX explicatum* (Herbipoli [Würzburg], 1668). For a general description of this instrument and those similar to it, Mara Miniati, "Les *cistæ mathematicæ* et l'organisation des connaissances au XVIIe siècle," in *Studies in the History of Scientific Instruments*, eds. Christine Blondel *et al.* (London: Rogers Turner Books Ltd, 1989), 43–51.

39 Archduke Joseph Karl to Kircher, 31 December 1661, in which he wrote: "Placet mihi sedulitas, ut magna cum facilitate capiam disciplinas Mathematicas, quas iure merito regias ac principe dignas appellasti. Futurium vero est, ut mihi magis magisque placeat organum." See also the draft of the letter Kircher sent to Karl Joseph on 7 July 1661, in which he described his instrument: "Mitto per manus P. Pizzoni Serenissimae Vestrae Celsitudini, Instrumentum quoddam quod Organum mathematicum appello; in quo summa facilitate traduntur illae Artes mathematicae, ad quas Serentitatem vestram inclinari audio, Regibus et Principibus uti par est, nullo non tempore summo in honore habitae?" Both letters are found in the Archivio della Pontificia università Gregoriana, Rome, APUG 555, f.075r and f.098r respectively. The letters were extracted from the Athanasius

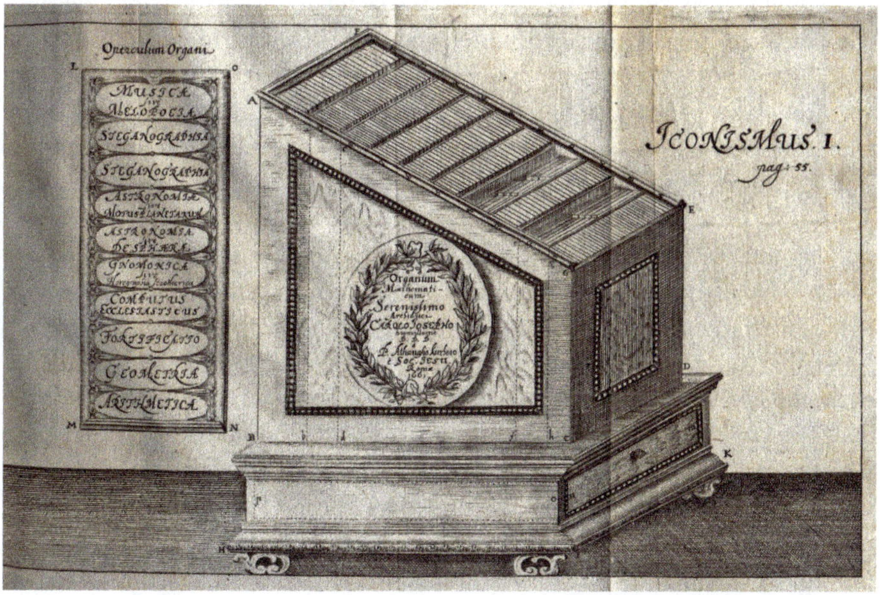

FIGURE 2 Kircher's *organum mathematicum* as depicted in Gaspard Schott's *Organum mathematicum libris IX explicatum* (1668).
COURTESY OF WIKIMEDIA COMMONS: HTTPS://COMMONS.WIKIMEDIA.ORG/WIKI/FILE:ORGANUMMATHEMATICUM.JPG

his Peripatetic followers, was to be the sum of all knowledge about logic—and a tool for all inquiry. Not only could it stand for a material device or an abstract notion, *organum* often connoted universal rather than specific knowledge about an instrument (in all the inferences of the word) or a theoretical conception. Aristotle's teaching is useful in understanding how an instrument came to signify so many things in early modern Europe.

In describing the reason we become more intelligent as we become older, Aristotle explained that "God has given us two instruments within ourselves: the *hand* for the body and *intelligence* for the soul." In the same way we perfect the use of our hands with time, intelligence attains its highest perfection as we get older. That said, Aristotle argued, there exists a natural time frame in acquiring the proper learning habits, "for it is by using the instruments that we acquire a disposition":

Kircher Correspondence Project established by Michael John Gorman and Nick Wilding http://193.206.220.68/kircher/index.html#0 (accessed 8 October 2022).

> Intelligence comes to us later than the faculty of using the hands, because the instruments used by the intelligence are posterior to those used by the hands. For science is an instrument of the intelligence (for it is useful to the intelligence just as flutes are useful to the flute player), and many things in nature are instruments of the hands, but nature itself and its creations are prior to science.[40]

The hammer is the instrument of the hand, which in turn is the instrument of the body; likewise, science is the instrument of intelligence, which in turn is the instrument of the soul. To Aristotle, therefore, an *organon* was something that mediated between the doer and the deed.[41] In his *Physics*, for example, he explained that movement rested on three things: the moved (the deed), the mover (the doer), and the instrument of motion (what mediated between the two). The moved was in motion; the mover caused motion yet was unmoved; and, finally, the instrument of motion had to move something else and be itself in motion—for, Aristotle said, "it changes together with the moved, with which it is in contact and continuous, as is clear in the case of things that move other things locally, in which case the two things must up to a certain point be in contact."[42] The element fire is an interesting case. It has many kinds of motion and one power—the power to cleave: it can condense, rarify, dissolve and harden, "for different results are achieved by fire in the work of the goldsmith, the coppersmith, the carpenter, and the cook." But, as Aristotle explained, "it is truer to say that the arts themselves achieve these different results, for that by using fire as an instrument they soften, liquefy, and desiccate substances, and some they temper."[43] As an instrument, however, fire is not a formal cause. Aristotle's predecessors, he emphasized, invested "the forces they assign to the simple bodies—the forces which enable these bodies to bring things into being—with too instrumental a character." They wrongly treat the saw (and other instruments of carpentry) "as the cause of the things

40 Aristotle, *The Complete Works of Aristotle*, ed. Jonathan Barnes, 2 vols., rev. ed. Oxford translation (Princeton: Princeton University Press, 1984), *Problems**, book 30, §955b22–956a10, vol. 2, 1503. Taken from the online resource *Past Masters* via Harvard University (accessed 9 May 2018).

41 It is important to note that the body itself is also an *organon*: "Now exactly in the same way the body, since it is an instrument—for both the body as a whole and its several parts individually are for the sake of something—if it is to do its work, must of necessity be of such and such a character, and made of such and such materials." Ibid., *Parts of Animals*, book 1, 642a1–642a13, vol. 1, 999.

42 Ibid., *Physics*, book 8, §256b13–256b27, vol. 1, 428.

43 Ibid., *On Breath***, §485a28–485b2, vol. 1, 772. A few paragraphs later, Aristotle mentions that "But while the crafts use the fire merely as an instrument, nature uses it as a material as well."

that come-to-be." Fire is no different: "however true it may be that fire is active, i.e., sets things moving" it is crucial to understand that fire is also "inferior to the tools or instruments in the manner in which it sets things moving."[44] In other words, the element fire is simply the instrument (artificial or natural) of another instrumental, formal cause.

In the early modern period, the Dutch chemist and physician Herman Boerhaave developed in his *Elementa Chemiae* a theory of chemistry revolving around six such unusual 'instruments':

> And thus in Chemistry we have Instruments by which we are able to perform the actions we desire; and these we usually reduce to six, as the principal, with those Chemists who have cultivated the Art with the greatest accuracy; and they are as follow. Fire, Water, Air, Earth, the Solvents which the Artists called *Menstruums*, and the common furniture of the Elaboratory.[45]

The core concept behind these 'Instruments of Chemistry'—*de Instrumentis Chemicorum*—does not come from Boerhaave himself. Carefully analyzed by John C. Powers, Boerhaave's principles originated primarily from the Leipzig medical professor Johannes Bohn, whose *Dissertationes chymico-physicae* (1685) became a model for the Dutch chemist's lectures at Leiden.[46] Even though fire, of all these instruments, was recognized as the most important and powerful one,[47] they were similarly defined:

44 Ibid., *On Generation and Corruption*, book 2, §336a13–336a23, vol. 1, 550.

45 Herman Boerhaave, *Elements of Chemistry* ..., 2 vols. (London, 1735), 78. Nicolas Poisson, in his commentary of Descartes's treatise on mechanics, talks about 'machines naturelles' regarding the basic elements of nature. Poisson, *Traité de la mécanique composé par Monsieur Descartes* ... (Paris, 1668), 19.

46 The best studies on this topic are from John C. Powers, "Chemistry Without Principles: Herman Boerhaave on Instruments and Elements," in *New Narrative in Eighteenth-Century Chemistry*, ed. Lawrence M. Principe (Dordrecht: Springer, 2007), 45–61. Powers, "Fire Analysis in the Eighteenth Century: Herman Boerhaave and Scepticism About the Elements," *Ambix* 61 (2014): 385–406. See his book as well, Powers, *Inventing Chemistry: Herman Boerhaave and the Reform of the Chemical Arts* (Chicago: The University of Chicago Press, 2012). On chemical instruments in general, Frederic L. Holmes and Trevor H. Levere, eds., *Instruments and Experimentation in the History of Chemistry* (Cambridge, MA: MIT Press, 2000).

47 The "wisest nations of old reverenced and worshipped it, as a supreme being" according to the Dutch chemist; some of the most famous chemists "have acknowledged it as the source of all their knowledge, and hence have professed themselves Philosophers of Fire." Boerhaave, *Elements of Chemistry*, 78. For Bohn, fire is "tanquam potissimi & universalis magis instrumenti chymici" and elsewhere: "Ignem instrumenti chymici titulum mereri, nullus unquam dubitavit, ut vel ex recensitorum Scriptorum collatione videre est, ejusque

> For in every Art that instructs us in the changing of Bodies, that particular thing is called an Instrument, which either has or may have given it a certain peculiar motion, which being applied to the Body to be changed will produce such an alteration in it, as the Art had before determined.[48]

Boerhaave did not consider these basic natural instruments as simple, or as elements in the traditional Aristotelian sense—they could, for instance, be combined by chemical operations with other substances. However, as Rosaleen Love explains, Boerhaave's instrument was a tool that could apply a 'peculiar motion' (*definitus motus imprimi potest*) and "that itself remained unchanged in the processes it initiated." In the case of combustion, the instrument fire was "the source of change while itself remaining unchanged, while remaining indeed beyond the possibility of change."[49] Boerhaave's 'instrument' was epistemologically akin to Aristotle's 'instrument of motion' species, which caused motion *and* was itself in motion, but remained unchanged.

In *Generation of animals*, in trying this time to explain "which part comes into being after which," Aristotle complicated matters further by classifying instruments into two separate classes: "that which can generate, and that which serves as an instrument to what is generated." Both, however, *existed for the sake of the end.*[50] Whether it was material (hammer, saw, lancet), immaterial (fire, breath) or living (bodily organ) an *organum*'s raison d'être was about the action leading toward a specific goal. Hence, "As every instrument and every bodily member is for the sake of something, viz. some action, so the whole body must evidently be for the sake of some complex action. Thus, the saw is made for sawing, for sawing is a function, and not sawing for the saw. Similarly, the body too must somehow or other be made for the soul, and each part of it for some subordinate function, to which it is adapted."[51] This follows directly from Plato, Aristotle's teacher, who affirmed that "actions are real as well as the

usus in operationibus ferè cunctis quàm amplissimus observatur". Bohn, *Dissertationes chymico-physicae ...* (Leipzig, 1696), 17 and 66.

48 Boerhaave, *Elements of Chemistry*, 78. This notion was perhaps not so unusual. According to a certain W. Simpson, "And as *Trismegistus* in *Asclepio*, saith, *Aer est organum vel machina omnium, per quam om|nia fiunt*, not as an Element, but as a Machine for the motion of all Bodies." Simpson, *Philosophical Dialogues Concerning the Principles of Natural Bodies Wherein the Principles of the Old and New Philosophy are Stated, and the New Demonstrated More Agreeable to Reason, from Mechanical Experiments and its Usefulness to the Benefit of Man-kind* (London, 1677), 109.

49 Rosaleen Love, "Herman Boerhaave and the Element-Instrument Concept of Fire," *Annals of Science* 31 (1974): 547–59, quotes on pp. 553 and 555.

50 Aristotle, *Generation of Animals*, book 2, §742a16–742b17, vol. 1, 1151.

51 Ibid., *Parts of Animals*, book 2, §645b15–645b20, vol. 1, 1005.

things." In the voice of Socrates, Plato explained that "the actions also are done according to their proper nature, and not according to our opinion of them. When, for example, we set out to cut something, can we do so in any way we please, and with any chance instrument? I think rather that if we cut with the proper instrument only, and according to the natural process of cutting, then we shall be successful in our cutting and perform this action rightly; but if we go against nature we shall fail and achieve nothing." Plato suggested the existence of a natural instrument for every natural action—all actions possessing distinct natures. Thus, in performing the action of burning something, "not every way is the right way; but the right way is the natural way, and the right instrument the natural instrument."[52] Following Plato, Aristotle emphasized in his physics, physiology, and metaphysics the effectiveness of instruments, not their materiality or lifelessness.[53] For this reason Aristotle did not see a difference between the human body and any other natural or artificial instrument. Epistemically (not ontologically) speaking, *anything* generating an action that led toward an end was an *organum*.[54] Aristotelian logic was one such abstract instrument.

According to most early modern savants and scholars, logic was *stricto sensu* an art—an instrument designed to think accurately, naturally.[55] Since Antiquity, the notion of *ars* was understood as the knowledge of doing things. Things here should not only be grasped as material objects, but also as

52 Plato, *The Dialogues of Plato*, trans. and ed. by Benjamin Jowett, 4th ed. (Oxford: Oxford University Press, 1953), vol. 3, *Cratylus*, §387a–387b, 45. Taken from the online resource *Past Masters* via Harvard University (accessed 10 May 2018).

53 In his *Politics*, Aristotle described slaves and servants as property, which were also 'instruments' for maintaining life (Aristotle, *The Complete Works of Aristotle*, *Politics*, book 1, §1253b24–1254a17, vol. 2, 1989).

54 Conceptually, whether the human hand (as an instrument) was used as a mnemonic device, such as the so-called Guidonian hand to teach the theory of music, or as a mathematical instrument, such as a nocturnal to tell the time at night as depicted on the title page of Peter Apianus's *Instrument Buch*, it was ontologically different from Ambroise Paré's artificial and mechanical hand. See, for instance, Ibid., *Parts of Animals*, book 1, §642a1–642a13, vol. 1, 999. On how Leibniz understood this problem, Smith, *Divine Machine*, esp. 106ff.

55 See, for instance, Leibniz: "Ut ita tam magnificae inscriptionis excusationem nullam inveniam, nisi hanc unam: Logicam veram non tantum instrumentum esse, sed et quodammodo principia ac veram philosophandi rationem continere, quia generales illas regulas tradit, ex quibus vera falsaque dijudicari, adhibitisque solis definitionibus et experimentis omnes conclusiones demonstrari possunt." Leibniz, *Die philosophischen Schriften von Gottfried Wilhelm Leibniz*, ed. by C. J. Gerhardt, 7 vols. (Berlin: Weidmannsche Buchhandlung, 1875), Vol. 4, 137. Taken from the online resource *Past Masters* via Harvard University (accessed 8 May 2018). Chauvin has a similar definition in his *Lexicon Philosophicum*, 363.

productions of the mind (*ars constat cognitione & effectione*). Art, more precisely, was acknowledged as a poietical or mechanical habitus, an organized and methodical disposition of body and mind toward the production of things and knowledge. Whether an art was defined as perfect or mediocre, it always entailed a system of principles (or actions) leading to an end. For our purpose, it is the second definition of *ars* in Rudolph Goclenius's *Lexicon philosophicum* (1613) that is especially interesting. Related to the idea of an end, logic (as a form of art) was said to be a *habitus organicus*, namely an instrumental habit that uses the notion of *organum* and disposition of the mind to achieve a predetermined goal.[56] It was against this Logic, this end-producing and purely rational instrumental habit at the foundation of medieval and Renaissance knowledge claims, this "moral instrument by which the mind is directed in the knowledge of things,"[57] that Bacon offered his *Novum Organum* (1620).

In his dedication to the English king, Lord Verulam pleaded for regeneration, a renewal of the sciences that was long overdue. He compared James I to Solomon, and entrusted his sovereign to "emulate that same king in another way, by taking steps to ensure that a Natural and Experimental History be built up and completed … So that at last, after so many ages of the world, philosophy and the sciences may no longer float in the air, but rest upon the solid foundations of every kind of experience properly considered." To achieve this goal, Bacon proclaimed he had "supplied the Instrument (*equidem organum probui*)."[58] The latter was perhaps not unlike the ones producing 'matters of fact' for a court of law, to which Bacon was familiar considering his training and profession.[59] Whether or not it was the case, Bacon likened his 'instrument' rather to a nautical compass of the mind, which could be used to "sail to the more remote and secret places of nature."[60] He made full use of the analogy to the material culture of tools and machines stemming from the mechanical arts to describe how best to renew the use of the human intellect. Aphorism two is famous in describing how one should think about this *novum organum*:

56 R. Goclenius, *Lexicon philosophicum* (New York: Georg Olms Verlag, 1980 [1613]), *s.v. ars*. We will discuss this further in chapter 3.

57 Chauvin, *Lexicon philosophicum*, 363. Micraelius associates *logica* to Aristotle and to art: "*Logica* est Aristoteli modus seu methodus sciendi talis nemper ars" *Lexicon philosophicum*, column 602. For Galileo, geometry was a better *strumento* than logic. Galileo, *Discorsi*, 135.

58 Francis Bacon, *The New Organon*, ed. Lisa Jardine and Michael Silverthorne (Cambridge: Cambridge University Press, 2000 [1620]), 4–5.

59 On this familiar topic, see Barbara Shapiro, *A Culture of Fact: England, 1550–1720* (Cornell: Cornell University Press, 2000). The role of witnessing is now familiar to readers owing to Steven Shapin's body of work.

60 Bacon, *The New Organon*, 10–11.

> Neither the bare hand nor the unaided intellect has much power; the work is done by tools and assistance [*instrumentis et auxiliis res*], and the intellect needs them as much as the hand. As the hand's tools [*instrumenta manus*] either prompt or guide its motions, so the mind's tools [*instrumenta mentis*] either prompt or warn the intellect.[61]

In the preface, Bacon even went so far as to say that "from the very start the mind should not be left to itself, but be constantly controlled; and the business done (if I may put it this way) by machines."[62] There is a constant reminder throughout the work of the importance of method and order, something evidently found in well-organized machines and skilled, tool-using craftsmen. To reform *scientia*, according to Bacon's rhetoric, one ought to step into a workshop rather than a library to discover heuristic practices, thus helping to secure a new instrument of knowledge.[63]

Bacon wanted his reader to understand how much the mechanical arts had and would continue to contribute to human civilization, and that these were based on a few axioms and "the patience and the subtle, ordered movement of hand and tool."[64] To drive the point home, Bacon also compared his *organum* to the simplest of mathematical instruments: "In drawing a straight line or a perfect circle, a good deal depends on the steadiness and practice of the hand, but little or nothing if a ruler or a compass is used. Our method is exactly the

61 Ibid., 33.

62 Ibid., 28. Bacon wanted to compile a whole "History mechanique" that would serve to synthesize, convert, and translate craft knowledge into useful natural philosophical knowledge. As he wrote in his private notebook, *Commentarius solutus*: "To procure an History mechanique to be compiled with care and diligence and to professe it that is of the experimts and observations of all Mechanicall Arts. The places or thinges to be inquyred are; first the materialls, and their quantities and proportions; Next the Instrumt and Engins requesite; then the use and adoperation of every Instrumt; then the woork it self and all the processe thereof wth the tymes and seasons of doing every part thereof. Then the Errors wch may be comytted, and agayn those things wch conduce to make the woorke in more perfection. Then all observacions, Axiomes, directions. Lastly all things collaterall incidt or intervenient." Quoted in Mark Thomas Young, "Manual Labor and 'Mean Mechaniks': Bacon's Mechanical History and the Deprecation of Craft Skills in Early Modern Science," *Perspectives on Science* 25 (2017): 521–50, on p. 529.

63 Bacon, *The New Organon*, 70. Mark Young argues that Bacon's and his successors' rhetoric concerning artisans and craftsmanship was only that: a way for the former to appropriate the hard-won work from the latter. Natural philosophers, in the end, did not wish to legitimize artisanal practice, but rather aimed at demonstrating its inferiority to their own intellectual work. The natural philosophical rhetoric did not serve to break the barrier between artisans and savants, but rather to reinforce it. Young, "Manual Labor and 'Mean Mechaniks'."

64 Bacon, *The New Organon*, 69.

same."[65] Bacon's *organum* was meant as a new logic of practice and discovery, a new instrumental habit (i.e., logic, defined as a *habitus organicus*) elaborated to reform the old peripatetic one based on purely deductive and syllogistic arguments.[66] Most importantly, this *new organon* was not created for the sake of generating predetermined—logical—end results, but was rather open-ended. The programmatic feature of Bacon's orderly method was strictly found in how individuals ought to train and use this instrument; it was not necessarily about the results it would generate. Acquiring Bacon's method—or Descartes's, for that matter, as we will see shortly—i.e., training oneself to operate this new instrument, was the key to knowledge claims and useful discoveries. It was no different than someone trying to master the use of a telescope or an air pump to uncover nature's secrets. Material or not, an *organum* did not work on its own. It needed a user who had to partake in an intensive and ever more specialized training to become proficient with the instrument. Embedded in every rational or material *organum*, therefore, was a set of practices—habits—that could produce, when correctly carried out, natural philosophical knowledge.[67]

René Descartes's *méthode* is not epistemically speaking much different than Bacon's *novum organum*—though, of course, the former is deductive whereas the latter is inductive. In a comment made in 1648 on the motion of individual particles with regard to the overall structure of celestial matter, Descartes declared to the young Dutchman Frans Burman that although "the entire system [of the universe] is in a state of equilibrium,"

---

65 Ibid., 50.

66 Peter Pesic demonstrates how Bacon was, in a very specific context, greatly influenced by Aristotelian physics, from which he drew his concept of the 'violence of impediments.' Pesic notes that "Bacon used the received texts of Aristotle and ancient myths as points of departure for his 'new philosophy,' which relied on those older sources even as it departed from them." Pesic, "Francis Bacon, Violence, and the Motion of Liberty: The Aristotelian Background," *Journal of the History of Ideas* 75 (2014): 69–90, quote on p. 70.

67 Though original and "aimed at providing detailed recipes suited for operational use," as Peter Dear puts it, Bacon's programmatic instrument did not "rock the boat of established understandings of natural philosophy." Dear, "What is the History of Science the History *of*? Early Modern Roots of the Ideology of Modern Science," *Isis* 96 (2005): 390–406, on p. 396. And, judging from Deborah Harkness's study of Elizabethan London, Bacon "disingenuously overlooked the instrumental mania that held London in its grip and ignored the instrument makers who were his neighbors in the parish of St. Dunstan in the West." Harkness, *The Jewel House: Elizabethan London and the Scientific Revolution* (New Haven: Yale University Press, 2007), 246. For a counterexample regarding Bacon's actual involvement with craftsmen, Cesare Pastorino, "The Mine and the Furnace: Francis Bacon, Thomas Russell and Early Stuart Mining Culture," *Early Science and Medicine* 14 (2009): 630–60.

> [it] is a very difficult thing to conceive of, because it is a mathematical and mechanical truth. We are not sufficiently accustomed to thinking of machines, and this has been the source of nearly all errors in philosophy.[68]

By then, and for close to two decades, Descartes had abolished artisans as the epistemic emblem of order and replaced them with the more visual and tactile mechanical order of machines.[69] This 'mechanical turn' was so compelling that, as Graham Burnett tentatively illustrates, a closer connection between the mechanically produced hyperbolic lenses of the *Dioptrique* and the mind's eyes' 'metaphysical lens'—or hyperbolic doubt—found in the *Meditations* could be entertained without too much of a stretch.[70] The machine not only supplanted the artisan; it ultimately incarnated Cartesian metaphysics and natural philosophy. The concept of machine became a thing knowledge.[71]

In *Sidereus nuncius*, Galileo did not mention the telescope by name, referring instead to "the instrument [*organum*] with the benefit of which [great things of nature] make themselves manifest to our sight."[72] It is not unusual here for Galileo to speak of a mechanical instrument as an *organum*—a thing that serves a specific purpose—following the traditional Aristotelian definition of the word (see above). Where Descartes distanced himself from Galileo and other natural philosophers, however, was in the explicit epistemic and

68 John Cottingham, ed., *Descartes' Conversation with Burman* (Oxford: Clarendon Press, 1976), 44 §73.

69 On this assertion, Jean-François Gauvin, "Artisans, Machines, and Descartes's *Organon*," *History of Science* 44 (2006): 187–216. These paragraphs on Descartes are based on this article.

70 D. Graham Burnett, *Descartes and the Hyperbolic Quest: Lens Making Machines and Their Significance in the Seventeenth Century* (Philadelphia: Transactions of the American Philosophical Society, new series, vol. 95, No. 3, 2005), 125–32.

71 Davis Baird, *Thing Knowledge: A Philosophy of Scientific Instruments* (Berkeley: University of California Press, 2004).

72 Galileo Galilei, *Sidereus nuncius*, in *Le opere di Galileo Galilei*, ed. Antonio Favaro, 20 vols (Florence: Tip. di G. Barbèra, 1890–1909), vol. 3, 59. For the English translation, Galileo, *Sidereus nuncius or The Sidereal Messenger*, trans. Albert van Helden (Chicago: The University of Chicago Press, 1989), 35. Before the word *telescopium* became the preferred one to identify such a device, several vernacular terms were simultaneously employed, such as *fistula dioptrica* and *perspicillum* in Latin, *lunettes* in French, *instrument om verre te sein* in Dutch, *perspective cylinder* and *perspective trunckes* in English, and *cannone dalla vista lunga* and *occhiale grande* or *di canna* in Italian. Massimo Bucciantini, Michele Camerota, and Franco Giudice, *Galileo's Telescope: A European Story*, trans. Catherine Bolton (Cambridge, MA: Harvard University Press, 2015), 10. On the naming of the telescope, Edward Rosen, *The Naming of the Telescope* (New York: Henry Schuman, 1947).

ontological connections he made between an artificial and a natural *organon*; between a mechanical instrument and a bodily organ.[73] In his *Dioptrique*, the correlation between mechanization and *organon* was firmly established. In this essay Descartes compared the same telescope to an 'organe exterieur,' an *organon* or instrument that could be put over another internal organ, namely the human eye.[74] For Descartes it was simply a matter of joining together two mechanical organs, the telescope and the human eye—the latter, as claimed by Kepler already, being nothing more than another machine, a *camera obscura*.[75] More mechanization here meant a sure path toward perfection, in this case the perfection of vision.[76]

Because it was simply unbearable to think that an invention *si vtile & si admirable* as the telescope was solely due to naked cunning and hands-on experience, Descartes's *Dioptrique*, which he began writing in the early 1630s, became the rational response to the artisan's lack of method.[77] Although the *Dioptrique* may look like a manual of technical instruction aimed at producing telescopes, it should be considered above all as a work edifying the method.[78] The essay was not intended for a hypothetical überartisan—as Bruce Eastwood argues—but was rather meant for artisans *tout court*, to those *âmes déréglées* populating the mechanical arts.[79] The *Dioptrique* was written in such a way that it established precisely how the mechanical arts should be carried out;

---

73 A very good analysis of *organon* qua instrument is given by Bates, "*Machina ex Deo.*" See also Dennis Des Chene, *Spirits and Clocks: Machine and Organism in Descartes* (Ithaca: Cornell University Press, 2001).

74 "Si bien qu'il ne nous reste a considerer que les organes exterieurs, entre lesquels ie comptens toutes les parties transparentes de l'œil, aussy bien que tous les autres cors qu'on peut mettre entre luy & l'obiet." Descartes, *Œuvres de Descartes*, eds. Charles Adam and Paul Tannery, new edn., 11 vols (Paris: Vrin, 1996), vol. 6, 148. [henceforth cited as AT vi, 148].

75 Descartes, *Dioptrique*, AT vi, 114–7.

76 For a detailed analysis of this viewpoint, Ofer Gal and Raz Chen-Morris, "Empiricism Without the Senses: How the Instrument Replaced the Eye," in *The Body as Object and Instrument of Knowledge: Embodied Empiricism in Early Modern Science*, eds. Charles T. Wolfe and Ofer Gal (Dordrecht: Springer, 2010), 121–47. Leonardo da Vinci already considered the human body the most perfect of instruments: "It does not seem to me that coarse men with lewd habits and little reasoning power deserve such a beautiful instrument [the human body] or such a variety of mechanisms." Quoted in Galluzzi, *The Italian Renaissance of Machines*, 118.

77 Descartes to Golius, [January 1632], AT i, 234–5 in which Descartes mentions he will send the first part of his *Dioptrique* that deals with refraction, without the philosophy.

78 Philippe Hamou, *La Mutation du visible. Essai sur la portée épistémologique des instruments d'optique au XVIIe siècle*, 2 vols. (Villeneuve D'Ascq (Nord): Presses universitaires du septentrion, 1999), vol. 1, 239–88.

79 Bruce Stansfield Eastwood, "Descartes on Refraction: Scientific Versus Rhetorical Method," *Isis* 75 (1984): 481–502.

it showed that building complex machines should no longer be a matter of one's cunning or experience. The *Dioptrique* was meant to inculcate into the practitioners of the mechanical arts Descartes's new and universal logic—his *organon*, or *méthode*. An orderly soul structured around the Cartesian method required above all an organized body, one that acted as a material instrument, a suitable *organon*. Through methodical order, Descartes sought to incorporate a new (absolute) way of knowing into early modern bodies. His *organon* as I would call it—his method or knowledge-producing instrument—generated a logic of practice in both natural philosophy and the mechanical arts, ordering the mental and manual skills of philosophers and artisans together toward the act of creating ideas and machines. The method as exhibited in the *Dioptrique* was the key in bringing forth a rational foundation for the mechanical arts, and therefore in building exact instruments and machines. Owing to the method, instruments (or 'bodily' parts) could be designed and built in such a way as to enhance natural abilities, skills, and *habitus* of living organs; instruments, *organa*, were mechanically upgrading the human body to a higher degree of perfection.[80] The method had a foot in both the realm of rational philosophy and the machine world. It was neither pure thought nor *vile méchanique* alone: it was both at the same time.

Descartes desired to reorganize thinking and hands-on practices, and consequently to oversee how natural philosophers created ideas and artisans manipulated matter. He wanted to ensure, more exactly to control, the acquisition of knowledge. Yet the method was not only beneficial to natural philosophy. Matthew Jones has convincingly shown that Descartes's *Géométrie*—another essay that epitomizes the Cartesian method—was a sophisticated system of 'spiritual exercises' aimed at cultivating one's self, at finding a better—and orderly—way of life through the practice of higher mathematics (to examine oneself geometrically Mersenne would say).[81] As a universal generator of orderly souls, the method thus suggested one fundamental sociocultural outcome: based on the inherent *bon sens* of mankind, Descartes's method became a constraining yet multifaceted pedagogical instrument, a pedagogical tool or *organon* that could shape any individual into a *honnête homme*.[82] According to

80 Neil M. Ribe pointedly argues that nature in the end "is not a source of standards but is itself subject to the higher standard of Cartesian rationality". Ribe, "Cartesian Optics and the Mastery of Nature," *Isis* 88 (1997): 42–61, quote on p. 53.

81 Matthew L. Jones, "Descartes's Geometry as Spiritual Exercise," *Critical Inquiry* 28 (2001): 40–71. Marin Mersenne, *Qvestions inovyes, ov recreation des sçavans* (Paris, 1634; facsimile Stuttgart, 1972), question xii, 45–6.

82 Descartes's universal *bon sens* is the very first assertion he makes in the *Discours de la méthode*. On Descartes's philosophy of education in general, Daniel Garber, "Descartes,

Peter Dear, Descartes used the notion of mechanization to establish the criteria of intelligibility in natural philosophy and from the latter, to observe how it constrained bodily behaviors. Domineering one's own passions came down to exerting control over one's own body. Descartes wanted, in short, to understand in mechanical terms the movement involved in the civilizing process of absolutism.[83] In this context, the man-machine concept was an attempt at embodying the Cartesian method into the emergent early modern social body.

Consonant with Francis Bacon's *novum organum*, Descartes's *organon* can ultimately be understood as an instrument that "more or less equalises intellects, and leaves little opportunity for superiority, since it achieves everything by most certain rules and forms of proof."[84] By fully ordering and constraining knowledge-making processes (both body and mind) anyone, embracing the right method, could become a lord and master of Nature. Descartes, writing in French and thus reaching to a wide-ranging audience, made sure no one would be left out of his natural philosophical civilizing process.[85] Descartes's most famous opus was, among many other things, a universal book of *civilité*.

2 Organization of Knowledge

In his *Critique of Pure Reason* (1781; second edition in 1787), Immanuel Kant explained that "[i]t is humiliating for human reason that it accomplishes nothing in its pure use, and even requires a discipline to check its extravagances

---

or the Cultivation of the Intellect," in *Descartes Embodied: Reading Cartesian Philosophy Through Cartesian Science* (Cambridge: Cambridge University Press, 2001), 277–95.

83 Peter Dear, "A Mechanical Microcosm: Bodily Passions, Good Manners, and Cartesian Mechanism," in *Science Incarnate: Historical Embodiments of Natural Knowledge*, ed. Christopher Lawrence and Steven Shapin (Chicago: The University of Chicago Press, 1998), 51–82.

84 Bacon, *The New Organon*, aphorism CXXII, 95. Guido Giglioni writes, for example, about Bacon's *novum organum*: "the project outlined in the *Novum Organum* is a program for taming the *ingenium*, mechanizing the mind and controlling the appetites. No methodological blueprint for a future mechanization of nature is involved. … Bacon is in favor of forcing the mind to follow a methodological path that has been successfully tested, where nothing is left to chance and everyone can reach the same results starting from the same mental capacities." Giglioni, "How Bacon Became Baconian," in *The Mechanization of Natural Philosophy*, eds. Sophie Roux and Daniel Garber (Dordrecht: Springer [Boston Studies in the Philosophy of Science], 2013), 27–54, quote on p. 32.

85 On Descartes's audience generally, Jean-Pierre Cavaillé, "Descartes stratège de la destination," *XVIIe siècle* 177 (1992): 551–9; Cavaillé, "'Le plus éloquent philosophe des derniers temps': les stratégies d'auteur de René Descartes," *Annales: histoire, sciences sociales* 49 (1994): 349–67.

and avoid the deceptions that come from them." Maybe, Kant believed, "[t]he greatest and perhaps only utility of all philosophy of pure reason is thus only negative, namely that it does not serve for expansion, as an organon, but rather, as a discipline, serves for the determination of boundaries, and instead of discovering truth it has only the silent merit of guarding against errors."[86] For Kant, an *organon* (as he called it himself in German) was an elemental, powerful cognitive and epistemic instrument one found in the toolbox of transcendental philosophy. As he explained a decade earlier, in what is now called the Vienna Logic, Kant referred to two kinds of cognition: "An *a priori* one, which is independent of experience; and an *a posteriori* one, which is grounded on empirical principles." General logic (or *logica artificialis*), for instance, rests only on *principia a priori*; it is among what are known as *scientiae rationales*. The aim of such a science, according to Kant, is "either to extend or to elucidate our cognitions." Since logic can only elucidate our cognition ("because it considers form, which cannot be extended"), and rests on rules *a priori*, it is called a canon: "Logic is a rational science, a canon for the understanding." Conversely, "A science that extends our cognitions [*Kenntniße*] is called an organon."[87] Only an organon, therefore, can secure a growth of our knowledge. The latter concept is in fact at the core of a 'special science' Kant named 'a critique of pure reason.' As he elucidated in the second edition of his masterpiece, "reason is the faculty that provides the principles of cognition *a priori*. Hence pure reason is that which contains the principles for cognizing something absolutely *a priori*." Given such a definition, Kant can illustrate what an *instrument* of pure reason is:

> An **organon** of pure reason would be a sum total of all those principles in accordance with which all pure *a priori* cognitions can be acquired and actually brought about. The exhaustive application of such an organon would create a system of pure reason. But since that requires a lot, and it is still an open question whether such an amplification of our knowledge is possible at all and in what cases it would be possible, we can regard a science of the mere estimation of pure reason, of its sources and boundaries, as the **propaedeutic** to the system of pure reason. Such a thing would not be a **doctrine**, but must be called only a **critique** of pure reason.[88]

86 Immanuel Kant, *Critique of Pure Reason*, trans. and eds. Paul Guyer and Allen W. Wood (Cambridge: Cambridge University Press, 1998), 672.

87 Immanuel Kant, *Lectures on Logic*, trans. and ed. J. Michael Young (Cambridge: Cambridge University Press, 1992), 252–3. Kant defines logic thusly: *Logica est scientia regularum universalium usus intellectus*. (p. 253) Mathematics is such an organon.

88 Kant, *Critique of Pure Reason*, 149. (bold emphases original)

Kant's object of cognition is not the nature of things, "which is inexhaustible," but rather "understanding, which judges about the nature of things, and this in turn only in regard to its *a priori* cognition."[89] The critique, therefore, is a preparation—a propaedeutic—"if possible, for an organon, and, if this cannot be accomplished, then at least for a canon," whether in the former circumstance it consists in the amplification or, in the latter, in the mere limitation of cognition of a complete system of the philosophy of pure reason.[90]

Compared to a general logic, "which is merely a **canon** for judging," Kant's notion of *organon* provides us with a clear epistemology and organization of scientific knowledge.[91] In the next chapter, I argue that the organ, as a musical instrument, played a similar—though material—role for Marin Mersenne in the 1630s. Like other types of instruments, the etymology of the pipe organ is derived from the Greek *organon* and its subsequent Latin translation. In Greek Antiquity, however, *organon* never referred to a pipe organ. The origins of *organum* as a definition for a pipe organ came from St. Augustine and the Latin Church Fathers. Only with the coming of the Christian era did one see a clear evolution from the general connotation of *organum* to the more specific musical term.[92] Once established, however, the committed attribution of the term organ to a musical instrument became so prevalent in Renaissance Europe that the Italian lutenist, composer, and music theorist Vincenzo Galilei felt the need to spell out its original meaning. An organ, he wrote, was an instrument, whether natural or artificial, that could best achieve its intended end:

> Although we read the term 'organ' countless times in the ancient writers on the subject of musical instruments and other subjects, this arises from their understanding by the term whatever pleases them. Its meaning is 'instrument' and 'rising to a height' [*ascendere in alto*], which is the

89 Ibid., 150. This is how Kant distinguishes between doctrine and critique: a doctrine is defined when "the rules of judgment precede the passing of judgment"; when the passing of judgment precedes the rules, it is a critique. Kant, *Lectures on Logic*, 253.

90 Kant, *Critique of Pure Reason*, 150.

91 Ibid., 198–9, where he writes: "Now one can take it as a certain and useful warning that general logic, **considered as an organon**, is always a logic of illusion, i.e., is dialectical. For since it teaches us nothing at all about the content of cognition, but only the formal conditions of agreement with the understanding, which are entirely indifferent with regard to the objects, the effrontery of using it as a tool (organon) for an expansion and extension of its information, or at least the pretension of so doing, comes down to nothing but idle chatter, asserting or impeaching whatever one wants with some plausibility." (bold emphasis original)

92 Barbara Owen, Peter Williams, and Stephen Bicknell, "Organ," *Grove Music Online* (accessed 17 August 2006).

> nature of every sound and voice. Ultimately, what remains is this particular name of that instrument which has the capacity to realize best the effect of its meaning.[93]

Galilei used the original Aristotelian definition of organum to argue against his rival music theorist Gioseffo Zarlino's dichotomy between art and nature. Whether a body part, a tool or a musical instrument, an 'organ' was for Galileo's father an instrument made toward achieving a specific end. Zarlino replied to Galilei in his *Sopplimenti* that, on the contrary, the pipe organ acquired the universal name of 'organ' only because it had been invented by imitation of nature, specifically the bodily organs that produced the human voice—the only organum that could produce just intonations—where the pipes stood for the throat, the bellows for the lungs, the keyboard keys for the teeth, and the player for the tongue.[94] Zarlino's analogy became a very common one in the seventeenth century, mentioned by music aficionados such as the physician from Sens Christophe de Villiers as well as engineers like Salomon de Caus and scholars like Athanasius Kircher.[95] (Descartes, who had read and mastered

93 Vincenzo Galilei, *Dialogue on Ancient and Modern Music*, trans. Claude V. Palisca (New Haven: Yale University Press, 2003), 361–2.

94 Gioseffo Zarlino, *Sopplimenti musicali del rev. M. Gioseffo Zarlino da Chioggia ...* (Venice, 1588), book VIII, 288: "Per la qual cosa dico, che l'Organo proposto s'acquistò questo nome uniuersale & commune d'Organo proprio & particolare, per una certa eccellenza dalle parti naturali, che formano la Voce, che si chiamano Istrumenti naturali: percioche fù fabricato alla guisa del Corpo humano, corrispondendo le Canne alla Gola, I Mantici al polmone, I Tasti à I Denti, & colui che sona alla Lingua, & cosi l'altre parti di esso à quella che sono nell'Huomo." Pierre Trichet mentions Zarlino's viewpoint, adding a parallel between the function of the wind within the organ and the human soul within the body. Trichet, *Traité des instruments de musique (vers 1640)*, ed. François Lesure (Neuilly-sur-Seine: Société de musique d'autrefois, 1957), 25.

95 Christophe de Villiers to Mersenne, [mid-November 1633], in *Correspondance du P. Marin Mersenne: religieux minime*, ed. and annotated by Cornélis de Waard (with the collaboration of René Pintard), 17 vols. (Paris: G. Beauchesne, 1933–1988), vol. 3, 547: "Et neantmoins de tous les instruments nul n'approche de si prez les organes de la voix de l'homme que l'orgue qui a, ce me semble, les souflets pour poulmon, le porte-vent pour trachee artere, et pour le larinx, glotte, epiglotte et cavité depuis iceux jusq'au palais, le tuyau de l'orgue et ses partyes, en sorte mesme que de cette analogie, je conclurois l'orgue plus antienne que pas un autre instrument, n'ayant esté fait sur d'autre prototype que celuy des partyes dediees à la voix humaine." Salomon de Caus, *Les Raisons des forces movvantes*, book 3, 4. Kircher, *Musurgia universalis*, book 1, chapter XIII, vol. 1, 24–5. A historical survey leading to the nineteenth century is found in Thomas L. Hankins and Robert J. Silverman, *Instruments and the Imagination* (Princeton: Princeton University Press, 1995), chapter 8, "Vox mechanica: The History of Speaking Machines."

Zarlino,[96] also made famous this analogy between the pipe organ and the human body in his *Treatise on Man*. But, contrary to what the Italian humanist proclaimed, the human body here imitated the mechanical organization of the pipe organ, not the opposite way around.[97])

Marin Mersenne's book on the voice in the *Harmonie universelle* (1636–37) leaned heavily on such an art-nature mechanistic interpretation. The ear, for example, was not endowed with special invisible powers, but was just a mechanical instrument to catch sounds: "I thus say the ear does not know sounds, and that it only serves as an instrument & organ to guide [these sounds] towards the mind, which study their nature and properties"[98] In describing the bodily organs producing sounds, Mersenne identified them as the "instrumens & organes de la voix," playing with the semantic and epistemic meaning of these two words. He also explained that instruments were "the other parts of harmony" since they put it into practice, "particularly if one includes the voice among the instruments." And to explain complex features of vocal sound production, Mersenne often turned to analogies derived from mechanical components of a pipe organ.[99] Peter Dear says that for Mersenne a "deep

96 Descartes, *Abrégé de musique*, trans. and ed. Frédéric de Buzon (Paris: Presses universitaires de France, 1987), 7–8. There is only one reference in Descartes's Compendium, and it is to Zarlino. See also Stephen Gaukroger, *Descartes: An Intellectual Biography* (Oxford: Clarendon Press, 1995), 74–80.

97 Descartes, *Traité de l'homme*, AT ix, 165–6. It is often said in fact that Descartes came up with the idea of the human body as a machine by contemplating the automated figures in the grottoes of the Royal gardens at Saint-Germain-en-Laye, designed by the Francini brothers. Interestingly enough, looking at de Caus's *Raisons des forces movvantes*, for instance, one sees that most grotto automata were actually powered by hydraulic organs. Perhaps more than anything else, I would argue, the collapse of art and nature onto each other in Descartes's thinking came from his mental picture of the pipe organ.

98 Mersenne, "De la voix, des parties qvi seruent à la former, de sa definition, de ses proprietez, & de l'oüye," in *Harmonie universelle, contenant la théorie et la pratique de la musique*, 3 vols. (Paris: Centre national de la recherche scientifique, 1963), book I, prop. LII, vol. 2, 79: "Ie dis donc premierement que l'oreille ne connoist pas les sons, & qu'elle ne sert que d'instrument & d'organe pour les faire passer dans l'esprit qui en considere la nature & les proprietez".

99 Mersenne, "De la voix, des parties qvi seruent à la former, de sa definition, de ses proprietez, & de l'oüye," in *Harmonie universelle*, book 1, prop. XIII, vol. 2, 13: "La grande varieté des sons que l'homme fait procede de la diuersité des organes, & des instrumens de la voix, ou de la differente maniere dont ils se peuuent mouuoir pour battre l'air" and prop. IV, vol. 2, 7: "Il n'y a rien qui puisse mieux seruir à l'explication de cette difficulté que l'anche des *regales*, que l'on appelle *voix humaines*; car à proportion que l'on ouure ceste anche en retirant le mouuement en haut, la voix deuient plus graue; & quand on le pousse plus bas pour fermer l'anche, elle deuient plus aiguë: De mesme quand la glotte

and reciprocal relationship" existed between the sciences of mechanics and music, insofar as universal harmony "functioned not merely as a metaphysical ornament to technical work in the mathematical sciences but as a vital heuristic and tacit demonstrative premise in specific pieces of mathematical natural philosophy."[100] Not only in natural philosophy, according to the minim Father, but also in politics, moral, religious contemplation, and the art of war. What has been much less investigated, however, is the place musical instruments take in this notion of *harmonie universelle.*

---

s'ouure dauantage, elle fait la voix plus graue, & quand elle se ferme, elle la fait plus aiguë." Mersenne, "Traité des instrvmens a chordes," in *Harmonie universelle*, book 1, préface au lecteur, vol. 3, n.p.). English translation in Mersenne, *Harmonie universelle: The Books on Instruments*, trans. Roger E. Chapman (The Hague: M. Nijhoff, 1957), 13.

100 Peter Dear, *Mersenne and the Learning of the Schools* (Ithaca: Cornell University Press, 1988), 116.

CHAPTER 2

Organ Making and Natural Philosophical Knowledge in Marin Mersenne's *Harmonie universelle*

The popularity of instrumental music during the Renaissance breached a few dogmatic walls regarding the use and virtue of musical instruments. These were ubiquitous in Renaissance Europe, and served several functions in sixteenth-century urban, parochial, and court cultures. Just consider Rabelais's descriptions of popular fêtes and festivals, elaborate French royal entries in cities, or even Girolamo Cardano's youthful habit of wandering at night through Milan, dripping with perspiration at daybreak from the physical exertion of serenading on his lute.[1] Not only did scholars begin to pay attention to the manufacture of musical instruments—like Henri Arnaut de Zwolle's famous treatise (ca. 1440) on the organ and lute[2]—European courts found renewed interest and opportunities for instrumental music. At the court of Mantua, for example, the Marchesa Isabella d'Este Gonzaga enrolled the service of the instrument maker and virtuoso instrumentalist Lorenzo da Pavia to make a variety of musical instruments, such as viols, clavichords, harpsichords, and lutes.[3] In France, François I had numerous *joueurs d'instruments* in ensembles called

1 On Rabelais, see his *Gargantua* and *Pantagruel*, and Mikhail Bakhtin, *Rabelais and his World*, trans. Helene Iswolsky (Bloomington: Indiana University Press, 1984). On French royal entries and other events in streets, cemeteries, and fields, or during hunting parties and military campaigns, Isabelle Cazeaux, *French Music in the Fifteenth and Sixteenth Centuries* (Oxford: Basil Blackwell, 1975), chapter 7. On Cardano's epic all-nighters, Clement A. Miller, trans. and ed., *Hieronymus Cardanus (1501–1576): Writings on Music* ([n. p.]: American Institute of Musicology, 1973), 17.

2 G. Le Cerf, ed. (with the collaboration of E. R. Labande), *Les Traités d'Henri-Arnaut de Zwolle et de divers anonymes (ms B. N. Latin 7295)* (Paris, 1932). Arnaut was a professor of medicine to Monseigneur le duc de Bourgogne. He was also interested in astronomy and even received monetary rewards for making astronomical instruments. His treatise on music is one of the first to study and describe the manufacture of organs, lutes, and other keyboard instruments like the *clavisimbalum*. See also Edward L. Kottick, "Building a 15th-Century Lute," *The Galpin Society Journal* 26 (1973): 72–83. Edwin M. Ripin, "The Early Clavichord," *The Musical Quarterly* 53 (1967): 518–38.

3 William F. Prizer, "Isabella d'Este and Lorenzo da Paria, 'Master Instrument-Maker'," *Early Music History* 2 (1982): 87–127. See also Iain Fenlon, "Isabella d'Este e i suoi contemporanei. Musica e mecenatismo presso le corti dell'Italia settentrionale," in *Bernardo Clesio e il suo tempo*, ed. Paolo Prodi (Rome: Bulzoni, 1986), 607–37. On women and musical instruments

 | DOI:10.1163/9789004504615_005

musique de l'Écurie and *musique de chambre* as well as *chantres* who sang in the king's Grande Chapelle.[4]

On a sociological scale, music had a striking impact on the early modern French court. It was not by any means fortuitous that the *ballet de cour, air de cour*, and *musique mesurée* were introduced at court during the Wars of Religion. Music and dance, according to Kate van Orden, participated in the "broad process of behavioral and cultural disciplining that the French aristocracy underwent beginning in the sixteenth century." Court music and dance performances were deployed not so much to quench violent behavior (as one might think) as to make sure instead that the notorious aristocratic excess of bloodshed was forced out in a controlled and orderly way—in the pyrrhic dances or *ballets à cheval*, for instance. Music became a royal instrument of social order on many levels, whether during royal balls, royal entries or sacred *Te Deum* ceremonies. "Through the action of music," argues van Orden, "social bodies cohered."[5]

Although music contributed to religious, courtly and popular activities, the value and significance of musical instruments continually led to artistic, epistemological, and theological disagreements in the Christian West. Up until the Renaissance, and through most of it actually, a distinctive religious and scholarly consensus considered the human voice as the purest of 'instruments,' allowing anyone well trained to sing following the strictest of principles (or consonances) derived from ancient music theory. This idea went back primarily to Boethius's early sixth-century *De institutione musica*, in which the Latin philosopher classified *vox* in three categories—reciting a prose oration (emphasis on words spoken), singing a sequence of intervals (emphasis on sung voices), and doing both at the same time, or the intermediate voice. For Boethius, the notion of *vox* meant the spoken or sung voice as well as the abstract tone. In contrast to the abstract or perfect mathematical tone, the human voice was limited, though it remained the greatest musical instrument of all. Boethius

in general, Lujza Tari, "Women, Musical Instruments and Instrumental Music," *Studia Musicologica Academiæ Scientiarum Hungaricæ* 40 (1999): 95–143.

4 Christelle Cazaux, *La Musique à la cour de François Ier* (Paris: École Nationale des Chartes, 2002), 69–160. On royal chapels in general, Juan José Carreras and Bernardo García García, eds., *The Royal Chapel in the Time of the Habsburgs: Music and Ceremony in Early Modern European court*, trans. Yolanda Acker with English ed. Tess Knighton (Woodbridge, UK: Boydell Press, 2005).

5 Kate van Orden, *Music, Discipline and Arms in Early Modern France* (Chicago: The University of Chicago Press, 2005), quotes on pp. 7 and 36 respectively. On dance, see Sarah R. Cohen, *Art, Dance, and the Body in French Culture of Ancien Régime* (Cambridge: Cambridge University Press, 2000). A similar approach is found in Robert M. Isherwood, *Music in the Service of the King: France in the Seventeenth Century* (Ithaca: Cornell University Press, 1973).

acknowledged that fact early in his treatise, insisting that human nature be the one limitation to the boundlessness of *vox*.[6]

Musical instruments, despite the rhetorical emphasis on *vox*, played an important role in the daily life of early modern Europeans. They were heard everywhere, in all circumstances and by all walks of life. They were also purchased in larger quantities than ever before by the end of the sixteenth century—especially the lute, given that it was still fairly inexpensive.[7] It was, however, different to assert that musical instruments could make original contributions to the study of music. The father minim Marin Mersenne (1588–1648), the well-known humanist and *secrétaire général* for a whole generation of European scholars, upheld such a position from the mid 1620s onward, taking the side of Vincenzo Galilei—Galileo's father—in the latter's heated debate against Gioseffo Zarlino. Galilei's claim that artificial instruments were as good, and sometimes even better than the human voice in producing knowledge about music was of fundamental importance to Mersenne. In fact, this stance became the foundation on which rested the father minim subsequent study of sound.[8]

Within this changing epistemic context, Marin Mersenne's books on musical instruments, featured in his well-known *Harmonie universelle* (1636–37),

---

6 "The voice [*vox*] which is continuous and that with which we run through song are inherently boundless. For by consideration generally agreed upon, no limit is placed either on flowing through words or on rising to high pitches or sinking to low ones. But human nature imposes its own limitation on both of these kinds of voice." Anicius Manlius Severinus Boethius, *Fundamentals of Music*, trans. Calvin M. Bower (New Haven: Yale University Press, 1989), §12–13, 20–21, quote on §21.

7 François Lesure, "La facture instrumentale à Paris au seizième siècle," *The Galpin Society Journal* 7 (1954): 11–52. For twenty *sous tournois* one could buy either a violin or a lute with its case in the middle of the sixteenth century. These prices for cheap instruments were found as well throughout the first half of the seventeenth century. Madeleine Jurgens, *Documents du Minutier central concernant l'histoire de la musique (1600–1650)*, 2 vols. (Paris: S.E.V.P.E.N., 1967; La Documentation Française, 1974), vol. 2, 84–93. Another good indication of the widespread distribution of lutes in the beginning of the seventeenth century is given by Jean Titelouze, a famous Rouen organist and one of Marin Mersenne's first correspondents. In Titelouze's youth (he was born in 1563), everyone admired "un homme qui touchoit le lut et assez mal pourtant; et maintenant j'en voy cent plus habilles gens que luy mille fois, que l'on ne daigne pas presque escouter." Titelouze to Mersenne, 2 March 1622, *Correspondance du p. Marin Mersenne: religieux minime*, ed. Cornelis de Waard, Paul Tannery, René Pintard, Bernard Rochot and Armand Beaulieu, 17 vols. (Paris: PUF and C.N.R.S., 1933–1988), t. 1, 76 [hereafter cited as CM I, 76]. On the rising influence of the lute in sixteenth-century France, see the excellent book by Jean-Michel Vaccaro, *La Musique de luth en France au XVIe siècle* (Paris: Éditions du Centre National de la Recherche Scientifique, 1981).

8 Vincenzo Galilei, *Dialogue on Ancient and Modern Music*, trans. Claude V. Palisca (New Haven: Yale University Press, 2003). On the traditional meaning of *vox* see Rossetti, *Libellus de*

became the best early modern illustration of the developing experimental and mechanistic approach to the study of sound. Musical instruments were not only data-gathering tools, but powerful symbols of the emergent mechanical philosophy. Since music was Mersenne's paradigmatic *scientia* of *harmonie universelle*, musical instruments became ipso facto the paradigmatic 'scientific' instruments of the mechanical sciences, and consequently of Mersenne's universal harmony and of the 'modern' natural philosopher.[9] This chapter is about the transformation of the theoretically inclined medieval *musicus* into the early modern 'parfait musicien,' an individual in touch not only with the theory and practice of music, but also with the producers of its material culture.[10]

---

rudimentis musices, ed. Albert Seay, Critical Texts, no. 12 (Colorado Springs: Colorado College Music Press, 1981), 1–60. Gioseffo Zarlino, *Sopplimenti musicali del rev. M. Gioseffo Zarlino da Chioggia ...* (Venice, 1588). Gioseffo Zarlino, *Le Istitutioni harmoniche* (Venice, 1561; facsimile of the second edition, Arnaldo Forni Editore, 1999). Philippe Vendrix, "La dichotomie vocal/instrumental dans la théorie musicale aux confins de la Renaissance et du Baroque," in *Le Concert des voix et des instruments à la Renaissance* (Paris: C.N.R.S., 1995), 71–81.

9 In the Minim's mind, musical instruments were actually linked to our understanding of the mechanical laws of art and nature. Peter Dear demonstrates that the concept of universal harmony emerged from Mersenne's study of mechanics, which itself was a science wholly dependent on material objects. "By integrating music with a mechanical account of sound," Dear argues, Mersenne "succeeded in creating a way of treating mechanics itself as an exemplification of the harmonious relationships for which music provided the prototype." Musical harmony was thus explained by the science of mechanics; yet, Mersenne strongly believed that the concept of harmony itself laid at a deeper level than mechanics. Harmony, according to Mersenne, was the presupposed and fundamental notion behind mechanics and consequently of all of Creation. Music, Dear continues, "provided the paradigm of harmony by which the rest could be developed and judged." Dear, *Mersenne and the Learning of the Schools* (Ithaca: Cornell University Press, 1988), quotes on pp. 117 and 139 respectively. A more recent, and excellent, account of Mersenne's universal harmony is Natacha Fabbri, *De l'utilité de l'harmonie: Filosofia, scienza e musica in Mersenne, Descartes e Galileo* (Pisa: Edizioni della Normale, 2008). On the use of material objects in the science of mechanics, Domenico Bertoloni Meli, *Thinking with Objects: The Transformation of Mechanics in the Seventeenth Century* (Baltimore: The Johns Hopkins University Press, 2006).

10 On the role of mathematics and natural philosophy in the development (and historical study) of music, H. Floris Cohen, *Quantifying Music: The Science of Music at the First Stage of the Scientific Revolution, 1580–1650* (Dordrecht: D. Reidel Publishing Company, 1984). H. Floris Cohen and Jacomien Prins, "Music, Philosophy, and Science," in Iain Fenlon and Richard Wistreich, *The Cambridge History of Sixteenth-Century Music* (Cambridge: Cambridge University Press, 2019), 439–71. Penelope Gouk, *Music, Science and Natural Magic in Seventeenth-Century England* (New Haven: Yale University Press, 1999). Myles W. Jackson, "Music and Science During the Scientific Revolution," *Perspectives on Science* 9 (2001): 106–15. Benjamin Wardhaugh, *Music, Experiments and Mathematics in England, 1653–1705* (Aldershot, UK: Ashgate, 2008). Peter Pesic, *Music and the Making of Modern*

Mersenne demonstrated better than most contemporary savants how and why doing natural philosophy was an enterprise involving everything but armchair philosophizing. Mersenne's books on musical instruments—and especially the organ—epitomized like none other in the first half of the seventeenth century the new mechanical natural philosophy, which was ultimately grounded on theory, experiments, and a thorough understanding of the mechanical arts.

1 Mersenne's Seven Books on Instruments in the *Harmonie universelle*

The scholarly dispute that erupted between Galilei and Zarlino concerning the epistemic value of musical instruments was so consequential to Mersenne's conceptual system that he mentioned it at the outset of his books on instruments. In the third proposition of book one dedicated to string instruments, Mersenne—siding all along with Galileo's father—stated that

> Actually, if one tunes instruments according to the perfection of the theory, there is no doubt that they will have no need of the perfection of the voice, which can be corrected and adjusted by their means, for one can show whether the voices are singing exactly only by showing that they conform to the perfect instrument. This Zarlino should have acknowledged, had he considered it attentively.[11]

Science (Cambridge, MA: MIT Press, 2014). On the role of mathematics in Mersenne's work, Patrice Bailhache, "L'Harmonie universelle: La musique entre les mathématiques, la physique, la métaphysique et la religion," *Les Études philosophiques* 1/2 (1994): 13–24. Justin Grosslight, "Small Skills, Big Networks: Marin Mersenne as Mathematical Intelligencer," *History of Science* 51 (2013): 337–74. Catherine Goldstein, "Routine Controversies: Mathematical Challenges in Mersenne's Correspondence," *Revue d'histoire des sciences* 66 (2013): 249–73.

11 Mersenne, *Harmonie universelle, contenant la théorie et la pratique de la musique*, 3 vols. (Paris: Centre national de la recherche scientifique, 1963), "Traité des instrvmens a chordes," book I, prop. III, 8. Hereafter, I will not cite the *Harmonie universelle* volumes, but only the book referred to within, adding a reference to the volume thusly: (HU3). English translation in Mersenne, *Harmonie universelle: The Books on Instruments*, trans. Roger E. Chapman (The Hague: M. Nijhoff, 1957), 22. (hereafter cited as Mersenne, *The Books on Instruments*). Musicians, Pierre Trichet heard, were "beaucoup plus ravis et satisfaicts de leur jeu que de leur chant: et la raison me semble estre fondée sur ce qu'ils ne peuvent démonstrer qu'ils chantent justement qu'en faisant apparoir que leurs voix sont conformes à un instrument parfaict." Trichet, *Traité des instruments de musique (vers 1640)*, ed. François Lesure (Neuilly-sur-Seine: Société de musique d'autrefois,

According to Mersenne (and, in a modified way, also Aristotle[12]), the craftsmanship of musical instruments—when perfected by theory—was superior to the naturally created organs of the human voice, and therefore the latter should be tuned to the former. Objects of art, in other words, were not always predestined to becoming poor imitations of nature—Aristotle's *ars imitatur naturam*.[13] Mersenne pushed even further the Renaissance argument that musical instruments played a significant role to our understanding of music. In his writings, the material culture of music became nothing short of a key element to the study of music—and consequently to our understanding of universal harmony.

Mersenne's books on instruments are well known in music history. Published in 1636 as part of the much larger *Harmonie universelle*, these books, along with book two of Michael Prætorius's 1619 *Syntagma musicum*, are the most important contributions to our knowledge of early modern musical instruments. It is difficult to identify exactly when Mersenne decided to add a treatise—a compendium—on musical instruments to his study of music. In his *Questiones in genesim* of 1623, Mersenne was interested primarily with Antiquity's musical thought, devoting several columns to the perfection of ancient Greek music and to the more contemporary *musique mesurée* of Jean-Antoine de Baïf's Académie de poésie et de musique. His approach was then humanist, established on the study of rhetoric, dialectic, classical writings and recent works by Pierre de Ronsard, Pontus de Tyard, and Baïf himself.[14] Four years later, when

1957), 14. Claude V. Palisca, "Mersenne *pro* Galilei *contra* Zarlino," *Nuova civiltà della macchine* 16 (1998): 74–80.

12 See Aristotle, *Problems**, book 19, §918a29–918a34, in *The Complete Works of Aristotle*, ed. Jonathan Barnes, 2 vols. (Princeton: Princeton University Press, 1984–1985), vol. 2, 1431: "Why, if the human voice is more pleasant than an instrument, is the voice of a man singing without words—as, for example, when singing nonny-noes—not so pleasant as a flute or lyre? Or is it true that even in the case of an instrument we get less pleasure if it is not expressive of meaning? The instrument, however, has an advantage even in its actual effect; for while the human voice is pleasanter, instruments strike the note better than the human mouth, which is why they are pleasanter to hear than nonny-noes." This reference was taken from the *Past Masters* internet resource http://library.nlx.com (accessed via Harvard College Libraries, 22 April 2007).

13 Mersenne, "Traité des instrvmens a chordes," book I, prop. III, 9 (HU3): "Certes il me semble que l'Art peut estre dit superieur à la Nature ou surpasser la Nature, lorsqu'il donne quelque degré de perfection à vn sujet, auquel elle ne le peut donner: ce qui n'empesche pas qu'elle ne surmonte l'Art en plusieurs autres choses." A good example related to acoustic instruments: Matteo Valleriani, "Galileo's Abandoned Project on Acoustic Instruments at the Medici Court," *History of Science* 50 (2012): 1–31.

14 David Allen Duncan, *The Tyranny of Opinions Undermined: Science, Pseudo-Science and Scepticism in the Musical Thought of Marin Mersenne* (PhD diss., Vanderbilt University,

Mersenne published the *Traité de l'Harmonie universelle*, music had become more than a pure intellectual pursuit derived exclusively from ancient learning. Music, through the idea of a universal harmony, could teach a new method of learning, one not faithfully restricted to the 'tyranny of opinions.' Ancient writing, Mersenne stressed, should not be blindly received before one had done his own sense experiments.[15] Although Mersenne's natural philosophy did not completely steer clear from the 'learning of the schools,' as convincingly shown by Peter Dear, there was nevertheless in the *Traité de l'Harmonie universelle* a fairly unusual emphasis given to experiments and the material culture of music.[16] From the mid 1620s–a full decade before he finally published his books on instruments—Mersenne deemed it significant to start investigating, ascertaining, and rationalizing the role of instruments and experiments with regard to his new natural philosophy.

The 1636 books on instruments, printed before any other from the *Harmonie universelle*, are divided according to the traditional Boethian system of string (books 1–4), wind (books 5–6), and percussion (book 7) instruments. One of their avowed and most apparent purposes was to serve as an encyclopedia of all instruments that existed in the history of humanity. To this end, Mersenne's vast network of correspondents proved to be of tremendous assistance. He received, for example, an account and a drawing of a South Asian instrument that had just arrived in London from China during the summer of 1634.[17] From

1981), esp. chapter 3. On the Académie de musique et de poésie the classic work remains Frances Yates, *The French Academies of the Sixteenth Century* (London: Warburg Institute, 1947; reprinted in London: Routledge, 1988).

15 Mersenne, *Traité de l'Harmonie universelle* (Paris, 1627), book II, préface au lecteur, n.p.

16 Mersenne, *Traité de l'Harmonie universelle*, book I, préface au lecteur, n.p., where he describes the content of book IX, never printed but heralding the later books on instruments.

17 It was an Indian *vina*, though Mersenne did not know its exact name and was not even sure whether it originally came from China or India. As soon as he could, he had an engraving made and included it with a short (secondhand) description to a section on Indian and Turk musical instruments. Mersenne to Peiresc, 2 July 1634, CM IV, 230; Peiresc to Mersenne, 16 July 1634, ibid., 244 in which Peiresc tells Mersenne he should get in contact with Sr. Langlois dit Chartres, wonderful musette and flute player, and someone who could discover the owner of this instrument and arrange for an engraving. Mersenne to Peiresc, 2 February 1635, CM V, 46, in which he says that Claude Hardy brought back a drawing of the said instrument. Mersenne, "Traité des instrvmens a chordes," book IV, prop. XX (sic, XVIII), 227–8 (HU3). Mersenne to Doni, 2 February 1635, CM V, 35. In the *Harmonicorum libri XII* (1636) Mersenne writes: "Placet autem hîc Sinense vel Indicum Instrumentum describere, cuius figuram beneficio atque diligentiâ summi viri domini Hardy ex Anglia accepi." *Harmonicorum libri XII*, vol. 2, book II, prop. XXI, 111. On the *vina*, Claudie Marcel-Dubois, *Les Instruments de musique de l'Inde ancienne* (Paris: Presses Universitaires de France, 1941), 79–80.

Italy, where most of Mersenne's research on ancient musical instruments was focused, 'Messieurs Gaffarel & Naudé' sent engravings depicting instruments used by Greeks, Romans, and Egyptians as found on several marble artworks and medals.[18] In Rome, moreover, the musical theorist and humanist Giambattista Doni was actively seeking information on ancient musical instruments toward the preparation of a book on Greek music. Not only did Doni share some of his discoveries with Mersenne (who waited in vain for the published work[19]), he also looked for drawings of unusual contemporary Italian instruments. In addition, he sent to Mersenne a detailed sketch and a brief account of his newly invented instrument, the Lyra Barberina (dedicated to Pope Urban VIII, Maffeo Barberini), a sort of lyra (though resembling a lute) on which gut strings were affixed to one side and metal strings to the other.[20] Closer to home, in France, Christophe de Villiers, a physician of Sens and one of Mersenne's most prolific correspondents on music and demonology, was the first to draw the Minim's attention toward an uncommon musical instrument: the *trompette marine*, a one-string instrument that sounded like a brass trumpet.[21] After de Villiers sent a drawing and a complete physical description

18 Jacques Gaffarel to Mersenne, (June 1633?), CM III, 443–4. Gabriel Naudé also sent to Mersenne an engraving taken from an Antiquity marble artwork representing Orpheus with his lyra, which was also added to the *Harmonie universelle* on the verso of the title page. See Naudé to Pierre Gassendi, 6 March 1632, CM III, 266–7. See also Naudé to Mersenne, 12 November 1633, CM III, 533–6 for more of Naudé's efforts in securing images from ancient instruments.

19 Mersenne to Peiresc, 14 May 1634, CM IV, 136, where Mersenne writes: "Il m'escrivit il y a tres longtemps qu'il y metteroit quantité de choses de la musique et des instruments antiques. Dieu veuille que je puise voir son livre avant que de mettre fin au mien, affin de luy donner la louange qu'il meritera, comme je luy ay mandé." On the topic of Greek music in Italy, Patrizio Barbieri, "Musiche e strumenti 'alla greca' nel seicento Italiano," in *Francesco Buti tra Roma e Parigi: Diplomazia, poesia, teatro*, ed. Francesco Luisi (Roma: Torre d'Orfeo Editrice, 2009), 349–64.

20 Doni to Mersenne, 15 October 1633, CM III, 508–9. See also Doni to Mersenne, 8 April 1634, CM IV, 90. Mersenne mentions Doni's lyra in his *Harmonie universelle*, "Traité des instrvmens a chordes," book II, prop. 6, 100 (HU3). However, Mersenne neglects to mention it in the Latin version, the *Harmonicorum libri XII*, which earned him a minor reproach from Doni (who, as an excuse for Mersenne, blamed the quality of the drawing), Doni to Mersenne, [May 1636], CM VI, 71–2. On the lyra barberina, see Claude V. Palisca, *G. B. Doni's* Lyra Barberina (Bologna: A.M.I.S., 1981), which is a facsimile with a commentary of Doni's book published posthumously.

21 On the history, music, and construction of this instrument, see Cecil Adkins, "Trumpet marine," *Grove Music Online* (accessed 9 August 2006). It was known since the fifteenth century. On the relationship between de Villiers and Mersenne, Pietro Biasiol, "Christophe de Villiers: Un *savant* di provincia nella corrispondenza di M. Mersenne," *Rivista Critica di Storia della Filosofia* 36 (1981): 379–400.

of the instrument he had examined, it still took Mersenne several more months before he heard it play in Paris.[22] Although all these correspondents and their exposés on ancient, foreign, strange, and modern musical instruments were useful, one person, owing to his own vast network of relationships, did more than anyone else in trying to secure musical knowledge for Mersenne: Nicolas-Claude Fabri de Peiresc, the celebrated gentleman from Provence and Mersenne's most important patron.

Toward the end of 1632, Peiresc—with the assistance of Pierre Gassendi—began to recognize the worth of Mersenne's music undertaking and instrument compendium. According to Peter Miller, Peiresc's project of retracing the music of Eastern Christians became one of his most audacious information-gathering plans in the 1630s.[23] Regarding uncommon musical instruments, Peiresc then only knew about one *Sanbucca* (or Sambuca), a harp of Greek origin built in Rome, though he had never received a good account of how it sounded. On foreign music from the Levant, however, Peiresc told Gassendi he would send a small *mémoire* to several non-Western locations to assist with Mersenne's research.[24] Peiresc had developed over the years a vast network of correspondents in the Middle East and North Africa, which he used at length, for instance, toward his study of geography and cartography.[25]

---

22 Christophe de Villiers to Mersenne, 3 March 1634, CM IV, 59–60. See also de Villiers to Mersenne, 17 August 1634, ibid., 319–22; de Villiers to Mersenne, 20 October 1634, ibid., 370–1, where he says that he is happy Mersenne could finally hear it, since it is the only way to believe what people say about it. Mersenne's description is found in "Traité des instrvmens a chordes," book IV, prop. XIV [sic, XII], 217–22 (HU3). The account Mersenne ultimately gave in the *Harmonie universelle* was completely different from de Villiers's basic physical description—reminiscent instead of Pierre Trichet's and Descartes's. On Trichet and Descartes, François Lesure, "Pierre Trichet's *Traité des instruments de musique:* supplément," *The Galpin Society Journal* 15 (1962), 70–81, on pp. 78–80; Descartes to Mersenne, [15 September 1634?], CM V, 360–1.

23 Peter N. Miller, *Peiresc's Mediterranean World* (Cambridge, MA: Harvard University Press, 2015), 89–142 for Peiresc's Levant network and the scope of the search on Arabic music.

24 Peiresc to Gassendi, 21 December 1632, CM III, 351–2.

25 Peiresc's correspondents depended overwhelmingly on French collaborators, including ambassadors, secretaries of French ambassadors, consuls and vice-consuls, merchants, ship captains travelling in the Mediterranean and missionaries. See Sonja Brentjes, *Peiresc's Interests in the Middle East and Northern Africa in Respect to Geography and Cartography* (Berlin: Max Planck Institute for the History of Science [preprint 269], 2004). See also Sydney Aufrère, *La Momie et la tempête: Nicolas-Claude Fabri de Peiresc et la curiosité égyptienne en Provence au début du XVIIe siècle* (Avignon: Editions A. Barthélemy, c1990). I would like to thank Sonja Brentjes for this last reference and for providing me with a copy of her preprint.

Mersenne was quick in thanking Gassendi for putting in a good word on his behalf, while reminding the Provence savant at the same time of his promise to deliver a small drawing of the type of cymbals played in that part of France.[26] After an early May 1633 letter to Peiresc, in which Mersenne explained his desire to know every detail of music theory and musical instruments from Arabs, Turks and Persians—so that "nothing was omitted to please the curious"—scores of letters followed between them and between Peiresc and several other correspondents in Tunis, Alep (Syria), Cairo, Jerusalem, and Constantinople.[27] One particular Arab manuscript appeared promising on those matters, until it turned into a little epistolary drama between Mersenne and Peiresc, the former trying to find someone in Paris with the language skills to translate it, and the latter striving a year later to pass it along to someone else who would make a better use of it.[28] Despite Peiresc's efforts, as the correspondences show, Mersenne ultimately abandoned the

26 Mersenne to Gassendi, 5 January 1633, CM IV, 355. Mersenne received the drawing more than a year later, after many other epistolary exchanges with Peiresc himself. Mersenne to Peiresc, 1 May 1633, CM III, 394; Peiresc to Mersenne, 1 May 1634, CM IV, 109 and Mersenne to Peiresc, 14 May 1634, ibid., 135. Mersenne described them in his *Harmonie universelle*, "Des instrumens de percvssion," book VII, prop. 26, 53 (HU3).

27 Peiresc to Thomas d'Arcos, (end of January or beginning of February 1633), CM III, 373–4; Peiresc to Father Théophile Minuti, 12 October 1633, ibid., 494–5; Peiresc to Father Jacques de Vendôme, [12 October 1633?], ibid, 496; Peiresc to Jean Magy, 15 May 1634, CM IV, 155; Peiresc to Henri de Gournay, comte de Marcheville, 19 March 1634, ibid., 80. For a summary of Peiresc's efforts on behalf of Mersenne, Peiresc to Mersenne, 13–15 October 1633, CM III, 497–505. A typical request read thusly, this one to d'Arcos: "Pour un ouvrage excellent de la musique qui se va mettre soubz presse, on desireroit d'avoir quelque cognoissance de la musique et façon de chanter dont se servent aujourd'huy les Grecs vulgaires, les Turcs, les Persans, les Aegyptiens, les Mores et autres peuples de ces pays là. Sy en tout ou en partie, vous pouvez obliger ce bon personnage, je vous asseure qu'il n'en sera pas ingrat et que c'est un homme qui le vaut bien."

28 At first, Mersenne and Peiresc thought this manuscript would be very useful: "[on] y trouvera des fondements et maximes principales de la plus excellente musique des anciens Grecs, à ce que nous en avons peu juger à l'ouverture du livre, et à la consideration des figures geometriquement representatives des proportions et rapportz des tons, avec des differentes couleurs pour les entredistinguer et desmesler plus commodement; ce que je n'avois jamais veu en tous les livres imprimés et mss qui m'estoient passez par les mains. Ce qui nous a faict recognoistre que, parmy ces peuples barbares, il fault qu'il y ait eu des esprits bien desliés." Peiresc to Father Gilles de Loches, 20 May 1634, CM IV, 158–9. Nothing useful really came out from this manuscript. See Peiresc to Jean Magy, 15 May 1634, CM IV, 155; Peiresc to Mersenne, 18 June 1634, ibid., 177–8; Claude Saumaise to Peiresc, 2 September 1634, ibid., 344–5; Mersenne to Peiresc, 15 January 1635, CM V, 28; Mersenne to Doni, 2 February 1635, ibid., 41; Peiresc to Mersenne, 20 March 1635, ibid., 107–8; Peiresc to Mersenne, 5 May 1635, ibid., 162–72; Peiresc to N. Aubery, sieur du Mesnil, 8 May 1635, ibid., 173–4; Mersenne to Peiresc, 23 May 1635, ibid., 208–10; Mersenne to Peiresc, 25 May

idea of integrating into his *Harmonie universelle* any thorough knowledge of Greek and Oriental music after several years of fruitless investigation.[29] This sort of 'Orientalism' was still fairly uncommon in the first third of the seventeenth century. Though always at the margins of early modern French scholarship, Oriental learning progressed and even received kingly patronage during the reign of Louis XIV a few decades later.[30]

Mersenne's books on instruments do not offer comprehensive historical descriptions of ancient and modern musical instruments—unlike Galilei's and Doni's in Italy, and Trichet's in France.[31] Mersenne's books on instruments are somewhat closer to Prætorius's *Theatrum instrumentorum* in regard to the kind of detailed physical descriptions and explanations of sound production. Prætorius's book would certainly have been helpful to Mersenne if only he had acquired a copy before the publication of the *Harmonie universelle*.[32] But

1635, ibid., 215; Peiresc to Jacques Dupuy, 5 June 1635, ibid., 226–7; Peiresc to Dupuy, 26 June 1635, ibid., 255–6; Peiresc to Mersenne, 3 July 1635, ibid., 274–8, and many others.

29 Mersenne to Peiresc, 20 March 1634, CM IV, 82, where he wrote: "Je n'espère maintenant plus rien de la musique des Grecs, ni des Orientaux, après avoir attendu 2 ou 3 ou 4 ans après sans aucun fruit, comme vous sçavez, quoyque j'aye tenté la voye de Rome, de Venise et de Constantinople. Je croy que nous les surpassons tous en ceste matiere. C'est pourquoy je ne veux plus m'en mettre en peine." After receiving news of the Arabic manuscript, Mersenne showed a little bit more optimism, saying to Peiresc he would organize his books, especially the one on instruments, so that he could easily add anything new from those countries. Mersenne to Peiresc, 14 May 1634, CM IV, 133–4. In the end, nothing came out of this research. Surprisingly, both Peiresc and Mersenne would have gathered more information on Arabic music and musical instruments by looking at medieval Latin manuscripts on medicine, astrology, and philosophy kept in European collections. Charles Burnett, "European Knowledge of Arabic Texts Referring to Music: Some New Material," *Early Music History* 12 (1993): 1–17.

30 Nicholas Dew, *Orientalism in Louis XIV's France* (Oxford: Oxford University Press, 2009).

31 Galilei, *Dialogue on Ancient and Modern music*. Galilei probably influenced Doni's approach to the historical research of ancient Greek musical instruments. In the case of Trichet, his well-known contemporary manuscript located at the Bibliothèque Sainte-Geneviève in Paris was with Mersenne's books on instruments the most comprehensive treatise on the material culture of music. However, as Trichet wrote to Mersenne, his objective was quite different from the Minim's: "Quant à moi, tout ce que je pretends en mon livre est seulement de traitter historiquement des instruments de musique sans m'amuser à des recerches qui surpassent la capacité de mon esprit. Sur quoi voulant avoir vostre jugement, je m'estois mis en chemin pour vous aller voir à Paris, durant que la contagion affligeoit la ville de Bourdeaux." Trichet to Mersenne, 9 January 1631, CM III, 4–5. Trichet, *Traité des instruments de musique (vers 1640)*. This is not a complete edition of Trichet's manuscript. Other excerpts were later published by Lesure in *The Galpin Society Journal*.

32 Not that Mersenne didn't try to get hold of a copy, looking in the Netherlands (Beeckman) and mostly in Italy (Doni). Mersenne to André Rivet, 30 October [1628], CM II, 107–8;

even more than Prætorius's treatise on instruments—like none other before really—Mersenne's is filled with technical details of fabrication and theoretical methods of improving their designs. Mersenne wanted his readers to realize that to fully understand the nature of sound, one had to not only be au fait with the 'natural' producers of sounds, but most importantly au fait with the 'artificial' ones—the man-made musical instruments. This exceptional emphasis on 'mechanics' and 'craftsmanship,' which to some extent intruded into the standard theoretical and practical study of music, held a key role in Mersenne's natural philosophical practice and rhetoric. Simply said, musical instruments were transformed into emblematic devices of the new mechanical natural philosophy. And of them all, none better than the pneumatic organ, the king of all instruments, articulated that perspective.[33]

2 The Organ and Mersenne's Epistemology of Natural Philosophical Knowledge

Of all the known early modern musical instruments, the most sublime and perfect was said to be the organ. It was for the Church as well as for most scholars the *organum organorum*, i.e., the instrument of instruments. Cardano, for instance, after conceiving nine rules by which to judge the perfection of musical instruments called the organ the "most simple of simple instruments and the most elaborate of the elaborate."[34] To Mersenne's contemporary Trichet, the mechanical structure of and the melody coming from the organ let any listener wonder whether such an invention was divine rather than secular.[35] Mersenne

Peiresc to Mersenne, 13–15 October 1633, CM III, 501–2; Doni to Mersenne, 8 November 1634, CM IV, 393; Mersenne to Doni, 2 February 1635, CM V, 39–40 where he says that "Peut-estre que Scapin qu'on a icy veu jouer de 40 ou 50 sortes d'instrumentz sur le theatre avoit veu ce livre"; Doni to Mersenne, 8 September 1635, CM V, 392. Mersenne briefly discussed Prætorius's work in his 1638 *Novvelles observations physiqves et mathematiqves*, XIV. Obseruation: De la multitude des Instrumens harmoniques, & particulierement des nouueaux, included at the end of HU3. Trichet could not find it either in Bordeaux. Trichet, *Traité des instruments de musique (vers 1640)*, 15.

33 For a useful complementary study to what follows, Brandon Konoval, "Pythagorean Pipe Dream? Vincenzo Galilei, Marin Mersenne, and the Pneumatic Mysteries of the Pipe Organ," *Perspectives on Science* 26 (2018): 1–51.

34 Miller, *Hieronymus Cardanus (1501–1576)*, 56. In a Vatican manuscript, however, Cardano says that the lyra is the most elegant and perfect of instruments. Ibid., 178–82 and 199–206.

35 Mersenne, "Traité de l'orgue," Epitre, n.p. (HU3). Prætorius, *Syntagma musicum: Textes relatifs à l'orgue, comprenant aussi la basse générale ou continue*, trans. and ed. Jacques Leguy (Chatenay Malabry: Éditions Ars Musicæ, 1999), 25. Trichet, *Traité des instruments de musique (vers 1640)*, 23–4.

himself considered the organ one of the most admirable pneumatic machines ever invented. Although Mersenne did not try to give his mechanical organ an allegorical meaning, as Athanasius Kircher promptly did, any reader could tell it represented more than a simple musical instrument. (Fig. 3) Mersenne's description of the organ gives incredible insights into the practice of natural philosophy, which in addition to a thorough knowledge of theory, now involved experimental data gathering and hands-on *savoir-faire* from the mechanical arts. Not that making an organ made a savant out of the maker. (Mersenne's description by itself was insufficient to manufacture such an intricate piece of machinery.[36]) Making the organ was less important to Mersenne than the account of the rigorous procedure entailing its construction—seldom carried out by instrument makers. The organ, in other words, reified the practices of the nascent mechanical philosophy into a respected and altogether Christian early modern material entity. The musical harmony produced by the organ within the Church context was not only a powerful religious—particularly Counter-Reformation—symbol, it also epitomized and materialized the role theory, experiment, and the mechanical arts played in Mersenne's notion of *harmonie universelle*.

2.1 *Pipe Experiments and the Production of Sounds*

Up until the late sixteenth century, Pythagoras's and Boethius's methods based on the mathematical theory of proportions were usually called upon in matters related to the theory of sound.[37] Numbers wholly dominated the science of music: they not only structured musical theory, they also defined music's quintessence. This sublime power of numbers over music was attacked by Galilei

36 Actually, toward the end of the book on organs, Mersenne explicitly instructed his readers to go and visit the instrument maker's workshops to find out more about the accurate manufacture of organs: "As to the practice which are most certain, it is fitting to consult the best makers, such as Valeran, le Pescheur, and many others, which have made the greatest part of the organs which are seen in the churches, and from whom can be learned everything that is missing in this treatise" Mersenne, "Traité de l'orgue," book VI, prop. XLV, 412 (HU3). Mersenne, *The Books on Instruments*, 493.

37 Sebastian Virdung, a German priest who published the first treatise in the West on musical instruments, explained that "To write about these ringing instruments and also about organ pipes, I would choose Boethius, because these have to do with the mensur, that is, [with] the measurement of the tubes and the weight of the metals (like the hammer), and that is expressed through the theory of the proportions. [I have] written nothing at all about these [here], but [I am] saving [this subject] for the complete work." Beth Bullard, trans. and ed., *Musica getutscht: A Treatise on Musical Instruments (1511) by Sebastian Virdung* (Cambridge: Cambridge University Press, 1993), 110–1.

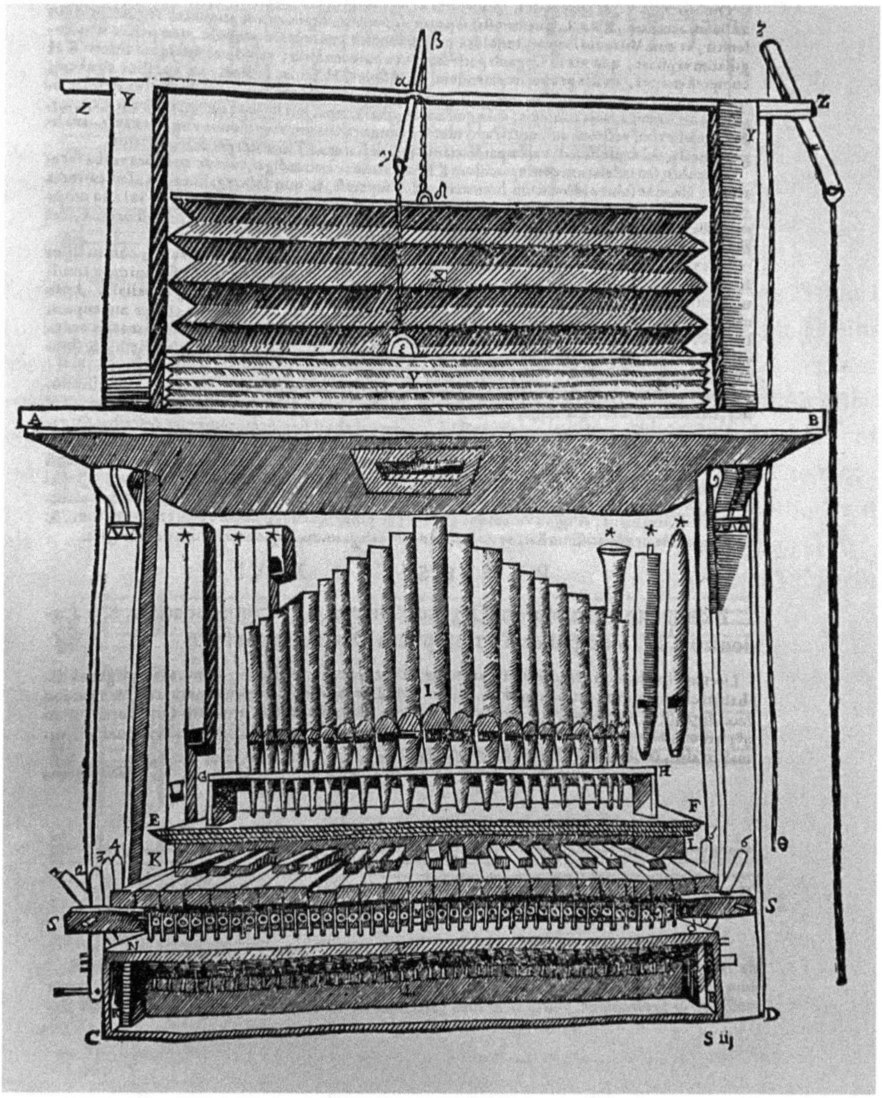

FIGURE 3 Mersenne's portative organ (positive).
MARIN MERSENNE, *HARMONICORUM LIBRI XII* (GENÈVE: MINKOFF REPRINT, 1972 [1636]), BOOK 3, PROP. XXV, 137

in his dispute with Zarlino. Guided by the Florentine humanist Girolamo Mei, Galilei took the path of experiments with musical instruments. Sense experience, for Galilei, had to replace the blind following of ancient authorities. Right from the beginning of his *Dialogue on Ancient and Modern Music*, Galilei—in the voice of Strozzi—fixed the new program of musical scholars:

> I desire in those things in which the sense is involved that we always set aside (as Aristotle says in the eighth [book] of the *Physics*) not only authority but seemingly plausible reasoning that may be contrary to any perception of truth. For it seems to me that those who, for the sake of proving some conclusion, want us to believe simply on the basis of authority without adducing valid arguments for it are doing something laughable, not to say (with the Philosopher) acting like silly fools.[38]

Galilei's anti-theory reaction led him later, in the *Discorso*, to become the first musician ever to reveal the falsity of Pythagoras's noted (and likely apocryphal) musical experiments in a blacksmith's workshop. Galilei destroyed all these with meticulous experiments of his own, which dealt over time a severe blow to number mysticism and ancient authorities in music.[39] Mersenne, drawing again on Galilei's investigations, could not believe that no one since Pythagoras had bothered doing these simple experiments to "discover the truth."[40]

Mersenne's experimental research with organ pipes is traceable to the early 1620s.[41] Thanks to Robert Cornier, the leader of a philosophical circle in Rouen,

38 Galilei, *Dialogue on Ancient and Modern Music*, 12.

39 Galilei, *Discorso*, 103–5. See also Galilei's unpublished manuscripts in Claude V. Palisca, *The Florentine Camerata: Documentary Studies and Translations* (New Haven and London: Yale University Press, 1989), 152–207. Palisca, *Humanism in Italian Renaissance Thought*, 269–77. Claude V. Palisca, "Scientific Empiricism in Musical Thought," in *Seventeenth-Century Science and the Arts*, ed. H. H. Rhys (Princeton: Princeton University Press, 1961), 91–137. On the influence of Pythagora's 'experiments,' Daniel Heller-Roazen, *The Fifth Hammer: Pythagoras and the Disharmony of the World* (New York: Zone Books, 2011). A recent comparison between the major works from the Galilei's father and son is found in Maurice A. Finocchiaro, "Vincenzo Galilei's Musicology and Galileo's Science: Methodological Comparison and Contrast," *Isis* 111 (2020): 740–58.

40 Mersenne, *Traité de l'Harmonie universelle*, book 2, théorème XIV, 447: All these great scholars "ont esté si negligens qu'ils n'ont pas fait vne seule experience pour découurir la verité, & pour desabuser le monde. Ie ne pense pas qu'il y ait vn homme de iugement qui vueille maintenant croire ce que disent tous ces Autheurs s'ils ne l'experimentent auparauant, puis qu'ils nous ont donné des fables pour des histoires en vne chose qui est si claire & si euidente".

41 For a good summary of the role of 'experiment' versus 'experience' in Mersenne's work, Peter Dear, "Mersenne et l'expérience scientifique," *Les Études philosophiques* 1/2 (1994): 53–67. Our analysis of organ pipes does not contradict Dear's analysis, but clearly emphasizes the role played by 'experiment' in seventeenth-century France. See also Michel Blay, "Mersenne expérimentateur: Les études sur le mouvement des fluides jusqu'en 1644," *Les Études philosophiques* 1/2 (1994): 69–86. For some perspective, Sigalia Dostrovsky, "Early Vibration Theory: Physics and Music in the Seventeenth Century," *Archive for the History of Exact Sciences* 14 (1975): 169–218. These early experiments could help explain how Mersenne learned to love Galileo, Daniel Garber, "On the Frontlines of the Scientific

Mersenne sought to have experiments done by other parties to confirm his results. To this end, Cornier employed an 'organiste' known within the circle of Rouen's *honnêtes hommes* concerned with musical matters. This organ maker, who also dabbled with pneumatics in Cornelis Drebbel-like submarine experiments, conducted a series of musical experiments to determine how sounds were modified when air went through pipes of equal length but different diameters. Although the organ maker told Cornier that the pipe's size mattered to the production of sound—which was well known to members of his profession—Cornier could not give Mersenne quantitative findings due to the organiste's prolonged absence from the city. He apparently did not leave any detailed written account. Cornier, however, heard from M. Le Febvre—another *curieux* on familiar terms with the said instrument maker (perhaps a member of the Rouen family of organ makers)—that the organiste had taken his measurements from rectangular tin plates, before they were rolled up into pipes, as Mersenne likely suggested. The size, therefore, was the width of the tin plate, or when rolled up the circumference; size was not measured as the diameter (or cross-section) of the pipe. Cornier assured Mersenne that when the organiste returned, everything would be settled to his satisfaction.[42]

Although this set of initial experiments may not have given Mersenne any significant information, these were the sort of evidence he was relying upon. Around the same time, Mersenne performed (or had people perform for him) other experiments in Paris, in which the size of several different pipes was kept constant while the length was adjusted according to musical intervals. Mersenne reported this set of experiments to Father Jean Chastelier, who replied he was no expert himself in these matters. He knew classical texts well enough, however, to quote chapter 10 of Censorinus's third century *De die natali*, in which the Latin author mentioned Pythagoras's own experiments with flutes. The quote Chastelier transcribed in the letter mentioned that flutes of identical size but of different lengths gave the consonances if, like strings, they were cut in the right proportions, i.e., a flute of twelve fingers (*digitorum*) long made a sound an octave inferior to a six-finger flute; and the same six-finger long flute was a fifth superior in tone to the one measuring nine fingers. To Chastelier, such writings were enough to convince him that Mersenne had erred somewhere during his experiments. Moreover, he did not see why he

Revolutions: How Mersenne Learned to Love Galileo," *Perspectives on Science* 12 (2004): 135–63. On the same topic, Renée Raphael, "Galileo's *Discorsi* and Mersenne's *Nouvelle pensées*: Mersenne as a Reader of Galilean 'Experience'," *Nuncius* 23 (2008): 7–36.

42 Cornier to Mersenne, 23 November 1625, CM I, 310; Cornier to Mersenne, 16 March [1626], ibid., 416–7; Cornier to Mersenne, 22 March 1626, ibid., 427–30.

would painstakingly search for the causes of these experimental inaccuracies when they were so remote from the real things—in other words, so far from the truth of the ancients.[43]

What had Mersenne discovered in this set of experiments? That if one used small diameter pipes, say of three *lignes* (roughly 6 mm) and a base length of half a foot, Pythagoras's explanation of consonances described by Chastelier was approximately verified—i.e., if you double the length of this small pipe, it will sound almost a perfect octave lower. But what Mersenne discovered, and Galilei before him, was that in bigger pipes this proposition did not stand anymore. In a series of numerical examples with larger pipes, Mersenne demonstrated that doubling the length of a pipe while at the same time keeping its diameter constant did not produce the required octave; the sound was off by half a tone, a tone, or even more.[44] (Mersenne sometimes defined size [*grosseur*] as the width of tin plates or diameter [cross-section] of pipes.) Similarly, though the Rouen organiste likely did not provide Mersenne with hard evidence, the Minim could report numerous experimental results proving that keeping the length of organ pipes constant while varying the size did not produce the required consonances either. Here he used five half-foot pipes of diameters ranging from three lines to four inches, always doubling in size following the geometrical series. The experiments showed that it was impossible to reach an octave when keeping a pipe's length constant while modifying its diameter. Mersenne wrote that to reach a sound an octave lower, one would have to add two inches in diameter and two feet of length to the biggest pipe. Mersenne's description of the pipes' dimension was precise to make sure that if "one encounters other intervals in pipes larger or smaller, he will have occasion for seeking the reason."[45]

Mersenne's experimental method and results were seriously undermining the tyranny of opinions some contemporaries upheld, e.g., Father Chastelier as seen above. After these two propositions, Mersenne could ascertain with good reason that "Since experiment has shown us that the pipes ought to be of different lengths and widths to make all the pitches of the organ, these two dimensions must be joined together, so as to have sounds which are proportional

43 Father Jean Chastelier to Mersenne, 11 July 1626, CM I, 478–9, where he writes in Latin: "Hæc Censorinus. Quo fit ut facile Antiquis assentiar et experimenta de quibus scribis credam non fuisse accurata. Unde non video cur me excruciem in quærendis causis experimentorum quæ rei veritati non congruunt."

44 Mersenne, "Traité de l'orgue," book 6, prop. XIII, 333–4 (HU3).

45 Mersenne, "Traité de l'orgue," book 6, prop. XII, 331–2 (HU3), quote on p. 331. Mersenne, *The Books on Instruments*, 416.

in their pitch, sweetness and harmony."[46] For someone like de Villiers, however, there were still many hidden things (*bien des choses cachées*) that reason could not account for regarding the ratio between length and width in pipes. De Villiers made his own experiments with rectangular and cylindrical pipes, though they were not organ pipes per se but simple pipes open at both ends, in which he blew in with his mouth. Using two identical tin plates, one shaped into a rectangular pipe and the other into a cylindrical one, he found out there was a minor tone between their respective sounds, the rectangular pipe making a *re* and the cylindrical one an *ut*. Both were four inches (*poulce du Roy*) in length minus one *ligne*, whereas the cylindrical pipe had a diameter of one inch and a half plus one *ligne* and the base diagonal of the rectangular pipe made two inches minus three *lignes*. Two similarly shaped pipes, but twice in length this time, sounded on the contrary in unison. De Villiers was astonished to find out these two pipes sounded the same although their volume was different (which he simply measured by filling them with water). It was strange, de Villiers explained, because the two smaller pipes sounded different, and their volumes were unequal. Why was it not also the case with the pipes twice in length? De Villiers was confused and, because of these results, concluded that a generalization regarding the length and size of organ pipes was almost certainly impossible.[47]

Mersenne knew, however, that the volume of pipes did not automatically determine how they would sound. This he probably owed to organ makers. Mersenne discovered that there was no standardization in the size of tin plates—plates that were later rolled up into cylinders. To make an eight-foot organ pipe, for instance, some organ makers used metallic sheets eight feet tall but with a width of two fifths, a third or a quarter its height. It thus meant that with some adjustments—using an *accordoir*,[48] or cutting a slice from one pipe

46 Mersenne, "Traité de l'orgue," book 6, prop. XIV, 334 (HU3). Mersenne, *The Books on Instruments*, 419. I changed *expérience* to experiment in this translation, which better reflects what Mersenne was doing.

47 De Villiers to Mersenne, 15 May 1635, CM V, 191–3. His conclusion: "Mais je ne trouve pas qu'on doive rien asseurer de ce[ci], parce que je crois autant de diversitez de hauteur et largeur ès tuyaux, il y aura toujours diversité de raison en leur tons, tellement que je crois qu'il est bien difficile de disposer la comparaison des hauteurs et largeurs des tuyaux pour en faire des proportions de tons et tirer des consequences par la reigle de troys." (p. 193) See also de Villiers to Mersenne, 25 February 1635, ibid., 63–4; de Villiers to Mersenne, 1 May 1635, ibid., 148–51.

48 A cone-shaped device used to widen the upper-end of organ pipes in order to adjust how they sound.

or another[49]—two slightly different pipes would sound the same, despite their volumes being different. On this point, artisanal practices were probably even more useful than experiments themselves. To Mersenne, therefore, "it is of little importance in what proportion they are made, so long as they speak well; but since that depends on the industry and opinion of the maker [*Facteur*], and since pipes are met with whose height is five or six times the width, it is not necessary to explain this more extensively."[50]

Mersenne's ultimate objective was to standardize and rationalize the production of organ pipes. As he said, it did not matter whether the width of the tin or lead plates was a third, two fifth or a quarter of its length. This proportion, however, had to remain the same throughout the building process. Mathematical ratios of musical intervals compelled it. From one template, therefore, it became possible to find the precise division of the octave. Regarding how the width and length should vary, Mersenne explained that

> the pipes make the desired interval exactly if their height and width have the same ratio as the said interval. For example, if the diameter of a pipe a foot in length is made 1⅚ of an inch wide [22 *lignes*], it will make the octave perfectly with the pipe half a foot in length, whose diameter is half the other. Thus, it happens that if the pipes are taken as solid cylinders, they ought to be in triple ratio with the ratio of the intervals, which represent the roots, and the simple length of the pipes, as the double ratio of that of the said intervals represent the cylindrical surfaces of the pipes.[51]

49 Mersenne, "Traité de l'orgue," book 6, prop. XXX, 368 (HU3), where he says about the organ makers: "C'est pourquoy il n'est pas necessaire que les Facteurs soient bien exacts à la taille de leurs tuyaux, puis qu'il est necessaire qu'ils y touchent soit en les roignant, ou en les eslargissant & restressissant auec leurs accordoirs : quoy qu'il soit bien à propos qu'ils les fassent en raison triplée de l'interualle des sons qu'ils doiuent faire, afin qu'ils facent des tons mieux proportionnez, & plus plains & nourris. Mais il n'est pas besoin de les aduertir de cecy, puis que nous experimentons qu'ils roignent souuent plusieurs tuyaux d'vn pied entier, ou de plusieurs pouces pour les mettre d'accord."

50 Mersenne, "Traité de l'orgue," book 6, prop. IIII, 319 (HU3). Mersenne, *The Books on Instruments*, 404. In another proposition, Mersenne mentions that "Mais il suffit de remarquer que l'on peut donner vne infinité de differentes figures aux tuyaux tant ouuerts que bouchez, suiuant les differentes inuentions de la Geometrie: par exemple, on les peut faire de parties de parabole, d'hyperbole ou d'ellipse: d'où les Facteurs peuuent tirer de nouuelles graces pour l'harmonie." Ibid., prop. VII, 323. Mersenne makes similar claims in prop. XLIII. See also prop. XVIII, 346 where Mersenne says that a surveyor named Cornu was able to make a rectangular pipe sound in unison with a cylindrical one.

51 Mersenne, "Traité de l'orgue," book 6, prop. XIV, 335 (HU3); Mersenne, *The Book on Instruments*, 419.

In other words, two proportional pipes, one half the other, would sound an octave between them. This 2:1 ratio was thus comparable to an 8:1 ratio in volume. Similarly, a 3:2 ratio in sound (or a fifth) between two pipes would have their ratio in volume changed in a proportion of 27:8. Mersenne produced a table, proposed by a clever surveyor named Cornu, which detailed the volumetric ratio of numerous other intervals.[52]

Such a rationalization of organ pipe making, based on artisanal practices and experiments, would ensure, according to Mersenne, the most accurate of sound production in organs. The internal volume of pipes was not as much a valuable tool of measurement as an a posteriori theoretical reasoning extrapolated from facts. Although experiments and artisanal practice gave Mersenne useful information on the relationship between the construction of organ pipes and how they sounded, simple mathematical order compelled a thorough orderliness once the fundamentals had been discovered. And it is this theoretical conclusion that could be useful to organ makers. Therefore, they "have need of no greater understanding to make their diapason and their very exact pipes, although all the pipes of which I have spoken may not be sufficient to give all the just consonances against each key." That is why, based on the same simple and experimentally determined principles, Mersenne added "another diapason as general as can be wished, as long as one does not go to infinity."[53]

"[T]he table that follows contains all that can be reasonably desired on the subject [the division of the octave], aside from which there is nothing for the makers to know."[54] (Fig. 4L–R) In this full-page table, Mersenne combined knowledge acquired from experiments, artisanal practices, and the theory of mathematical proportions. This table—drawn to scale, the height being one foot (*pied de roy*)—contains eleven columns showing the length and width of organ pipes according to several divisions of the diapason (or octave). Column one shows the thirteen degrees of the equally tempered diapason (twelve equal semitones). Columns two and three show another tempered diapason invented by Francisco Salinas, the famous Spanish music theorist. Columns four and five show thirteen pipes in their perfection, according to a meantone temperament, following the previous proposition. Moreover, "the fourth column

52 Ibid., 335–6.

53 Ibid., 337. Mersenne, *The Book on Instruments*, 421.

54 Mersenne, "Traité de l'orgue," book 6, prop. XV, 338 (HU3); Mersenne, *The Book on Instruments*, 422.

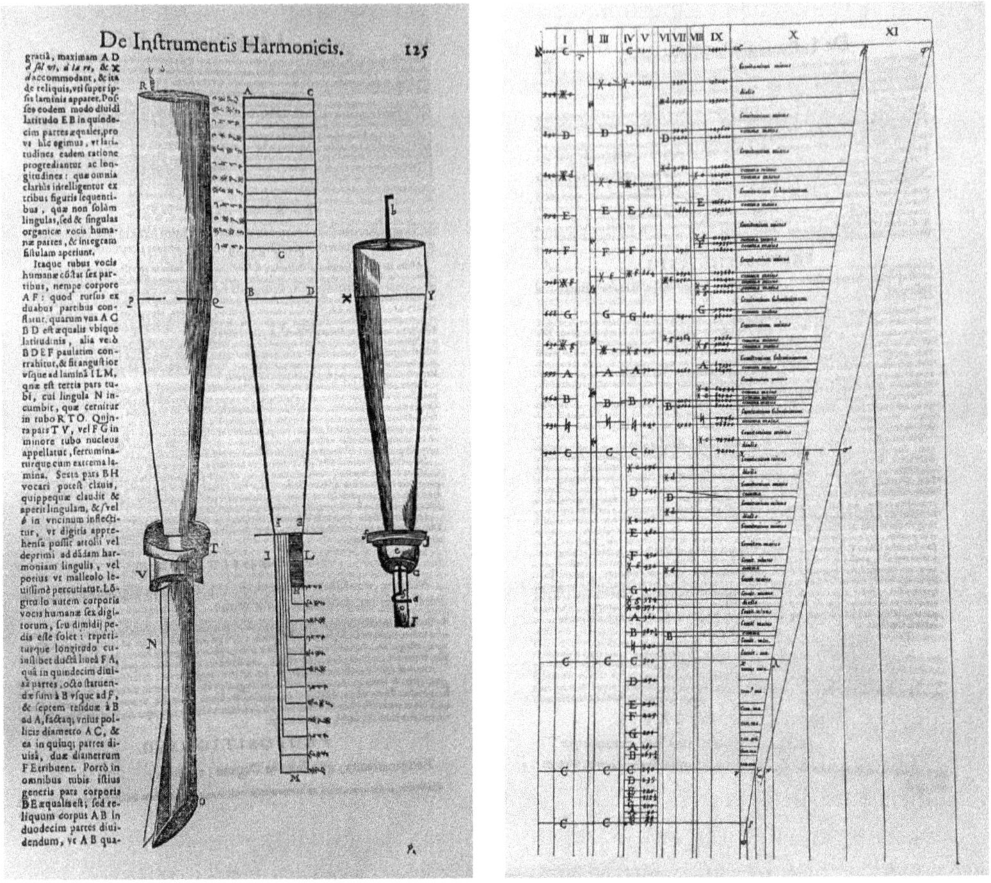

FIGURE 4.L–R Mersenne's flue pipe and table of organ pipe diapasons. Left, a flue pipe showing how it should be cut to obtain the desired musical interval. On the right, the universal table with which organ makers can build the most precise and just set of organ pipes for various diapasons.
MARIN MERSENNE, *HARMONICORUM LIBRI XII* (GENÈVE: MINKOFF REPRINT, 1972 [1636]), BOOK 3, PROP. XIV, 125; BOOK 3, PROP. XVIII, 128

contains the entire four octaves, that is to say the whole keyboard of the organ; and it has thirteen degrees in the first octave, and fourteen in the second, so that there is seen the degree or key which is missing in the said ordinary organ keyboard."[55] Columns six and seven show the nineteen pipes derived from the perfect diapason, which comprise the three genera of music, i.e., diatonic, chromatic, and enharmonic. Next, and

55 Ibid.

> because the seventh column does not comprise all of the perfect diapason of Salinas, which has twenty-five degrees or twenty-five pipes, and since it still lacks two degrees in this one, as I have shown elsewhere, I wish here to propose the most perfect diapason of all those which have been given up to now, that is to say, the one contained in the eighth and ninth columns, for the eighth contains thirteen degrees, which are proper to it (in which the octave of the sixth and seventh columns is excluded) since it has thirty-two degrees, pitches, keys, or pipes in its octave, each of which follows the justness of the harmonic numbers.[56]

Column ten contains all thirty-two intervals of the diapason "from which the makers can take the just measure of all sorts of one-foot pipes, or that of pipes which are smaller or larger. This can be done similarly with the other columns."[57] The last column mentions the width to length proportions (which is somewhat arbitrary, as said earlier) given to the two rectangular templates on which mathematical ratios were based upon: one is one fourth and the other one sixth of the one-foot-long template.

This table systematizes what Mersenne learned from organ makers, experiments conducted by himself or others on organ pipes, and from the pure mathematical theory of music. To Mersenne, such a table could (and should) become the organ maker's fundamental paper *organon*, guaranteeing—in principle at least—the perfection of organ pipes. It was a powerful rhetorical and didactic tool illustrating as well to natural philosophers how important experiments and artisanal knowledge really were in formulating theoretical generalizations.[58]

Mersenne, however, was fully aware of the material shortcomings of his rationalization. He knew, for instance, that there was often no better instrument than the ear to tune an organ. The difficulty did not reside in his theoretical generalization, but rather in actual organ-pipe making. His table would be perfectly fine except for the industry of men (of its lack thereof):

---

56 Ibid., 340. Mersenne, *The Books on Instruments*, 423–4.

57 Ibid.

58 On tables in general, Domenico Bertoloni Meli, "The Role of Numerical Tables in Galileo and Mersenne," *Perspectives on Science* 12 (2004): 164–90. Bertoloni Meli describes many of Mersenne's tables found in the *Harmonie universelle*, which have mostly didactic, philosophical, and esthetic purposes. In our case, however, I believe experimental data are embedded in this table and the way Mersenne constructed it is not solely rooted in theory.

> Unless one is able to trim all the pipes exactly enough and adjust the wind so equally and with so much skill that all the pipes are found in tune without its being necessary to make use of the ears to tune them. This would always happen if one observed all the proportions and circumstances of which I have spoken in this book, if the material should follow the exactness of the mind and if the manual operation corresponded perfectly with science. But since this is beyond the industry of men, who cannot anticipate the great multitude of occurrences which accompany lead, tin, wood, and the other materials with which the pipes are made, and which are met even in the air, I assume that the ears are entirely necessary to tune the pipes.[59]

Reason was not enough. To achieve perfection, countless experiments on organ pipes needed to be performed, and it was where musical instrument makers became so indispensable. Why, for example, did organ pipes sing different intervals when air pressure varied? What was the relationship between air pressure, musical intervals, and the material components of pipes? For Mersenne the "manufacturers can help out Philosophy by preparing a catalogue of the pipes which rise only a semitone, or a third, or a fourth, or a fifth, etc., for it will be easier to find the reason when one understands the qualities of the pipes which cause the difference of these pitches."[60] Francis Bacon's empirical method of data accumulation was perhaps an influence here. (Although critical of Lord Verulam's brute fact gathering, Mersenne was

59 Mersenne, "Traité de l'orgue," book 6, prop. XXIX, 363 (HU3). Mersenne, *The Book on Instruments*, 446. He repeats something similar in the subsequent proposition: "Or si l'on pouuuoit tailler les tuyaux si iustes, & leur donner le vent si esgal, qu'ils se trouuassent d'accord en les mettant sur les registres, sans qu'il fust besoin de les roigner, ou de les toucher de l'accordoir, l'oreille ne seroit pas necessaire pour accorder, mais il est tres-difficile de faire les tuyaux si iustes qu'il n'y faille nullement toucher pour les eslargir, estressir, accourcir, ou alterer, soit que l'on vse de temperament, ou que l'on les tienne iustes suiuant les raisons harmoniques, d'autant qu'il est trop difficile d'obseruer toutes les proportions des bouches, des languettes & de l'ouuerture des pieds, & que la matière des differents tuyaux n'est pas semblable: ioint que les petits tuyaux proportionnez comme les grans, ne font pas les mesmes interualles, comme i'ay remarqué en parlant de ceux qui sont d'esgale hauteur & de differente largeur." (pp. 367–8) On the type of materials and their use in making organ pipes over time, Patrizio Barbieri, *Physics of Wind Instruments and Organ Pipes 1100–2010* (Rome: Il Levante Libreria Editrice, 2013), 1–27.

60 Mersenne, "Traité de l'orgue," book 6, prop. XXIX [sic, XIX], 346–7 (HU3). Mersenne, *The Book on Instruments*, 430. Mersenne was interested in this topic since the early 1620s. See Cornier to Mersenne, 18 August [1625], CM I, 263–7.

nonetheless working on a translation of *Sylvan sylvarum*.[61]) And to accomplish such a monumental and difficult labor, who would be better qualified than the people actually building and designing organs?

2.2 *Organology or the Art of Musical Instrument Making*

Organology is the "study of musical instruments in terms of their history and social function, design, construction and relation to performance." It is a modern historiographical concept that deals with the scientific study of musical instruments and the symbolism and folklore functions of the material culture of music as investigated, for example, by ethnomusicologists.[62] I want to use it here in a particular way, not unlike John Tresch's and Emily Dolan's Foucauldian new organology concept, as an 'ethics of instruments' put forward by Mersenne to enlighten the role and value of artisans and musical instruments (and material culture in general) toward the aim of perfecting the mechanical arts themselves.[63] Not only were artisans useful in performing experiments, as described in the previous section, they were also the sole source of hands-on knowledge, which according to most natural philosophers needed to be rationalized. Studying and describing the art of musical instrument making meticulously gave Mersenne compelling arguments against armchair philosophers. Experiments went hand-in-hand with instrument making, and answers to the improvement of natural philosophy were unearthed in the

---

61 Cornier to Mersenne, 24 December [1627], 611–2: "Si vous pouvés achever la traduction du *Sylva Sylvarum* à l'ayde de vostre Anglois et le donner au public, je croy que vous feriés une chose fort aggreable à beaucoup de monde. Pour moy je vous diray que je n'estime pas tant en Bacon la curiosité de ses experiences comme les consequences qu'il en tire, et la methode, avec laquelle il s'en sert. C'est pourquoy (encor que ses observations soient fort ordinaires), je pense que ce seroit une chose fort aggreable à beaucoup de monde de congnoistre ses procedés." On Mersenne's criticism, found mostly in his *Vérité des sciences* (1625) and focused on Bacon's *Novum organum*, see Robert Lenoble, *Mersenne, ou la naissance du mécanisme* (Paris: Vrin, 1943), 325–35. See also Claudio Buccolini, "Mersenne et la philosophie baconienne en France à l'époque de Descartes," in *Bacon et Descartes. Genèse de la modernité philosophique*, ed. Elodie Cassan (Paris: ENS Editions, 2014), chapter 6.

62 Laurence Libin, "Organology," *Grove Music Online* (accessed 12 September 2006). On general issues, Geneviève Dournon, "Instrumentariums et classifications," *Revue de musicologie* 79 (1993): 197–207. More specifically, Conny Restle, "Organology: The Study of Musical Instruments in the 17th Century," in *Instruments in Art and Science. On the Architectonics of Cultural Boundaries in the 17th Century*, eds. Helmar Schramm, Ludger Schwarte, and Jan Lazardzig (Berlin: Walter de Gruyter, 2008), 257–68.

63 John Tresch and Emily Dolan, "Toward a New Organology: Instruments of Music and Science," in *Music, Sound, and the Laboratory from 1750–1980*, eds. Alexandra Hui, Julia Kursell, and Myles Jackson, *Osiris* 28 (2013): 278–98.

workshops. This argument is not novel, yet Mersenne's books on instruments are a significant case in point, which has never been fully explored.[64]

Michael Prætorius held a genuine admiration toward musical instrument makers: "When we come to describe musical instruments, we should treat them as the art-works of outstanding, intelligent craftsmen, who have brought them into being by manual and intellectual effort. By applying precise plans to suitable materials, they have skillfully fashioned instruments which can be used to publish the glory of God, or—which is perfectly legitimate—to give pleasure to mankind, with their sweet, harmonious sounds."[65] Mersenne did not oppose the common *joueurs d'instruments*, who delighted simple folks in the street with their sweet harmonies, yet he believed they were less valuable than bona fide artisans.[66] Indeed, musical instrument makers were truly central to Mersenne's work. As he reminded the readers of the book on organ, due to the great and multifaceted complexity of this machine, "whatever one may say and whatever figures one can give to explain everything that concerns the construction of the organ, it is very difficult to have it understood when one has not seen one made, or has not considered the pieces in the large as well as in detail."[67] To fully understand how an organ works, therefore, one had to observe how it was actually put together. This can mean only one thing: Mersenne regularly conferred with organ makers—and as a rule with musical instrument makers. There is no other way he could have collected all the

64 The classic literature on this topic is Paolo Rossi, *Philosophy, Technology, and the Arts in the Early Modern Era*, trans. Salvator Attanasio (New York: Harper & Row, 1970). Jim Bennett, "The Mechanic's Philosophy and the Mechanical Philosophy," *History of Science* 24 (1986): 1–28. Pamela O. Long, "Power, Patronage, and the Authorship of *Ars*: From Mechanical Know-How to Mechanical Knowledge in the Last Scribal Age," *Isis* 87 (1997): 1–41; Paula Findlen, *Possessing Nature: Museums, Collecting, and Scientific Culture in Early Modern Italy* (Berkeley: University of California Press, 1994). Lorraine Daston and Katharine Park, *Wonders and the Order of Nature, 1150–1750* (New York: Zone Books, 1998). Deborah Harkness, *The Jewel House: Elizabethan London and the Scientific Revolution* (New Haven: Yale University Press, 2007). Pamela Smith, *The Body of the Artisan. Art and Experience in the Scientific Revolution* (Chicago: The University of Chicago Press, 2004). Paola Bertucci, *Artisanal Enlightenment. Science and the Mechanical Arts in Old Regime France* (New Haven: Yale University Press, 2017). Alexander Marr, *Between Raphael and Galileo. Mutio Oddi and the Mathematical Culture of Late Renaissance Italy* (Chicago: The University of Chicago Press, 2011).

65 Michael Prætorius, *Syntagma musicum II: De organographia*, trans. and ed. David Z. Crookes (Oxford: Clarendon Press, 1986), 21.

66 On Mersenne's view of *ménestriers*, see *Les Prelvdes de l'Harmonie vniverselle, ov qvestions curievses Vtiles aux Predicateurs, aux Theologiens, aux Astrologues, aux Medecins & aux Philosophes* (Paris: Chez Henry Gvenon, 1634), 185.

67 Mersenne, "Traité de l'orgue," book 6, préface au lecteur, n.p. (HU3). Mersenne, *The Books on Instruments*, 392.

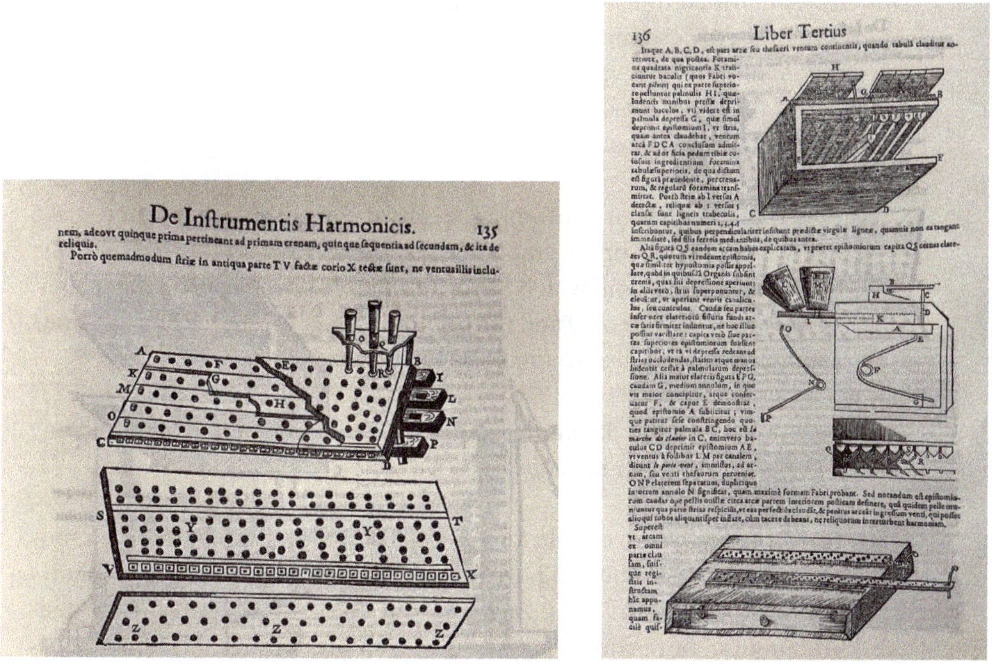

FIGURE 5.L–R Mersenne's illustrations of an organ wind-chest and clavier mechanisms. Both engravings show how the organ's wind chest and clavier are linked together. Though these engravings were copied from Salomon de Caus, their descriptions were based on actual artisanal knowledge. De Caus's *Raison des forces mouuantes*, book 3, 10–12.
MARIN MERSENNE, *HARMONICORUM LIBRI XII* (GENÈVE: MINKOFF REPRINT, 1972 [1636]), BOOK 3, PROP. XXV, 135–6

facts on the choice of woods, metals, tools, glue, etc. and could explain how to resolve tricky technical obstacles such as building the organ's pipes, wind chest and keyboard. (Fig. 5L–R)

In a few instances, Mersenne did borrow from and refer to published works, such as Salomon de Caus's *Raisons des forces mouuantes* to describe the clavier, the wind-chest, and how organ pipes were molded and shaped. (Even though some of Mersenne's friends thought de Caus was a rather poor music theorist and organ maker.[68]) He also studied Vannoccio Biringuccio's *De la pirotechnia*,

68 Cornier to Mersenne, 27 January 1626, CM I, 350–1: "A la premiere [lettre de Mersenne reçue par Cornier] donc je vous diray qu'entre tous ceux que je congnois qui ont quelque intelligence de la musique, De Caux est tenu pour fort peu intelligent en ce qui concerne cete science et ay-je oüi tenir que, pour son honneur, il eust fort bien faict de ne publier pas son livre. Il me semble aussi que lors que vous estiez icy [son voyage à Rouen en mai 1625], Le Vasseur vous dict quelque chose de cela mesmes à propos de ces tuyaux

originally published in 1540 but translated into French and printed in Rouen in 1627, in order to describe the fabrication of bellows and most importantly to expound on how bells of various shapes and sizes were cast-ironed and tuned. Yet most of the time book knowledge was insufficient. On metalworking, for example, Mersenne writes:

> Now if one wishes to know more, he must consult the casters and make many experiments on this matter, with sulfur, which makes red iron cast by its touch alone, and which has an admirable power over all the metals, as well as with antinomy and the other minerals. From this one can draw enough understanding to establish a particular science.[69]

But Mersenne was also fully aware that bell makers were not always utilizing reason in making and tuning large bells:

> Now this *brochette*, or rule, of the bells so follows the harmonic ratios of the tones, that one will have perfect tuning of the bells if one follows the thicknesses marked on each line. But I wish to explain here the general method of making whatever diapason one wishes, for the founders go only by guessing when they have bells to cast weightier, thicker, and larger than those which are marked on their *brochettes*, which they cannot increase because they do not know the construction through a certain and infallible science which I explain here.[70]

---

d'orgues et vous tesmoigna avoir une opinion contraire à ce que dict de Caux." De Caus's book on music is *Institutions harmoniques* (Frankfurt, 1615) and on organ making, *Raisons des forces mouuantes* (Frankfurt, 1615). See also Cornier to Mersenne, 21 September 1625, ibid., 294.

69 Mersenne, "Des instrumens de percvssion," book 7, prop. VI, 8 (HU3). Mersenne, *The Books on Instruments*, 507. On the generation of metals in the earth, Mersenne mentions Palissy and Agricola. On Biringuccio, Cornier to Mersenne, 16 March 1626, CM I, 417–8; Cornier to Mersenne, 29 March [1628], CM II, 50–51.

70 Mersenne, "Des instrumens de percvssion," book 7, prop. VII, 11 (HU3). Mersenne, *The Books on Instruments*, 510. Mersenne's correspondent in Nevers mentioned how makers also went 'à taston' while making bells: "Il n'y a aucun maistre fondeur en cette ville [Nevers], duquel on puisse sçavoir de combien il amoindrit chaque cloche pour chaque ton, ny combien il donne d'espesseur au diamètre et à la hauteur de la cloche, ny en quelle proportion il mesle ses matieres pour faire son metail. Bien ay-je ouy dire que les fondeurs permettent à chacun de jetter dans la fonte, qui une piece d'argent, qui un sol ou autre métail. Ce qui me faict croire que communement ils n'y regardent de si prez, quoy que je sçache qu'en plusieurs lieux les choches y sont fort bien d'accord, et en autres elles sonnent musicalement." Bredeau to Mersenne, 13 July 1628, CM II, 99.

Mersenne no doubt wanted to rationalize artisanal practices when possible—perhaps more than Descartes, who told Mersenne he had never considered what would be the most perfect shape of a bell;[71] and, according to Doni, musical instrument makers lacked so much in theoretical knowledge that the Italian scholar found no harpsichord maker in Rome who had ever heard of the monochord.[72] Theory was one thing, which Mersenne could describe from other books or deduce from experiment. With regard to the precise description of artisanal practices, however, Mersenne could not have done it without engaging in an authentic workshop experience.

All in all, information gathered from printed works—or manuscripts[73]—is thin in Mersenne's books on instruments, and thus cannot explain the incredible number of statements about artisanal knowledge found in the book on organs (or in any of the others). Take, for instance, Mersenne's description on how to shape and weld an organ pipe (quoted at length here):

> As to the solder and the method of soldering the pipes, it must be remarked that the mixture is composed of a twelfth part of tin to the lead, and that it must be forged quite evenly on a polished anvil like that of the tinsmiths. Now after the pipes are proportioned, one takes each of them aside, and before rolling it on the mill, one rubs the sides of the tube with chalk diluted with a little water and gum, and for this one heats the mixture a little. This done, one begins to bend the side of the tube, which ought to be along the length of the mould on which it is rolled. Afterwards, one strikes all around the said mould with a rule that is completely flat and rather long. And then the body of the tube is removed from the mould, so that there remains the width of a feather between the ends which are to be joined together, so as to rub with the preceding mixture. And after it is dry, one takes a small knife, which is so adjusted in the hand that the length of the blade passes along the thumb and its shaft is between the little finger and the fourth. And then one places the point,

71 Descartes to Mersenne, January 1630, *Œuvres de Descartes*, eds. Charles Adam and Paul Tannery, new edn., 11 vols. (Paris: Vrin, 1996), vol. 1, 111.

72 Doni to Mersenne, 8 April 1634, CM IV, 88–9: "Vous ne sçauriez croire combien regne l'ignorance par deça en faict de teorie et vous le pouriez juger de cecy, que voulant faire un monochorde (encores que de plusieurs cordes), je n'ay trouvé personne de ces faiseurs de clavicimbles, dont il y en a de fort experts, qui en ayent ouy parler; et je suis fort estonné qu'un estude si beau et si plaisant soit si fort delaissé."

73 This is to be expected for artisanal knowledge. For example, I found one manuscript destined for engravers, sculptors, painters, gilders, and draughtsmen in the Bibliothèque nationale de France, département des manuscrits français, 9155. Folio 46r ff. that deals with how to decorate wood for furniture, cabinets, luths, and violins.

> pressed by the thumb, lightly upon the edge of each side, and by running along the edge of each side from one end to the other, one presses them so that being joined together they form a small channel or gutter. Now having grated it, one takes the end of a candle and rubs on it so that the solder will run along it better. This is made of a point of lead, a pound of fine tin, and a quarter of tin-glass for the lead pipes. But one makes use of two pounds of pure tin, one pound of lead and a quarter of tin-glass of those of tin. Yet this mixture depends on the judgement and pleasure of the manufacturers. To this must be added that to solder the pipes well, it is necessary to place a little solder at the two ends before having it run the length, so as to fix the two sides of the body and to straighten and adjust one with the other.[74]

Neither Prætorius's celebrated *organographia* nor de Caus's engineer-like works described in print, in so many circumstances, the steps and techniques required to fashion an organ pipe. (Mersenne made numerous highly-detailed descriptions, e.g., the wind chest.[75]) Contrary to what might be expected, Mersenne was accustomed to call on artisans and converse with them. He was, for instance, so thrilled about a Spanish monk visiting Paris, who apparently built organs as well as played them, that two of his correspondents warned him against his own enthusiasm.[76] Although few such examples are available to us, one is thoroughly interesting and can serve to prove this point.

It starts with a letter from Peiresc dated early May 1634. In it, the Provence gentilhomme told Mersenne that a young lawyer from Aix, named Mr. Gailhard, would visit him on a mission: to gather every imaginable piece of information on public and private fountains. Peiresc wanted to know—presumably for one or both of his estates—the 'vulgar' names of all the parts (tools, machines, pipes, conduits, faucets, basins, etc.) needed for the manufacture and ornamentation of fountains; the correct proportions of water conduits depending on the space available, and how they should be organized to vary the flow

74 Mersenne, "Traité de l'orgue," book 6, prop. VI, 321–2 (HU3). Mersenne, *The Books on Instruments*, 406. See also prop. XVII, 344–5. For a general description of organ-pipe making, Barbieri, *Physics of Wind Instruments and Organ Pipes 1100–2010*, 29–92. When possible and deemed pertinent, Mersenne quoted ancient knowledge. In his *Harmonicorum libri XII* Latin edition of 1636, for instance, Mersenne refers to Vitruvius in discussing the *étoffe* used to make organ pipes and other metallic parts instead of artisans, or *Facteurs*—which was obviously meant to serve a purpose other than exposing artisanal epistemology. Mersenne, *Harmonicorum libri XII*, prop. II, 115.

75 Mersenne, "Traité de l'orgue," book 6, prop. XLIV, 400–04 (HU3).

76 Cornier to Mersenne, 16 March [1626], CM I, 417; Father Chastelier to Mersenne, 11 July 1626, ibid., 479.

of water; and finally all the prices and the names of the artisans who performed this kind of work. Peiresc believed that Mersenne would have more patience than the young messenger to accomplish such a task. This should be a simple and enjoyable undertaking—Peiresc indicated—spent browsing within the workshops of the most intelligent and capable artisans. Peiresc, moreover, thought that knowing these specific matters would help to rectify (*redresser*) poor artisanal practices.[77] (Peiresc himself, in his native South, seemed accustomed to interacting with artisans, calling them his 'best friends' when requiring specific information about types of leather, for instance, for binding books.[78])

Whether Mersenne actually wanted to rectify the work of *fontainiers*, he nevertheless complied with Peiresc's request. To do so, he consulted the overseer of the Rougi and Belleville-sur-Sablon fountains in the suburbs of Paris. The large number of potential difficulties in building such an extensive work had Mersenne write to Peiresc that the latter should send someone on site to study all there was to learn. Mersenne saw all the tools of the trade, and he told Peiresc that besides a manager, carpenters, stonecutters, plumbers, and bricklayers would be required. Mersenne even told Peiresc he could make experiments to better understand the flow of water in pipes—which remained one of the major problems at the time—and to determine why the wind caused loud noises in conduits.[79]

Peiresc was happy with Mersenne's first account. He also made sure the commission did not do violence to the Minim's natural inclination—implying here that perhaps Mersenne might be uncomfortable in the company of artisans. Mersenne obliged gracefully, though, never complaining—which he was otherwise prone to do. On a fête day, he even went 1,500 *pas* underground to carefully look at water conduits; the noise of the villagers, however, was so loud that he could not observe anything. He told Peiresc he would go back on a working day, when fewer people were in the streets. And when the Parisian *fontainier* became less cooperative with time, Mersenne was forced to call

77 Peiresc to Mersenne, 1 May 1634, CM IV, 110–11; Peiresc to Mersenne, 18 June 1634, ibid., 182–3. Peiresc was in the habit of asking such detailed memoirs. Around the same time, he asked Father Cœlestinus to spend around ten days in the vicinity of Mount Kasios (North-East of Alep) to look at the possible remnants of the ancient cult of Jupiter Cassius, observe the lunar eclipse of August 1635, investigate various natural properties of the mountain, transcribe any type of inscription he could find and back on the plain, and ask questions to the nearby peasants on vapors and fumes that ascended the mountain. Brentjes, *Peiresc's Interest in the Middle East and Northern Africa*, 2–3.

78 Miller, *Peiresc's Mediterranean World*, 181.

79 On these experiments, Michel Blay, "Mersenne expérimentateur," 72–73.

on another who worked for a gentleman in Liencourt, in the Pas-de-Calais.[80] Peiresc fancied a complete and comprehensive description on fountain making, similar perhaps to the ones Mersenne gave on musical instruments.

The information Peiresc sought was simply not found in books like Salomon de Caus's hydraulic opus. Exact techniques and artisanal practices, unsurprisingly, were available only from those who performed them. But what about secrecy? Artisans and engineers like de Caus, or the well-known Girard Desargues, often petitioned for lucrative contracts with the city of Paris to take care of waterworks. Hence, were Peiresc and Mersenne hindered in their fact-gathering project on fountains due to trade secrets? Mersenne alluded to this problem, observing that the *gens mecaniques* were afraid people like him desired to strip them of their *gaigne-pain*. It was (and still is) the usual trope, also acknowledged by de Villiers's hired *organiste* pertaining to the tuning of his organ; or by Peiresc, who told Mersenne about a bell maker that could give him helpful guidance, if it was not from the fact that, as all artisans, this one was jealous of his art.[81]

Was Mersenne a threat to artisans? Musical instrument makers (*maîtres faiseurs d'instruments de musique*) were incorporated by a letters patent signed in 1599 by Henry IV, perhaps a mark of their ever-growing numbers toward the end of the sixteenth century. Before the guild was established, most musicians crafted their musical instruments. Within the new regulation, musical instrument makers were not as free as before, but they were still allowed some leniency regarding what they could build. For instance, they were permitted by the letters patent to manufacture and decorate cases for their instruments, which stepped on the exclusive rights of leather makers (*gainiers*); likewise for instrument ornamentation, inlaid work of wood and precious metals, infringing on cabinetmakers (*ébénistes*) prerogatives.[82] Only individuals who were

80 Mersenne to Peiresc, 2 July 1634, CM IV, 230–31; Peiresc to Mersenne, 16 July 1634, ibid., 247–9; Mersenne to Peiresc, 26 July 1634, ibid., 268; Mersenne to Peiresc, 24 August 1634, ibid., 328; Mersenne to Peiresc, 2 February 1635, CM V, 47; Peiresc to Mersenne, 8 May 1635, ibid., 175–6; Mersenne to Peiresc, 17 May 1635, ibid., 203.

81 On fountains in general and how 'ordinary' they were in early modern France, Daniel Roche, *Histoire des choses banales. Naissance de la consommation, XVIIe–XIXe siècle* (Paris: Fayard, 1997), 169–74. See also Simon Werrett, "Wonders Never Cease: Descartes's *Météores* and the Rainbow Fountain," *British Journal for the History of Science* 34 (2001): 129–47. On de Caus's and Desargues's appointments with the city of Paris, *Registres des délibérations du bureau de la ville de Paris, publiés par les soins du Service historique*, 20 vols. (Paris: Imprimerie nationale, 1883–1984), vol. 18, 58–9; vol. 19, 275–6. De Villiers to Mersenne, 25 February 1635, CM V, 53–4; Peiresc to Mersenne, [6 September 1633], CM III, 474–5; Mersenne to Peiresc, 2 February 1635, CM V, 47.

82 René de Lespinasse, *Les Métiers et corporations de la ville de Paris*, 3 vols. (Paris: Imprimerie Nationale, 1886–1897), vol. 3, 593–6.

incorporated—namely those who had done their six-year apprenticeship in *bonne et due forme*—were allowed to build and sell musical instruments in Paris. Some artisans may have strongly defended their right to secrecy, but scores of *faiseurs d'instruments* lived in Paris in the first half of the seventeenth century, and several of them were prosperous—and educated—enough to partake in natural philosophical exchanges with an erudite like Mersenne.[83] The harpsichord maker Jean Denis wrote that all sorts of people came to visit him; some to discuss his work—as Mersenne likely did numerous times—and others to play on his harpsichords, during which time he would detect all the awkward gestures (*simagrées*) younger and older customers displayed in public.[84]

Mersenne, of course, wanted neither to build nor sell musical instruments. And even though the information he gathered from artisans was sometimes exhaustive, the question remains: would someone be capable of building an organ (or any other instrument for that matter) solely based on Mersenne's written descriptions? It is very doubtful. Mersenne often referred to the artisan's expertise in making such and such instrument's parts. For the spinet, its excellence depended upon "many conditions and particulars, but especially on the braces one places under the table, inasmuch as it is difficult to brace the spinets perfectly, and it is one of the greatest secrets of the art, the research of which I leave to the manufacturers."[85] In another proposition on the harpsichord, Mersenne even mentioned that at least 1,500 separate pieces were

83 The number of musical instrument makers in Paris can be approximated from the official records of the *minutier central*. Not all of them were as well-off as Jean Desmoulins, but a few seem to have been part of the bourgeois society. Take Thomas Le Vacher, 'facteur d'instruments de musique et bourgeois de Paris,' who inhabited a house on rue de la Pelleterie. After he died in 1624, furniture worth 175 *livres*, cloths worth 48 *livres*, a library containing ten books, and musical instruments worth 264 *livres* were found in his house. Jurgens, *Documents du Minutier central concernant l'histoire de la musique*, vol. 1, 772. The documents show that these instrument makers had a much nicer signature than other artisans, reflecting a better education. (Ibid., 26–8). See also Catherine Massip, "Facteurs d'instruments et maîtres à danser parisiens au XVIIe siècle," in *Instrumentistes et luthiers parisiens, XVIIe–XIXe siècles*, ed. Florence Gétreau (Paris: Délégation à l'action artistique de la Ville de Paris, 1988), 17–34.

84 Jean Denis, *Traité de l'accord de l'espinette*, new introduction by Alan Curtis (New York: Da Capo Press, 1969 [second ed. of 1650]), 39. Mersenne mentioned Denis in the "Traité des instrvmens a chordes," book 3, prop. XX, 159 (HU3): "Il faut seulement remarquer que l'vn des principaux secrets de l'Epinette consiste à barrer la table, dont la bonté depend de l'excellente barrure, qui a esté pratiquée en perfection par Anthoine Potin, & Emery ou Mederic, que l'on recognoist auoir esté les meilleurs Facteurs de France, ausquels les meilleurs Facteurs de maintenant, à sçauoir Iean Iacquet, le Breton, & Iean Denys ont succedé, lesquels sont excellents en leur art".

85 Mersenne, "Traité des instrvmens a chordes," book 3, prop. I, 107 (HU3). Mersenne, *The Books on Instruments*, 159–60.

necessary to build one of these clavier instruments. He did not describe all the pieces in detail, which would make it virtually impossible for anyone without previous knowledge, experience, or access to a harpsichord maker to build a working model.[86] Moreover, it appears that some knowledge-sharing between different guilds may have been more common than is usually believed. Organ makers, for instance, needed to call upon artisans from other specialized crafts to learn the secrets of glue making, in order to experiment with how organ pipes sounded when made from different materials.[87] The six-year apprenticeship to enter the *faiseurs d'instruments'* guild, therefore, was not for nothing: musical instruments, most importantly the organ, were some of the most difficult early modern pieces of technology to manufacture. The concept of secrecy may be valid between artisans, but it probably lacked any significance or consequence between an artisan and a savant: both knew that a written text could never replace hands-on experience.[88] Not unlike Peiresc's fountains, Mersenne probably believed it was essential to grasp as many aspects as possible of the mechanical art of musical instrument making to understand it and, essentially, improve it further.

Though substantial and often drawn from life, Mersenne's descriptions of musical instruments were nonetheless akin to Renaissance theaters of machines, *theatrum mechanicorum* illustrations: they were convincing, but largely insufficient to create working models, as Jean Ferrier tried to tell Descartes in their famous epistolary exchange regarding his lens-grinding machine.[89] Mersenne understood well enough the craftsmanship of organs

86 Mersenne, "Traité des instrvmens a chordes," book 3, prop. XX, 159 (HU3).

87 Mersenne did not describe in detail all the possible glues "d'autant que cela ne sert de rien à nostre suiet, & que les Facteurs d'Orgues le peuuent apprendre des artisans qui s'en seruent, s'ils sont assez curieux pour experimenter la diuersité des sons, qui se peuuent faire par des tuyaux de toutes sortes de matiere, dont ils peuuent tirer beaucoup de secrets pour l'harmonie." Mersenne, "Traité de l'orgue," book 6, prop. XVII, 345 (HU3).

88 Several scholars have discussed this topic over the last twenty years. See, for instance, Pamela Long, *Openness, Secrecy, Authorship: Technical Arts and the Culture of Knowledge from Antiquity to the Renaissance* (Baltimore: The Johns Hopkins University Press, 2001). Pascal Dubourg Glatigny and Hélène Vérin, eds., *Réduire en art: La technologie de la Renaissance aux Lumières* (Paris: Éditions de la Maison des sciences de l'Homme, 2008). Koen Vermeir and Dániel Margócsy, eds., *States of Secrecy*, special issue of *The British Journal for the History of Science* 45, no.2 (2012): 153–280. Pamela H. Smith, *From Lived Experience to the Written Word: Reconstructing Practical Knowledge in the Early Modern World* (Chicago: The University of Chicago Press, 2022).

89 On the topic of machine representations, Wolfgang Lefèvre, ed., *Picturing Machines, 1400–1700* (Cambridge, Mass.: The MIT Press, 2004). On Descartes and Ferrier, Gauvin, "Artisans, Machines, and Descartes's *Organon*," *History of Science* 44 (2006): 187–216.

to recognize the important features of this technology, explaining in one particular proposition how to examine and judge the excellence of an organ.[90] The interesting thing here is that Mersenne's explanation was not unlike the contract an organ maker would put in writing to build or repair an organ. Everything from the bellows, wind chest, organ pipes and the choice of registration (*jeux*) gave indications of the quality and purpose of an organ.[91] As those contracts show, moreover, there was no unique way to build an organ. Each organ was different and personalized, depending on the quality of the material used and the allocated space to set it in. There was thus no single template. Organ makers always had to adapt the dimensions of the wind chest and bellows to fit them to a specific location, whether in a church or a private home. Mersenne was fully aware of this matter, mentioning that a typical wind chest for a principal 8' (an eight-foot pitch),

> such as ordinarily those are in churches where there is only one organ cabinet, one can make the wind chest five or six feet in length, according to the judgment and preference of the maker, for there are some who more easily find place for their stops on a wind chest of four feet, than others do on one of five feet. That is why I do not in any way limit the size so that the cleverness of the manufacturers can decrease them [*que la subtilité des Facteurs n'en puise diminuer*].[92]

Although there were probably as many design variations of musical instruments as there were *faiseurs d'instruments*, one still has the feeling reading the books on instruments that Mersenne sought to standardize, to harmonize the different artisanal practices; each instrument was described as a coherent and theoretical entity, an unproblematic ideal-type model based on a rationalization

90 Mersenne, "Traité de l'orgue," book 6, prop. XXXVII, 382–4 (HU3).

91 Dufourcq, *Le Livre de l'orgue français*, vol. 1, for the variety of contracts. Jurgens, *Documents du Minutier central concernant l'histoire de la musique* also gives good examples of contracts during the first half of the seventeenth century. For instance, the one signed on 23 December 1635, *Marché pour la réparation des orgues de l'église Saint-Jean-en Grève, par Valeran de Heman* (vol. 1, 814–6), one of the two organ makers mentioned by Mersenne, and from whom he could have taken most of his information on organ building.

92 Mersenne, "Traité de l'orgue," book 6, prop. XLIV, 400 (HU3). Mersenne, *The Books on Instruments*, 481. For the general description of the bellows, Ibid., prop. XXXIV, 378. Mersenne's descriptions are in agreement with Dufourcq's comprehensive analysis of the French organ. Dufourcq, *Le Livre de l'orgue français*, esp. vol. 3.

of craftsmanship.[93] Mersenne's seven books on instruments were more than an illustrated compendium: they were a rhetorical and epistemic tool aimed at *honnêtes hommes* and natural philosophers, which cogently demonstrated that improving the mechanical arts meant one had first to study them carefully, in every possible detail.

Artisans had a crucial role toward the understanding and perfecting of the mechanical arts. The instruments they crafted and their expertise with various kinds of materials were essential for the natural philosophical practice of experiments. As was demonstrated, earlier experiments with organ pipes led to theoretical generalizations about sound production. The reverse process, however, was not at all excluded—and of obvious significance. Not every aspect of musical harmony required artisanal help and practical experiments.

2.3 *The Theory of Organ Claviers*

Mersenne strongly believed that the mathematical foundation of music theory could greatly improve the practice of organ playing as well. He demonstrated this by studying the 'science of organ claviers.' In fact, Mersenne explained that "Zarlino would not have taken so much pains in explaining the syntone of Ptolemy, which misses many degrees, if he had had an understanding of the keyboards that I propose in the treatise on the spinet and the organ."[94] Mersenne, in brief, tried to relocate the complete knowledge of musical genres into a mechanical device, i.e., the organ clavier.[95]

Organ claviers—all clavier instruments actually—had an obvious problem: "Since I [Mersenne] have shown that the keyboard and the diapason, which contains the diatonic genre, which is used now in its perfection, have 32, 27, 25, or at least 19 keys or degrees in each octave, and since the ordinary keyboards, organ as well as spinet, have only thirteen, it follows that they cannot be just, if one wants to find in them everything that is in the nineteen degrees of the

93 What Mersenne did was similar to the eighteenth-century pre-objectivity "truth-to-nature" atlases, in which what the "image represented, or ought to represent, was not the actual individual specimen before them but an idealized, perfected, or at least characteristic exemplar of a species or other natural kind." Lorraine Daston and Peter Galison, *Objectivity* (New York: Zone Books, 2007), 42.

94 Mersenne, "Traité des instrvmens a chordes," book 1, prop. 3, 9 (HU3). Mersenne, *The Books on Instruments*, 22–3 (translation slightly modified).

95 Mersenne, "Traité de la voix, et des chants," book 3 entitled "Des genres, des especes, des Systemes, & des Modes de la Musique." (HU2) A very good case study on the role of mathematics (irrational numbers), music and keyboard manufacture comes a century before Mersenne, from Nicola Vicentino. Peter Pesic, "Hearing the Irrational: Music and the Development of the Modern Concept of Number," *Isis* 101 (2010): 501–30.

perfect keyboard [for the diatonic genre]."[96] Experiments revealed how to build organ pipes that would produce the sounds of the perfect diapason; now a clavier had to match this same perfect diapason, so that theory and practice could ultimately work hand-in-hand.

The traditional organ clavier (or keyboard) had (and still has today) thirteen keys (*marches*), which were tuned according to the meantone temperament, i.e., a tuning made of pure major thirds and in which the fifths are smaller and the fourths larger by a quarter of the syntonic comma—hence the label ¼-comma meantone. As quoted above by Mersenne, this sort of tuning does not give pure consonances, but tempered ones. To Mersenne, a good tuning had to follow the just intonation of consonances, not simply for musical or mathematical reasons, but more importantly to fight skepticism. In *La Vérité des sciences* (1625) and *Questions harmoniques* (1634) Mersenne declared that skeptics argued against music being a science because there existed no certain and evident principles to provide concord based on simple mathematical ratios. Tempered musical intervals, found on virtually all musical instruments, only conveyed a sense that music was established on the vagueness and ambiguity of music theorists, musicians, and musical instrument makers. For it was well known that string instruments were usually tuned to an equal temperament (based on twelve equal semitones) whereas for clavier instruments, as said above, the ¼-comma meantone was favored (though several other irregular temperaments existed and were in use in Renaissance and Baroque music).[97] To Mersenne, therefore, a new kind of keyboard (whether for organs, spinet or harpsichord) was needed to embed the theory of just intonation into the mechanical arts and thus, hopefully, fight against skepticism.

Proceeding methodically, Mersenne started with two thirteen-key claviers differently tempered, neither of which, however, displayed perfect major and minor thirds and sixths.[98] To produce all the just intoned consonances, these

---

96 Mersenne, "Traité de l'orgue," book 6, prop. XVI, 341 (HU3). Mersenne, *The Books on Instruments*, 424–5.

97 Mark Lindley, "Mersenne on Keyboard Tuning," *Journal of Music Theory* 24 (1980): 166–203. On temperament in general, see Lindley, "Temperaments," *Grove Music Online* (accessed 12 September 2006). On temperaments, Heller-Roazen, *The Fifth Hammer*, chapter 6.

98 "Or le clauier ordinaire tant des grandes Orgues que seruent aux Eglises, que des cabinets dont on vse dans les chambres particulieres, a treize marches sur chaque Octaue, & n'est nullement different de celuy des Epinettes, dont i'ay parlé dans le liure des instrumens à chordes: c'est pourquoy il n'est pas besoin de le mettre icy, si ce n'est pour faire plaisir aux Organistes & aux Facteurs, qui pourront plus aysément comparer les nouueaux clauiers que ie donne, auec le clavier ordinaire, afin de voir & de suppleer ce qui y manque pour auoir tous les accords & les interualles dans leur iuste proportion." Mersenne, "Traité de l'orgue," book 6, prop. XXII, 349 (HU3).

two claviers had to be combined into a seventeen-key clavier. Yet even this keyboard was insufficient to exhibit the just intonation of the complete diatonic genre, which needs at least eighteen tones (hence nineteen keys). The latter, although exhibiting the three musical genres, did not do so perfectly for the chromatic and enharmonic ones, yet would be the best tempered nineteen-key organ keyboard one could imagine, matching the third column of the organ-pipe table presented in figure 4 above. To fully render the perfect harmonic diapason, a clavier would need twenty-seven keys, the first row of keys for the diatonic genre, the second row for the chromatic and the last row for the enharmonic. (Fig. 6L–R) The table that accompanies the clavier's drawing was the real thing, though, displaying at a glance the perfection of the harmonic diapason, such that one (says Mersenne) could straightforwardly extract from it this twenty-seven-key clavier. (The table was supposedly so self-explanatory that "it would be unnecessary to add the figure of the keyboard, except that I desire to make its understanding so easy for the organists and the makers that they may find no difficulty."[99]) Mersenne did not explain how to manufacture these very special claviers (the only account by Mersenne on keyboard making was a general one on the movement mechanism of the keys, as seen in figure 5 above). Such distinctive claviers were not utterly uncommon, especially in the harpsichord family of instruments, for example the nineteen-key clavier built in 1639 by Joan Albert Ban based on Mersenne's own design, Zarlino's nineteen-key clavier built in 1548 by Maestro Comenico Pesarese, Doni's three-manual sixty keys per octave organ clavier described in his *Compendio del trattato de' generi, e de' modi* (Rome, 1635), and the two-manual thirty-six keys per octave clavier of Nicola Vicentino's *archicembalo*, "the foremost and perfect instrument, in that none of the keys lacks any consonances."[100] Drawn from the most exact theory of music—which could be confirmed by the monochord—Mersenne's twenty-seven-key clavier had great advantages:

> As for the usefulness that can be made of this keyboard, it is very great, for it shows exactly the intervals of the three *genres* of music, and gives a

99 Mersenne, "Traité de l'orgue," book 6, prop. XXIII, 356 (HU3). Mersenne, *The Books on Instruments*, 437.

100 Lindley, "Mersenne on Keyboard Tuning," 168–70. Ban's keyboard is found in his *Kort Zangh-Bericht* (Amsterdam, 1643) and information is given in Mersenne's correspondence. On Ban, Daniel P. Walker, "Joan Albert Ban and Mersenne's Musical Competition of 1640," *Music and Letters* 57 (1976): 233–55. On Vicentino, see his *Ancient Music Adapted to Modern Practice*, trans. Maria Rika Maniates, ed. Claude V. Palisca (New Haven: Yale University Press, 1996), quotation on p. 315. Stuart Isacoff, *Temperament: The Idea that Solved Music's Greatest Riddle* (New York: Alfred A. Knopf, 2001), 181. J. Murray Barbour, *Tuning and Temperament: A Historical Survey* (New York: Da Capo Press, 1972), 111.

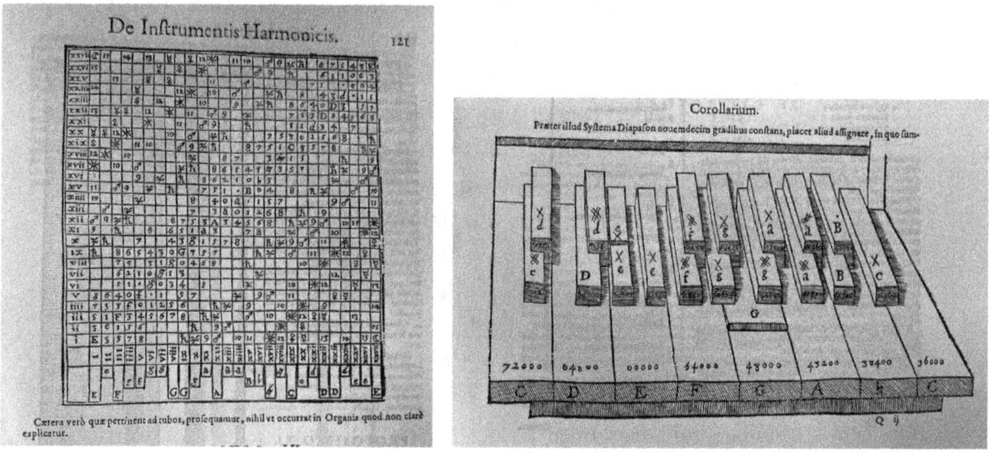

FIGURE 6.L–R Mersenne's perfect twenty-seven-key organ clavier. Left, a table that explains how to achieve the perfect twenty-seven-key organ clavier encompassing all three musical genres. Right, the clavier itself.
MARIN MERSENNE, *HARMONICORUM LIBRI XII* (GENÈVE: MINKOFF REPRINT, 1972 [1636]), BOOK 3, PROP. VIII, 119 AND 121

> greater light on the harmony that the Greeks [and the Latins] have written of. Thus, if one teaches music and the method of singing to children by means of its keys, they would be able to understand the most subtle ratios of all sorts of compositions and concerts in very little time, and to sing enharmonic airs as easily as the chromatic and diatonic ones. I am omitting many other uses that the organists can advise, if they use the keyboard on which they will make a quantity of beautiful passages and pretty touches which cannot be found on ordinary keyboards.[101]

And because these claviers were so perfect, nothing should stop organists from using them, even if it meant learning anew how to play the organ:

> Now it is certain that these keyboards ought to be preferred to the old ones, since they contain a greater number of consonances and other intervals in their justness … For it is of no importance that the difficulty of playing them is greater, inasmuch as it is not necessary to feel pity for

101 Mersenne, "Traité de l'orgue," book 6, prop. XXIII, 357 (HU3). Mersenne, *The Books on Instruments*, 439. On how the organ can help one learn to sing just without a *maître*, "Traité des chants," book 2, prop. XXXVII, 46–7 (HU2). Yet a *maître de chant* is needed to give more charm to the voice, since instruments cannot teach "certains charmes que l'on inuente tous les iours pour embellir les chants, & pour enrichir les Concerts."

> the pains nor to avoid the work which leads to perfection. To this I add that they will be played as easily as the others when the hands become accustomed to them, because they follow the infallible rule of reason. And there is no need for working to hide their imperfection, as happens in ordinary keyboards since they do not have them, as is seen in the two last keyboards of the preceding proposition in which all the consonances are quite just and without any temperament.[102]

In this case, musicians and the mechanical arts had to meet the terms of the music theorist, for only through the latter's *scientia* would a better musical instrument be designed and built, and consequently would music approach the long-lost perfection of Antiquity.

Mersenne argued elsewhere in his *Harmonie universelle* that music performed on ordinary (meaning tempered) organs, or other similar keyboard instruments, was as barbarous, rude, and unpleasant as orations *mal ordonnées*.[103] This appears in perfect agreement with the pure theory of musical harmony and the organ claviers he devised. Yet Mersenne was not always adamant about the best type of tuning for instruments. There is evidence in some of Mersenne's writings that he might have favored a universal diapason for all instruments, established on equal temperament, if it were not for the strong disapproval he received from Descartes, Doni, and the harpsichord maker Jean Denis, who was perhaps Mersenne's best source in Paris on keyboard instrument tuning.[104] In his *Traité de l'Harmonie universelle*, for

---

102 Mersenne, "Traité de l'orgue," book 6, prop. XXIII, 354 (HU3). Mersenne, *The Books on Instruments*, 437. Playing skillfully was a matter of practice, not of knowledge according to Trichet, *Traité des instruments de musique (vers 1640)*, 84–5.

103 Mersenne, "Traitez des consonances, des dissonances, des Genres, des Modes, & de la Composition," book 3, prop. 7, 161 (HU2): "Et si les Compositions que l'on ioüe sur l'Orgue ou sur les autres Instrumens à Clauier, ou à touches, peuuent estre comparees aux harangues des Orateurs : l'on peut dire que les pieces que l'on joüe sur les Instrumens ordinaires sont en comparaison de celles qui se ioûroient sur des Instrumens graduez selon lesdits Systemes parfaits, ce que sont les Oraisons mal ordonnees, fort rudes, & dont la locution est barbare & malplaisante, en comparaison des Harangues tres-polies, & si excellentes, qu'on n'y peut ajoûter, ny en oster aucune lettre sans en estropier le discours, & sans le rendre plus imparfait qu'il n'estoit deuant."

104 Lindley, "Mersenne on Keyboard Tuning," esp. 179–193. Dominique Devie, *Le Tempérament musical. Philosophie, histoire, théorie et pratique* (Béziers: Société de musicologie de Languedoc, 1990), 79–94. Barbour, *Tuning and Temperament*, where he says on p. 98: "Although Marin Mersenne was a zealous advocate of equal temperament in practice, he took pains to present literally dozens of tables in just intonation." Franz Josef Ratte, *Die Temperatur der Clavierinstrumente. Quellenstudien zu den theoretischen Grundlagen und praktischen Anwendungen von der Antike bis ins 17. Jahrhundert* (Kassel: Bärenreiter, 1991),

instance, although Mersenne mentioned that all instruments were imperfect due to their tempered tuning, he said that it would be impossible to tune equal tempered instruments such as the lute and the viola in concord with meantone tempered keyboard instruments "unless we modify the temperament of the ones or the others."[105] In the last proposition of the book on organ, Mersenne even described a geometrical method, devised by Roberval, to calculate the proportional means associated with the twelve equal semitones of equal temperament.[106] A theoretical ambiguity or indecision is present here that can also be easily connected to the crafting of organ pipes. Looking back again at the organ-pipe table depicted above in figure 4, the first column actually displays how to cut a pipe into twelve equal semitones, as an equal-tempered organ would require to play in tune with a string instrument. To favor equal temperament would certainly necessitate a completely new rationale, one that could still be established on mathematical ratios (as Mersenne also makes evident), but likely would not find support in musical practice and cultural taste. Mersenne's twenty-seven-key organ clavier became a true mechanical representation—a reification—of the most perfect musical harmony attainable by any of God's creation.

3 Musical Instruments and the 'parfait musicien'

The Baroque era saw many transformations and novelties in music: the Italian opera, the rebirth of monody, the spread of chamber cantata, and new genres of instrumental ensemble music. Regarding the latter, its renovation during the Renaissance found a natural setting within natural philosophy. Due in large part to the influence of instrumental music, the late sixteenth- and

---

228–241. Jean Denis did favor a tempered clavier for the harpsichord, but not equal temperament. To him, the most perfect temperament for a clavier was the meantone. Denis, *Traité de l'accord de l'espinette*, 9.

105 Mersenne, *Traité de l'Harmonie universelle*, book 1, théorème XXX, 301. "Il faut neantmoins remarquer que le temperament qui se fait par la seule distribution du *comma* est plus propre pour les instrumens à clauier, que pour ceux qu'on pince, ou dont on ioüe auec l'archet, qui supposent le temperament precedent, c'est à dire l'égalité des tons, & y ajoustent l'égalité des demy-tons par la distribution de la Diese, comme ie montreray en parlant des luths & des violes, qui ont tous leurs tons & leurs demy-tons égaux, & par consequent leur Octaue diuisées en trois Tierces majeures, ou en six tons, ou douze demy-tons. De la vient que … tout ce qu'on ioüe sur ces instrumens est imparfait, & qu'on ne les peut accorder iustement auec les orgues, ou auec l'épinette, si on ne change le temperament des uns ou des autres."

106 Mersenne, "Traité de l'orgue," book 6, prop. XLV, 408–12 (HU3).

early seventeenth-century *musica scientia* was amended from being a purely speculative mathematical science to becoming a more pragmatic analysis of sound—the nascent acoustics.[107]

Musical instruments, however, were not always considered indispensable to *musica scientia*. Toward the end of book one of *De institutione musica* Boethius, echoing Aristotle, explained *quid musicus est:*

> Now one should bear in mind that every art and also every discipline considers reason inherently more honorable than a skill which is practiced by the hand and the labor of an artisan. For it is much better and nobler to know about what someone else fashions than to execute that about which someone else knows; in fact, physical skill serves as slave, while reason rules like a mistress. Unless the hand acts according to the will of reason, it acts in vain. How much nobler, then, is the study of music as a rational discipline than as composition and performance! It is as much nobler as the mind is superior to the body; for devoid of reason, one remains in servitude ... It follows, then, that rational speculation is not dependent on the act of making, whereas manual works are nothing unless they are guided by reason. Just how great the splendor and merit of reason are can be perceived from the fact that those people—the so-called men of physical skill—take their names not from a discipline, but rather from instruments; for instance, the kitharist is named after the kithara, the aulete after the aulos, and the others after the names of their instruments. But a musician is one who has gained knowledge of making music by weighing with the reason, not through the servitude of work, but through the sovereignty of speculation.[108]

Episteme over techne, here you find the conventional Greek dichotomy between pure reason and mere manual work. Music, as episteme, was made an integral part of the scholastic *quadrivium*. To know music meant only one

---

107 John Walter Hill, *Baroque Music: Music in Western Europe, 1580–1750* (New York: W. W. Norton & Company, 2005). Tim Carter and John Butt, eds., *The Cambridge History of Seventeenth-Century Music* (Cambridge: Cambridge University Press, 2005), esp. Penelope Gouk's chapter "Music and the Sciences," pp. 132–57. Ann E. Moyer, *Musica scientia: Musical Scholarship in the Italian Renaissance* (Ithaca: Cornell University Press, 1992) is a very good overview of the science of music before the "experimental turn." See also Cristle Collins Judd, *Reading Renaissance Music Theory: Hearing with the Eyes* (Cambridge: Cambridge University Press, 2000). On the role of musical instruments in the transformation of instrumental music, Jonathan Wainwright and Peter Holman, eds., *From Renaissance to Baroque: Change in Instruments and Instrumental Music in the Seventeenth Century* (Aldershot, UK: Ashgate, 2005).

108 Boethius, *Fundamentals of Music*, §34, 50–51.

thing: to understand numbers' *ratio*. Mathematical intervals characterized everything about music; numbers gave music its power over the world (*musica mundana*) and humanity (*musica humana*).

Mersenne displayed a more humanistic approach, arguing instead that the *parfait musicien* needed all types of knowledge, not simply the mathematical culture of the *quadrivium*. In the *Traité de l'Harmonie universelle*, he described how the liberal arts and the other sciences were valuable to fully understand and practice music. Grammar, Mersenne explained, was used to go back to Greek roots of musical language (consonances, diapason, diapente, etc.) in order to grasp its true meaning, as it did for languages. Verbal rhetorical figures taught how to arrange the subject of music so it pleased listeners the most. Poetry was even more necessary than rhetoric, since it was tailor-made for singing. Logic taught how to reason (and thus how to understand music theory)—music depended upon arithmetic and geometry; astronomy showed how to find the consonances in the movements, sizes, intervals and distances of stars and planets. Music was subordinated to physics, which taught the nature of sounds and its properties. Metaphysics conveyed the relationships between sounds and all natural beings. Medicine taught how to find the temperament of listeners, and how to adapt songs accordingly. Morality explained how to behave at concerts, and politics demonstrated the modes and movements required to prevent the corruption and to preserve the virtues that made empires and republics flourish. Lastly, the mechanical arts (*mechaniques*) provided not only musical instruments to musicians, but were deemed necessary for the perfection of music.[109] Mersenne's concept of the perfect musician as defined here was not exactly original: it was in fact akin to the one found in Zarlino's *Istitutioni harmoniche*.[110] What makes Mersenne's description of the *parfait musicien* genuinely innovative was the addition of the mechanical arts at the end of the list. Musical instruments and musical instrument making entered *musica scientia* through Mersenne's notion that music was the best representation of universal harmony *tout court*.[111]

109 Mersenne, *Traité de l'Harmonie universelle*, Livre 1, théorème V, 20–22, entitled *Le parfait Musicien a besoin des autres sciences, des arts liberaux, & de quelques-uns des Mechaniques*. On rhetoric, David Allen Duncan, "Persuading the Affections: Rhetorical Theory and Mersenne's Advice to Harmonic Orators," in *French Musical Thought, 1600–1800*, ed. Georgia Cowart (Ann Arbor: U.M.I Research Press, 1989), 148–75.

110 Zarlino, *Institutioni harmoniche*, book 4, chapter 35, "Quel, che debbe hauere ciascuno, che desidera di venire a qualche perfettione nella musica."

111 A good summary of Mersenne's concept of universal harmony is found in Peter Pesic, *Music and the Making of Modern Science*, chapter 7.

In the *Préludes de l'Harmonie universelle* Mersenne explained, echoing Descartes, that the sciences should not be considered as separate fields of knowledge, but rather as a unity. In Mersenne's words, all the sciences "have created an unbreakable society [*ont iuré entr'elles vne inuiolable societé*]."[112] One, however, cannot study them all at once; but since they all 'hold hands,' one should rather focus on a particular science and from it, ascertain the others. Mersenne, of course, concentrated on music, where he reiterated what he had said in the *Traité de l'Harmonie universelle*, namely that studying music meant exploring all the rational and practical knowledge available to mankind, including the design of new musical instruments.[113] Mersenne said it best in the *Traité*, in which he explained that the making of musical instruments belonged to the *parfait musicien*, who must know all the materials engaged in the craftsmanship of all instruments, "otherwise he will find neither the pleasure nor the honor that comes from this knowledge, and thus will be deprived of the great utility he could gain from it."[114] To Mersenne, the material culture of music was as significant as the seven liberal arts and all the other sciences in achieving the status of the perfect musician. This, I believe, is one of Mersenne's foremost contributions to the study of music and, by extension, to the whole field of early modern natural philosophy—since music served as Mersenne's paradigmatic knowledge-producing *scientia*.

112 Mersenne, *Les Préludes de l'Harmonie universelle*, Question v, 155–6: "Les sciences ont iuré entr'elles vne inuiolable societé, il est quasi impossible de les separer, car elles souffrent plustost que l'on les déchire; & si quelqu'vn s'y opiniastre, son trauail ne luy en arrache que des lambeaux imparfaicts & confus. Elles ne viennent pourtant pas toutes ensemble, mais elles se tiennent tellement par la main, qu'elles se suiuent d'vn ordre naturel qu'il est dangereux de changer, parce qu'elles refusent d'entrer autrement où elles sont appellees."

113 Ibid., 138–43. Here Mersenne puts under the art of drawing the creation of musical instruments and other technological feats: "& par consequent à la Pourtraicture, tant pour cela, que pour desseigner les nouueaux instrumens que le Musicien peut inuenter en corrigeant les vns, & adioustant aux autres, & pour ordonner des grottes, & des machines hydraulique, & pneumatiques, qu'il rendra capables de toute sorte d'harmonie." (p. 143)

114 Mersenne, *Traité de l'harmonie universelle*, book 1, théorème 11, 10–11: "I'ajouste seulement *qu'elle* [*la musique*] *considere la nature des corps, & leurs proprietez*, parce que la fabrique des instrumens de Musique appartient au Musicien, qui doit connoistre le bois & les autres matieres qui seruent pour faire les instrumens de Musique, comme sont les peaux, les intestins & les metaux dont on fait les orgues, les luths, les violes, les cistres, les harpes, les flûtes, les trompettes, les hauts-bois, & les autres instrumens, s'il veut estre parfait Musicien, autrement il n'aura pas le plaisir ny l'honneur qu'il peut receuoir de cette connoissance, & sera priué de la grande vtilité qu'il en pourroit tirer."

From Mersenne onward, the role and purpose of musical instruments in *musica scientia* became indisputable. A 'Complete Musitian,' according to the author of a reader's epistle included in the 1653 English edition of Descartes's *Compendium musicæ*

> is required a more than superficial insight into all kinds of Humane Learning. For, He must be a Physiologist; that He may demonstrate the Creation, Nature, Proprieties, and Effects of a Natural Sound. A Philologer, to inquire into the first Invention, Institution, and succeeding Propagation of an Artificial Sound, or Musick. An Arithmetician, to be able to explaine the Causes of Motions Harmonical, by Numbers, and declare the Mysteries of the new Algebraical Musick. A Geometrician; to evince, in great variety, the Original of Intervalls Consono-dissonant, by the Geometrical, Algebraical, Mechanical Division of a Monochord. A Poet; to conform his Thoughts, and Words to the Lawes of præcise Numbers, and distinguish the Euphonie of Vowells and Syllables. A Mechanique; to know the exquisite Structure of Fabrick of all Musical Instruments, Winde, Stringed, or Tympanous aliàs Pulsatile. A Metallist; to explore the different Contemperations of Barytonous and Oxytonous, or Grave and Acute toned Metalls, in order to the Casting of tuneable Bells, for Chimes, &c. An Anatomist; to satisfie concerning the Manner, and Organs of the Sense of Hearing. A Melothetick; to lay down a demonstrative method for the Composing, or Setting of all Tunes, and Ayres. And, lastly, He must be so far a Magician, as to excite Wonder, with reducing into Practice the Thaumaturgical, or admirable Secrets of Musick.[115]

This long passage, which includes artisanal knowledge and the material culture of music, describes very well the new *scientia* of music by the mid seventeenth century.[116] Kircher's 1650 *Musurgia universalis*, the next great music

115 Cited from Benjamin Wardhaugh, *The* Compendium Musicae *of René Descartes: Early English Responses* (Brepols: Turnhout, 2013), 6–7. The stationer who likely wrote this epistle is Humphrey Moseley, though he must have received assistance from the work's translator, Walter Charleton, and from Lord William Brouncker, who composed a 'Necessary and judicious Animadversions' usually found at the end of the English translation. For an analysis of Descartes's work, Descartes, *Abrégé de musique*, trans. and ed. Frédéric de Buzon (Paris: Presses universitaires de France, 1987), 37–40. For more on the English context, Benjamin Wardhaugh, *Music, Experiment and Mathematics in England, 1653–1705* (Aldershot, UK: Ashgate, 2008).

116 Natural magic, since Bacon, has been more relevant in seventeenth-century England than on the Continent as regards the science of music. See Gouk, *Music, Science and Natural*

compendium after Prætorius's and Mersenne's, could not achieve its universal goal without introducing long descriptions of musical instruments—which borrowed heavily in fact from Mersenne's own, engravings included. This, I believe, is an argument too often forgotten when looking into the meaning of *harmonia universalis*. Mersenne's musical instrumentarium is perhaps the best demonstration that the mechanical arts and the material culture of music had finally crossed an important early modern intellectual threshold: they were finally included within the 'universal' of the universal harmony. Mersenne's musical instruments, considered one and all as unique *organum*, opened the door to a complete redefinition of what artisans, instruments, and experiments meant to natural philosophy.

Magic in Seventeenth-Century England. Another very good discussion on the same topic is Linda Phyllis Austern, "'Tis Nature's Voice': Music, Natural Philosophy and the Hidden World in Seventeenth-Century England," in *Music Theory and Natural Order from the Renaissance to the Early Twentieth Century*, ed. S. Clark and A. Rehding (Cambridge: Cambridge University Press, 2001), 30–67. This fits with the recent analysis of Matteo Valleriani, "From the Quadrivium to Modern Science," *Journal of History of Science and Technology* 16, no. 1 (2022): 121–32.

PART 2

Habitus

Introduction to Part 2

Otto Sibum's study of James Prescott Joule's determination of the mechanical value of heat convincingly established the foundational role the brewing industry—to which the Victorian physicist's family belonged—played in Joule's scientific practice. Combined with his replication of Joule's experiments, another field he helped develop and secure its *lettres de noblesse*, Sibum proposed the concept of 'gestural knowledge' to explain his and Joule's acquired skills in precision measurement. Sibum understood gestural knowledge as "knowledge united with the actor's performance of work, and it changes according to the specific kinds of performance ... and in ever new historical circumstances."[1] Though construed independently of Sibum's own interpretation in disciplines such as cognitive science and psychology,[2] 'gestural knowledge' has been for the past three decades a useful concept to think with in the history of science. Physicist turned historian and sociologist of science Andrew Pickering, for instance, employed Sibum's notion to further support his approach to scientific practice, an approach he called the 'mangle of practice.' To him, humans and machines interact in a continuous 'dance of agency,' taking the form of a 'dialectic of resistance and accommodation.' When investigating nature, humans perform actions with a machine (I would say, more generally, with an *organum*) that become the "balancing point, liminal between the human and the nonhuman worlds." To make machines work, humans require repetitive and specific gestures and skills, practices that need disciplining. These gestures and skills are answers (accommodations) to the unremitting resistance offered by both nature and machines. For Pickering, interestingly, Sibum's disciplined gestural knowledge is "*interactively stabilized* alongside the material (and conceptual) elements with which he worked": it is comprised within the mangle of practice. As an inclusive and comprehensive notion, the mangle ultimately "determines in time what scientific machines will look like, what scientists will believe, how they will conduct themselves

1 Otto Sibum, "Reworking the Mechanical Value of Heat: Instruments of Precision and Gestures of Accuracy in Early Victorian England," *Studies in the History and Philosophy of Science* 26 (1995): 73–106, quote on p. 76, n. 8. For a more general overview of the concept, Sibum, "Working Experiments: A History of Gestural Knowledge," *The Cambridge Review* 116 (1995): 25–37. On how quickly this concept was applied by historians of science, see Simon Schaffer, "Experimenters' Techniques, Dyers's Hands, and the Electric Planetarium," *Isis* 88 (1997): 456–83.

2 See, for instance, Daniel Bub and Michael Masson, "Gestural Knowledge Evoked by Objects as Part of Conceptual Representations," *Aphasiology* 20, no. 9 (2006): 1112–24.

 | DOI:10.1163/9789004504615_006

around those machines, and how (if at all) these pieces and others will hang together and relate to one another."[3] I find such hermeneutic constructions (many others exist) useful to think with, but I've asked myself one question for a long time, especially as it concerns early modern Europe: Can we examine *praxis* conceptually and more effectively utilizing an accurate and historically distinctive term?

Part 2 of this book looks at the early modern notion of *habitus* and how it relates to *organum*. It attempts to demonstrate that the production of scientific knowledge was understood as fully dependent on *habitus*, not just as a rhetorical and philosophical apparatus (chapter 3) but as an indicator of the effort, struggle, and investment required by artisans to manufacture these instruments of knowledge production (chapter 4). As demonstrated in Part 1, which examined the concept of *organum*, *habitus* presents itself as a comprehensive and historically fitting depiction of natural philosophical practices—and social customs. Before jumping into the early modern period, it is important to say a few preliminary words about *habitus* and how it is viewed today.

Several decades ago, one of the many things the French philosopher Maurice Merleau-Ponty did in his well-known study of the phenomenology of perception was to reestablish the authority of the human body as a genuine source of knowledge. He referred to the notion of *schéma corporel*—Paul Schilder's *Körperschema*—which he used to explain the total awareness of the body's movements (*gestes*) in the surrounding world. It is only through the action of the body in space that one grasps how the former occupies the latter.[4] Understanding the world is not simply an act of reason: it involves the whole body. The body, Merleau-Ponty argued, is the vehicle of a being into the world. To have a body is to be incorporated into a delineated milieu and to engage continuously with its substance.[5] A body is not *in* space but *inhabits* space. A knowledge of this space (*lieu*) *exists* that is unexplainable through a rational description or a silent *geste*. Accordingly, it is not the objective body—the

3 Andrew Pickering, *The Mangle of Practice. Time, Agency & Science* (Chicago: The University of Chicago Press, 1995), 104–9 (on Sibum), quote on p. 108 (emphasis original). Last quote is from p. 112. See also Pickering, "Material Culture and the Dance of Agency," in *The Oxford Handbook of Material Culture Studies*, eds. Dan Hicks and Mary C. Beaudry (Oxford: Oxford University Press, 2010), 191–208. Another approach based on what the authors call 'ethnomethodology of human-technology relations' is discussed in Bas de Boer, Hedwig te Molder, and Peter-Paul Verbeek, "Understanding Science-in-the-Making by Letting Scientific Instruments Speak: From Semiotics to Postphenomenology," *Social Studies of Science* 51, no. 3 (2021): 392–413.

4 Maurice Merleau-Ponty, *Phénoménologie de la perception* (Paris: Gallimard, 1945), 119.

5 Merleau-Ponty, *Phénoménologie de la perception*, 97: "Le corps est le véhicule de l'être au monde, et avoir un corps c'est pour un vivant se joindre à un milieu défini, se confondre avec certains projets et s'y engager continuellement."

mental image of one's own body into a Cartesian space grid—but the phenomenological body that moves, the body mobilized by the familiar objects, *gestes* and *lieu* in which it exists.[6] The body's motricity is the original intention of consciousness: the latter is no longer an 'I think' but rather an 'I can.' Motricity, therefore, is not a slave of consciousness. Because a body incorporates its own world, it has no need of a rational representation. It is not subordinated to a symbolic or objective function.[7] Though a rational perception of space can eventually liberate itself from the body's motricity in space, one has first to inhabit that space before objectivizing it. This non-objectivity of the inhabited space by a body in motion is what Merleau-Ponty calls habit (*habitude*).

Habit is thus motion (a notion not lost to early modern scholars as we will discover). To acquire habit is to master a dynamic incorporated into the *schéma corporel*. The notion of habit laid out here is good for the body as well as for an instrument binded to it. A walking stick, for instance, is no longer a simple object for the visually impaired who wields it. The stick's extremity becomes a sensory point improving the radius of touch. In a sense, it becomes the visually impaired individual's eyes. The perceived position of an object is not just given by the length of the stick, but rather by the extent of the motion created by the arm and the stick as one. This habituation to the stick means it is a part of oneself and thus participates in the voluminousness of the body.[8] It becomes an *organum*, in the universal sense of the word. The world of the visually impaired is thereby fully understood, fully 'observed' by means of the stick's motion. Habit is no longer simply motion: it is transformed into genuine perception. As Descartes claimed centuries ago, the blind's stick emerges as a fully-fledged appendix to the body, an instrumental extension to the *synthèse corporelle*.[9]

6 Merleau-Ponty, *Phénoménologie de la perception*, 122–3.

7 Merleau-Ponty, *Phénoménologie de la perception*, 161–4.

8 Merleau-Ponty, *Phénoménologie de la perception*, 166–8: "S'habituer à un chapeau, à une automobile ou à un bâton, c'est s'installer en eux, ou inversement, les faire participer à la voluminosité du corps propre. L'habitude exprime le pouvoir que nous avons de dilater notre être au monde, ou de changer d'existence en nous annexant de nouveaux instruments." The example of the stick used by visually impaired individuals is taken from Paul Schilder. Merleau-Ponty provides also a very nice example involving a pipe organ and an organist (pp. 169–70), taken from Jacques Chevalier.

9 Merleau-Ponty, *Phénoménologie de la perception*, 177–8. Descartes, in the first discourse of the *Dioptrique*, proposes a similar argument apropos the stick of a blind person. In explaining how we perceive colors (resulting from the reflection and refraction of light on matter), Descartes maintains it is analogous to the blind touching different types of matter (trees, rocks, water, etc.). What is perceived by someone who is visually impaired, via its stick, are the different types of motion or resistance offered by various materials. The blind 'sees' the material bodies surrounding him or her by the action of these bodies when they are sensed by the stick. It is in fact comparable to seeing with one's eyes. A visually impaired individual

Marcel Mauss stressed a decade before Merleau-Ponty that the "body is man's first and most natural instrument. Or more accurately … man's first and most natural technical object."[10] His description of the techniques of the body—the bodily movements involved in child rearing, sleeping, running, swimming, jumping, dancing, drinking, eating, washing and soaping, etc.—relied on habits that, according to Mauss, "do not just vary with individuals and their imitations, they vary especially between societies, educations, properties and fashions, prestiges."[11] The habits forming the techniques of the body are utterly dependent on the physiology, psychology, and sociology of its practitioner. Mauss endeavored in this famous article to develop a conceptual scheme while disregarding the material culture of habit. This is strange since his article is filled with references to objects, or instruments—the examples of the French spades during World War I or the Kabyle going downstairs with slippers (*babouches*) come to mind. Body and object cannot be separated that easily.

That is Jean-Pierre Warnier's take on Mauss. Starting from the techniques of the body Warnier, an anthropologist, emphasizes the 'natural' integration of technological objects to the *schéma corporel*. He does not claim, however, that technical objects are 'incorporated' to one's body—the objects' materiality always remains exterior to the human body. What is incorporated instead is the *dynamic* of the object. An object's embodiment (*faire corps avec l'objet*) means that the body has memorized the exact and appropriate *gestes*, which can then be repeated without any effort nor special attention.[12] Had Mauss understood this notion, his example of the Kabyle "man-with-slippers" would have suggested to him that "'the body' he was interested in is not the anatomo-physiological sum total of all the human organs, it is a dynamic synthesis of

---

perceives his or her surroundings as would a 'normal' person with his or her eyes: through the action of an instrument, or organ. The stick and the human eye are analogous. Descartes, *Dioptrique*, in *Œuvres de Descartes*, ed. Charles Adam and Paul Tannery, 11 vols (Paris: J. Vrin, 1996), vol. 6, 85–6.

10 Marcel Mauss, "Techniques of the Body," *Economy and Society* 2 (1973): 70–88, quote on p. 75.

11 Mauss, "Techniques of the Body," 73.

12 Jean-Pierre Warnier, *Construire la culture matérielle. L'Homme qui pensait avec ses doigts* (Paris: Presses Universitaires de France, 1999), 9–35. For a similar conceptual approach to material culture, see Marie-Pierre Julien and Jean-Pierre Warnier, eds., *Approches de la culture matérielle. Corps à corps avec l'objet* (Paris: L'Harmattan, 1999) and Bernard Conein, Nicolas Dodier, and Laurent Thévenot, eds., *Les Objets dans l'action. De la maison au laboratoire* (Paris: Éditions de l'École des Hautes Études en Sciences Sociales, 1993).

sensori-motricity in a given materiality."[13] In the case of Pascal's arithmetical machine discussed in chapter 3, it was the dynamic of the machine that was embodied, not the material object itself. I argue that memory shifted from the mind to the body. Once the *gestes* involved in handling the pascaline were fully memorized by the body, performing an arithmetical operation was no longer a feat of the mind, but rather one entirely executed by the body. The machine itself did not embody mathematics, no more than it was embodied by its user. Without motricity—gear motion and bodily *gestes*—the pascaline was merely a dry and precious piece of craftsmanship. Arithmetical operations, in other words, were the result of a technical object finely tuned and properly manipulated by techniques of the body. Mathematics, in short, moved from the absolute sphere of the mind to the down-to-earth materiality of human existence.

From making an instrument to achieving objectivity the scientific self goes through a necessary introspection where subjectivity, codified by one's own body and local systems of meaning, participates directly in the production of knowledge.[14] Behind this subjective codification, we recover what Michel Foucault called technologies of the self. They allow individuals, "to effect by their own means or with the help of others a certain number of operations on their own bodies and souls, thoughts, conduct, and way of being, so as to transform themselves in order to attain a certain state of happiness, purity, wisdom, perfection, or immortality." These technologies are not only instruments of domination in a Foucauldian sense but actual forms of production. They call into being a *praxis*, a practical reason aimed at transforming the natural world and reproducing social institutions. And such technologies of the self, subjected to a specific matrix of *praxis* generate *habitus*.[15] Habit is perhaps best understood nowadays as a *structure structurée et structurante*, following Pierre Bourdieu's salient definition, a restraining system of intellectual, social, and bodily prescriptions that nonetheless allow for an open-ended and

---

13 Jean-Pierre Warnier, "A Praxeological Approach to Subjectivation in a Material World," *Journal of Material Culture* 6 (2001): 5–24, quote on p. 7. See as well the important book edited by Christopher Lawrence and Steven Shapin, *Science Incarnate. Historical Embodiments of Natural Philosophy* (Chicago: The University of Chicago Press, 1998).

14 Lorraine Daston and Peter Galison, *Objectivity* (New York: Zone Books, 2007), chap. 4. For a case study based on the 'scientific self' concept developed in this book, Ian Hesketh, "Technologies of the Scientific Self: John Tyndall and His Journal," *Isis* 110 (2019): 460–82.

15 Luther H. Martin, Huck Gutman, and Patrick H. Hutton, eds., *Technologies of the Self. A Seminar with Michel Foucault* (Amherst: The University of Massachusetts Press, 1988), 18. Ian Burkitt, "Technologies of the Self: Habitus and Capacities," *Journal for the Theory of Social Behaviour* 32 (2002): 219–37. On practical reason, Pierre Bourdieu, *Raisons pratiques. Sur la théorie de l'action* (Paris: Seuil, 1994).

limitless set of outcomes.[16] As "organized forms of movement," Jonathan Sterne argues, technologies generally, "are essentially subsets of habitus." They organize social practices like other ways of making and knowing do.[17] For Chandra Mukerji, however, Bourdieu's habitus built around "cognitive predispositions and logics of practice" is too stringent and prevents inarticulate knowledge to take hold of possible cultural imaginaries. Only the inarticulacy of tacit knowledge, the "repetitive actions that on the surface seem mindless," what Gilles Deleuze calls 'repetitions,' offers "escape routes from discursive regimes and make possible new social constructions of reality."[18]

This too short and somewhat idiosyncratic commentary does not instantiate the richness behind a historically, philosophically, and sociologically telling notion such as *habitus*. It nevertheless compels us to focus on and perhaps reevaluate the value of studying scientific *praxis*. The goal in Part 2 of this book is thus to provide an original outline of how we should understand early modern natural philosophical practices through the pivotal concept of *habitus*. Together with *organum*, *habitus* forms a dyad that would appear not only relevant but faithful to how savants (and sometimes artisans) themselves discerned and formulated early modern science.

---

16 Pierre Bourdieu, *Le sens pratique* (Paris: Les Éditions de Minuit, 1980). For an interesting historiography of the term, especially in the US social sciences, going back to Émile Durkheim's and Max Weber's work, see Charles Camic, "The Matter of Habit," *The American Journal of Sociology* 91 (1986): 1039–87. See all the recent work by André Mary, *Le travail de l'habitus dans l'oeuvre de Pierre Bourdieu* (Paris: Karthala, 2022).

17 Jonathan Sterne, "Bourdieu, Technique and Technology," *Cultural Studies* 17 (2003): 367–89. He uses this interesting framework to say elsewhere, in a well-received book, that "the history of sound implies a history of the body" and that it "is an artifact of the messy and political human sphere." Sterne, *The Audible Past. Cultural Origins of Sound Reproduction* (Durham: Duke University Press, 2003), 12–13.

18 Chandra Mukerji, "The Cultural Power of Tacit Knowledge: Inarticulacy and Bourdieu's Habitus," *American Journal of Cultural Sociology* 2 (2014): 348–75, quotes on p. 348. On tacit knowledge, Harry Collins, *Tacit & Explicit Knowledge* (Chicago: The University of Chicago Press, 2010). On Collins's collective tacit knowledge and Bourdieu's habitus, Stephen P. Turner, *Understanding the Tacit* (New York: Routledge, 2014).

CHAPTER 3

Habitus in corpore, habitus in anima: Making and Thinking in Early Modern Europe

The instrument—or *organum*, as it was commonly labeled—was central to the early modern practice of natural philosophy, whether one dealt with theoretical, mathematical, or the so-called philosophical instruments. What is often left undiscussed, however, is the operation of an instrument, the instrumental procedures leading to experiments. What did one need to know—and do—to properly use an instrument so it generated knowledge? What training did the body or mind require? Behind each instrument of science was attached a special training, a distinct practice—a *habitus*. An *organum* was not an instrument restricted to a *terminus ad quem*, as I mentioned in the introduction to Part 1 of this book. Instead, an *organum* in the hands of natural philosophers became an instrument with a *terminus a quo*—a starting point if manipulated rightly and creatively.[1] Though instruments were essential to the practice of early modern science, they became useless if the required *habitus* was improperly carried out. The notions of *organum* and *habitus* are, I believe, the most fitting (and historically appropriate) concepts one can investigate with when approaching the role played by instruments and machines in relation to the *theorica et practica* of early modern natural philosophy.

Mathematical instruments easily bridged the scholastic chasm between theory and practice.[2] By the time Francis Bacon laid the foundation for a reformation of knowledge, the mathematical instrument-making trade had already developed into an important artisanal industry, especially in Nuremberg and London—and to a lesser extent in Paris and the rest of Europe. It provided tools for fields such as surveying, navigation, horology and clockmaking, ship building, drawing, map making, accounting, and the military. These instruments and their practitioners, according to Jim Bennett, transformed the practical

1 An original approach to this topic is Simon Werrett, *Thrifty Science. Making the Most of Materials in the History of Experiment* (Chicago: The University of Chicago Press, 2019).

2 (Al)chemical laboratories, apothecary shops, assaying shops, foundries, perfumeries, and such other *laboratorium chymicum*, where distillation, combustion, smelting and various related chemical operations were performed every day have been historically described as sites where *theorica* and *practica* were bridged. Due to the lack of space, I will not say much about them. On this topic, see the article by Ursula Klein, "The Laboratory Challenge: Some Revisions of the Standard View of Early Modern Experimentation," *Isis* 99 (2008): 769–82.

 | DOI:10.1163/9789004504615_007

mathematical sciences into legitimate natural philosophical knowledge.[3] Bennett also draws upon the notion of instrumental virtue to account for the relationship between an instrument, its operator, and the mathematical sciences in sixteenth-century Europe. Moral issues, he claims, "are relevant to both the instrument and the operator, while [the mathematical] forms of knowledge do not stand independent of action." The dichotomy between things made and things done, or between making and doing, collapsed in the mathematical practices of the Renaissance. According to Bennett, early modern geometry was 'embodied' in instruments and operators. Virtue was found not only in the *scientia* of geometry, but in the instrument and its operator as well. Both acquired and displayed virtue, and were as dependent on geometry's virtue as the latter was on the instrument's and operator's virtues.[4]

Not too far afield, and to reclaim an intellectual space for an ontology of observation, Lorraine Daston contends that it is "habit that makes perception of a world possible." Going beyond epistemology and neo-Kantian ideas about observation and theories, she turns her attention to ontology, or "how scientists furnish the universe with objects that are amenable to sustained and probing investigation but that rarely correspond to the objects of everyday perception." Citing Ludwik Fleck's observations of microscopic bacteria, Daston explains that the "novice sees only blurs and blobs under the microscope; experience and training are required in order to make sense of this visual chaos, in order to be able to see *things*." Daston adds that for Fleck, "learning to see like a scientist was a matter of accumulated experience—not only of an individual but of a well-trained collective."[5] Such experience and training have unique consequences on those who do acquire them. As Daston and Peter Galison explain elsewhere, "[r]adically different practices are *prima facie* evidence of different selves." Mind and body are at work here, deeply changing one's self.[6]

Almost three centuries before Fleck, a similar viewpoint emerged with the first microscopic observations. Robert Hooke's *Micrographia* (1665) was a

---

3 Jim Bennett, "The Mechanics' Philosophy and the Mechanical Philosophy," *History of Science* 24 (1986): 1–28. Idem, "The Mechanical Arts," in *The Cambridge History of Science. Vol. 3 Early Modern Science*, eds. Katharine Park and Lorraine Daston (Cambridge: Cambridge University Press, 2008), 673–95.

4 Jim Bennett, "Geometry in Context in the Sixteenth Century: The View from the Museum," *Early Science and Medicine* 7 (2002): 214–30.

5 Lorraine Daston, "On Scientific Observation," *Isis* 99 (2008): 97–110, quotes on pp. 99–100 (italics original). See also Lorraine Daston, ed., *Biographies of Scientific Objects* (Chicago: The University of Chicago Press, 2000). Lorraine Daston and Elizabeth Lunbeck, eds., *Histories of Scientific Observation* (Chicago: The University of Chicago Press, 2011).

6 Lorraine Daston and Peter Galison, *Objectivity* (New York: Zone Books, 2007), 199 (emphasis original). We will return to *habitus* and the self in chapter 4.

famous demonstration of the complexity involved in instrument-based knowledge. Microscopes were essentially of two types: simple (one lens or glass bead) and compound (two or more lenses). Though the latter was likely invented first (closely tied to the invention of the telescope), it was the simple microscope that provided natural philosophers with the most astonishing anatomical discoveries beginning in the last third of the seventeenth century. Individuals such as Marcello Malpighi, Nehemiah Grew, Jan Swammerdam, and especially Antoni van Leeuwenhoek not only increased the body of knowledge on living things, they also perfected the microscope and the observational practices related to its use.[7] Hooke recognized this fact about the compound microscope, explaining that seeing with such an instrument required a "sincere Hand and a faithful Eye, to examine, to record, the things themselves, as they appear." One hand manipulated the microscope, the other faithfully reproduced what was seen. In Catherine Wilson's words:

> When Hooke speaks, for example, of the coordination of hand and eye, one remembers that hand and eye are *not* coordinated in microscopical work, that a new series of coordinated movements must be learned, with some difficulty. And involvement with 'things' turns out to involve not only delight and gratification, movement and progress, but the resistance offered by everything real, and the boredom, frustration, and even disgust associated with every manifestation of facticity.[8]

Such a 'disciplined seeing,' Michael Aaron Dennis argued, was essential to communicate individual standard practices, to validate particular data *and* expand the community of practitioners.[9] Citing Daston again, without "these acquired habits of perception cultivated by observation, there would not only be no

---

7 Gerard L'Estrange Turner, *Essays on the History of the Microscope* (Oxford: Senecio Publication Co., c1980). Marc J. Ratcliff, *The Quest for the Invisible: Microscopy in the Enlightenment* (Farnham, UK: Ashgate Publishing Ltd, 2009). Domenico Bertoloni Meli, *Mechanism, Experiment, Disease: Marcello Malpighi and Seventeenth-Century Anatomy* (Baltimore: The Johns Hopkins University Press, 2011).

8 Catherine Wilson, *The Invisible World. Early Modern Philosophy and the Invention of the Microscope* (Princeton, N.J.: Princeton University Press, 1995), 102. A deep analysis of the 'resistance of things' in the sciences is provided by Andrew Pickering, *The Mangle of Practice. Time, Agency, & Science* (Chicago: The University of Chicago Press, 1995), where the production of knowledge results from a 'dance of agency' taking the form of a 'dialectic of resistance and accommodation' between human and non-human agencies.

9 Michael Aaron Dennis, "Graphic Understanding: Instruments and Interpretation in Robert Hooke's *Micrographia*," *Science in Context* 3 (1989): 309–64.

science; there would be no articulated visible (or auditory or tactile) world at all. This is the way perception furnishes the universe."[10]

The instrumentality of natural philosophy, the new instrument-based knowledge of the Scientific Revolution as demonstrated by Hooke's microscope, required usually singular habits, targeted at a class of instruments or, commonly, at a single instrument. The epistemic, sociological, and technical problems associated with Robert Boyle's air-pump experiments—and their replications—are among the best known and documented instances of the inherent technical savoir-faire embedded in machines and their use.[11] These so-called 'philosophical' instruments were constantly caught in the crossfire of theory v. practice discourses. In the *Sidereus nuncius* (1610), Galileo did not describe the telescope (the *fabrica* part) because his experience told him that once one knew how to make a low-powered telescope, a few months of hands-on practice would allow such an individual to make the high-powered instrument from which Galileo made his fortune. What Galileo emphasized, however, was the *usus* part of his instrument. Training others how to use this *organum* (as he called it) was crucial to Galileo's authority as a natural philosopher, because he needed credible witnesses to validate his discoveries. The features of early telescopes, such as a narrow field of vision, double images and color fringes, made it difficult for the untrained eye to see the moons of Jupiter, for instance. Without such practical observational training—one could say habit here following Daston—"people who looked through a telescope for only a few minutes could legitimately believe that Galileo's claims were artifactual, as numerous spurious objects could be seen through a telescope's eyepiece at any given time."[12] Galileo did not disclose craft practices—he kept a monopoly in place for a long time—but understood that to be taken seriously, dignitaries, patrons, and (eventually) natural philosophers had to acquire the observational habit of using a telescope in order to validate his observations. Practice, not theoretical assumptions, was at the core of Galileo's enterprise of self-fashioning. "In contrast to languages," Daston emphasized, "which may be learned either by ear or by grammatical rules, there seems to be only one route to competent perception, and that royal road is habit."[13]

---

10 Daston, "On Scientific Observation," 100.

11 Steven Shapin and Simon Schaffer, *Leviathan and the Air-Pump. Hobbes, Boyle, and the Experimental Life*. (Princeton: Princeton University Press, 1985). One of the most original studies on air pumps and the instrumental and experimental habitus of its maker is Terje Brundtland, "From Medicine to Natural Philosophy: Francis Hauksbee's Way to the Air-Pump," *The British Journal for the History of Science* 41 (2008): 209–40.

12 Mario Biagioli, *Galileo's Instruments of Credit: Telescopes, Images, Secrecy* (Chicago: The University of Chicago Press, 2006), 102.

13 Daston, "On Scientific Observation," 100.

This chapter investigates in more detail the core concept of early modern *habitus* and attempts to go beyond what Daston examined in her short article. The main objective is to figure out what natural philosophers meant and referred to by *habitus* and how they apprehended the notion regarding the production of natural philosophical knowledge. What I am not doing is rewriting the history and philosophy of scientific practice. Rather, by looking at philosophers such as Descartes, Locke, Hobbes, Hume, and Pascal, I argue that these scholars could not conceptualize their work without first considering and tightly framing what habit was and what it might (or might not) do for them. Enough evidence points to the fact that *habitus in anima* and *habitus in corpore* (the latter, in a broad sense, as bodily gesture) did not escape the natural philosophers' intellectual attention toward the establishment of a theory—and a practice—of knowledge production.

1 Defining *habitus* in Early Modern Europe

Habit—or *habitus*—following the standard scholastic definition was a quality of the first species associated with both the body and the mind—at the core of several key philosophical, physiological, and artisanal notions. Per Goclenius's *Lexicon philosophicum* (1613), we learn that *ars* was divided into three constituents: it was first a *habitus ποιητικος* (*factiius*) *seu μηχανικος*, a poietical or mechanical habitus that could be noble or ignoble (the knowledge of making things—*ars est ratio faciendi*); second, it was described as a *habitus organicus*, an instrumental habit such as logic;[14] third, as *effectus seu opus*, εξεως (*hexis*), i.e., a bodily disposition, *habitus in corpore*. Moreover, through the habit of thinking (*pro habitu cognitionis*) we can grasp difficult concepts in fields such as ethics and philosophy[15] and, as we will explain in a section below, it characterized the very concept and action of *philosophia*. In short, habitus was understood as an organized and methodical disposition of body and mind for the production of things and knowledge. Goclenius summarized in a full

14 The concept of *organicus* meant theoretical soundness, and indeed the principles of logic led to rigorous judgment and analysis. But first and foremost, logic was a *habitus* because it helped accomplish operations that nature alone could not make easily or with consistency (more on this *infra*). Theodor Zwinger's second volume of his 1571 expanded *Theatrum vitæ humanæ* (first edition, 1565) is dedicated to "Habitus Organicos siue Facultates Logicas uulgò dictas, Philosophiæ scilicet utriusque instrumenta," i.e., grammar, rhetoric, poetic, and logic.

15 Goclenius, *Lexicon philosophicum quo tanquam clave philosophiae fores aperiuntur* (Hildesheim: Georg Olms Verlag., 1964 [Frankfurt, 1613]), *s.v. intellectus*: "Secundo pro habitu cognitionis indemonstrabilium & formarum immaterialium in Eth. & Philos. sumitur."

lexical entry the meaning of habitus, which mainly followed what Aristotle and his most influential commentator, Aquinas, described before him.[16] It is interesting to note as well that Theodor Zwinger's monumental *Theatrum vitæ humanæ* of 1565 is organized (*dispositio*) as a book showcasing a branching structure akin to the one later given by Goclenius in describing habitus, using instead *Ethica* as a chief header—underlining again the significance of this concept to all human knowledge, activities, and behaviors.[17]

Habitus was thus divided into two parts by Goclenius: *habitus in anima* and *habitus in corpore* (Fig. 7). *Habitus in corpore* was connected to the body's health, its constitution: in other words, to medicine.[18] *Habitus in anima* was associated, on the one hand, to the passions, *Voluntate* (*sensitivo appetitu*, the appetitive faculties), including moral virtues, and, on the other hand, to the intellect.[19] The latter, moreover, had two elements: *Speculativus* for the sciences and *Operativus* for intellectual virtues (*practicus*, i.e., acting or things done, such as *prudentia*) and art (*poeticus sev factiuus*, i.e., making or things made).[20]

A further division is introduced for the *habitus intellectus*, under *Hypolepses*, or how certain or uncertain knowledge is ascertained (*apprehensiones rerum in intellectu*). (Fig. 8) When uncertain, knowledge is but a doxa, an opinion,

16 A detailed description of *habitus* in Aquinas's thought is found in Roy J. Deferrari, *et al.*, *A Lexicon of St. Thomas Aquinas Based on the* Summa Theological *and Selected Passages of his Other Works* (Washington, D.C.: Catholic University of America Press, ca 1948), *s.v. habitus*.

17 Theodor Zwinger, *Theatrum vitæ humanæ* (Basel, 1565), 32 for the branching diagram explaining the *dispositio* of the whole work. On the development of Zwinger's book, Ann Blair, *Too Much to Know. Managing Scholarly Information Before the Modern Age* (New Haven: Yale University Press, 2010), 193–202.

18 Goclenius, *Lexicon philosophicum, s.v. habitus*: "pro corporis constitutione … Haec significatio Medicis frequens est." The last definition is linked to what we wear: "pro vestitu seu habitudine corporis ad ambiens ipsum."

19 Goclenius, *Lexicon philosophicum, s.v. habitus*: "Habitu *primum* sumitur Categorice pro Qualitatis prima specie, id est, pro qua litate, qua quis habilis est ad operationes, quas per naturam solam edere facile & constanter nequit."

20 Goclenius, *Lexicon philosophicum, s.v. habitus*: "Ars est Habitus cum recta ratione ϖοητικος (Factiuus)." Aquinas explains the difference between Prudence and Art in his *Summa theologiae*, prima secundæ partis, Question 57, art. 4: "The reason for this difference is that art is the 'right reason of things to be made'; whereas prudence is the 'right reason of things to be done.' Now 'making' (*facere*) and 'doing' (*agere*) differ, as stated in Metaph. ix, text. 16, in that 'making' is an action passing into outward matter, e.g. 'to build,' 'to saw,' and so forth; whereas 'doing' is an action abiding in the agent, e.g. 'to see,' 'to will,' and the like. Accordingly, prudence stands in the same relation to such like human actions, consisting in the use of powers and habits, as art does to outward making: since each is the perfect reason about the things with which it is concerned." (See note 25 for the reference.)

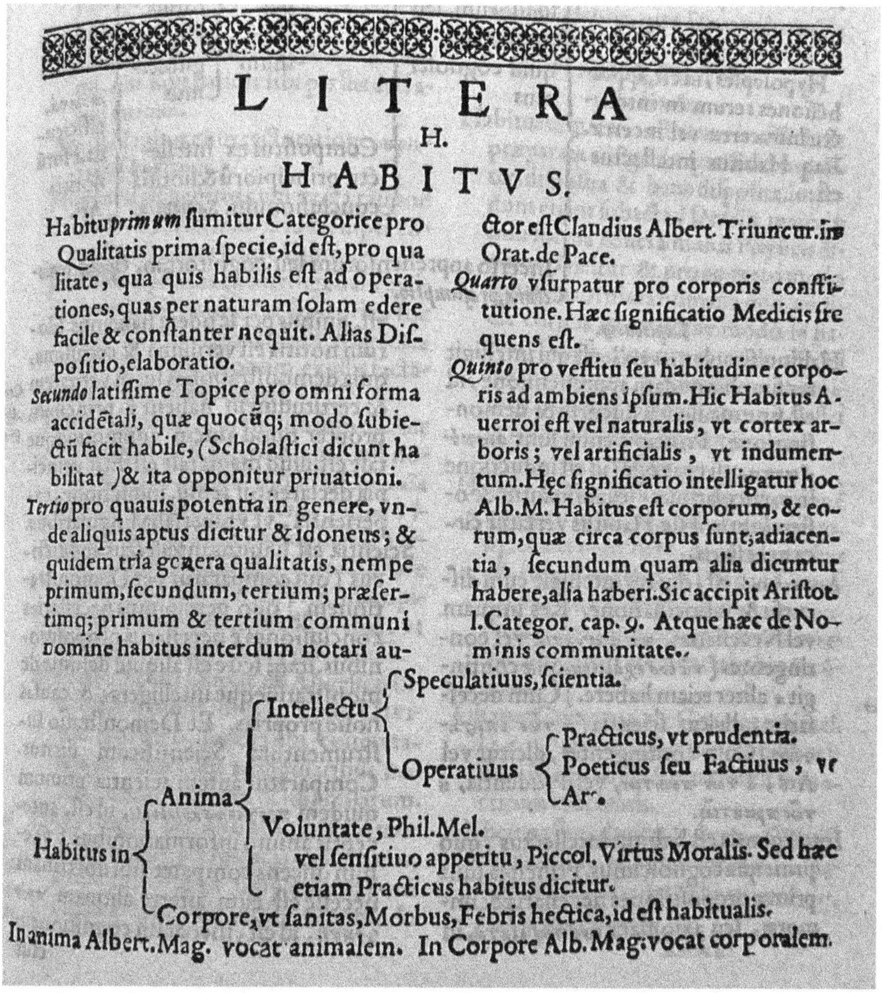

LITERA
H.
HABITVS.

Habitu primum ſumitur Categorice pro Qualitatis prima ſpecie, id eſt, pro qualitate, qua quis habilis eſt ad operationes, quas per naturam ſolam edere facile & conſtanter nequit. Alias Diſpoſitio, elaboratio.

Secundo latiſſime Topice pro omni forma accidẽtali, quæ quocũq; modo ſubiectũ facit habile, (Scholaſtici dicunt habilitat) & ita opponitur priuationi.

Tertio pro quauis potentia in genere, vnde aliquis aptus dicitur & idoneus; & quidem tria genera qualitatis, nempe primum, ſecundum, tertium; præſertimq; primum & tertium communi nomine habitus interdum notari auctor eſt Claudius Albert. Triuncur. in Orat. de Pace.

Quarto vſurpatur pro corporis conſtitutione. Hæc ſignificatio Medicis frequens eſt.

Quinto pro veſtitu ſeu habitudine corporis ad ambiens ipſum. Hic Habitus Auerroi eſt vel naturalis, vt cortex arboris; vel artificialis, vt indumentum. Hęc ſignificatio intelligatur hoc Alb. M. Habitus eſt corporum, & eorum, quæ circa corpus ſunt, adiacentia, ſecundum quam alia dicuntur habere, alia haberi. Sic accipit Ariſtot. l. Categor. cap. 9. Atque hæc de Nominis communitate.

Habitus in
- Anima
  - Intellectu
    - Speculatiuus, ſcientia.
    - Operatiuus
      - Practicus, vt prudentia.
      - Poeticus ſeu Factiuus, vt Ars.
  - Voluntate, Phil. Mel. vel ſenſitiuo appetitu, Piccol. Virtus Moralis. Sed hæc etiam Practicus habitus dicitur.
- Corpore, vt ſanitas, Morbus, Febris hectica, id eſt habitualis.

In anima Albert. Mag. vocat animalem. In Corpore Alb. Mag. vocat corporalem.

FIGURE 7 Goclenius's schematic tree summarizing the concept of *habitus*.
RUDOLPH GOCLENIUS, *LEXICON PHILOSOPHICUM QUO TANQUAM CLAVE PHILOSOPHIAE FORES APERIUNTUR* (HILDESHEIM: GEORG OLMS VERLAG., 1964 [FRANKFURT, 1613]), *S. V. HABITUS*

conjecture, or presumption.[21] Regarding the things one can know for certain, the *habitus intellectus* was *simplex* or *compositus*. If *simplex*, it was either νοητικὸς (formed from), i.e., the simple and naked things understood immediately (αμεσως [*amésos*]) without the help of discourse or demonstration, or διανοητικὸς (for thinking), i.e., things that were understood through discourse

21 Goclenius, *Lexicon philosophicum*, *s.v. habitus*: "Habitus incerto apprehensorum dicitur ∇οξα [*doxa*] opinio quae modo est (quieta, infirma, imbecilla)."

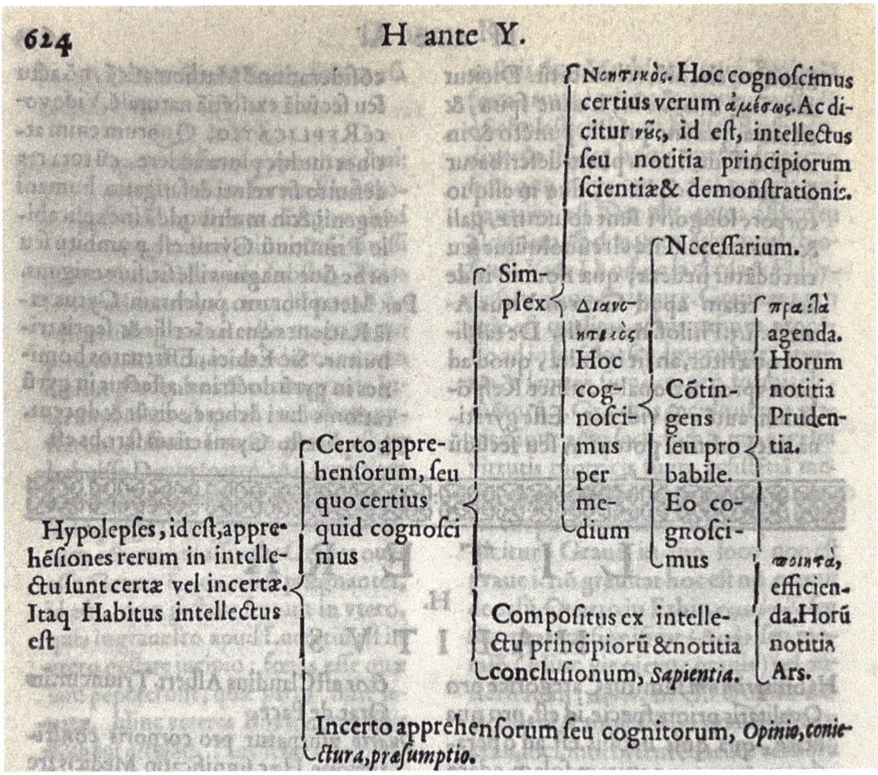

FIGURE 8 Goclenius's schematic tree summarizing the concept of *habitus intellectus*. RUDOLPH GOCLENIUS, *LEXICON PHILOSOPHICUM QUO TANQUAM CLAVE PHILOSOPHIAE FORES APERIUNTUR* (HILDESHEIM: GEORG OLMS VERLAG., 1964 [FRANKFURT, 1613]), *S. V. HABITUS*

and reasoning.[22] The former was associated with *intelligentia*, or the *habitus intellectus* from which first principles were believed to be true. The latter was associated with *scientia* when the conclusion reached was unavoidable (*necessarium*) or with art and prudentia when there was some accident or probability (*contingens*).[23] Lastly, the *habitus intellectus compositus* was the combination

22 Goclenius, *Lexicon philosophicum, s.v. habitus*: "Habitus simplex νοητικὸς est qui intelligit res simplici nudaq[ue]; apprehensione, id est, sine medio seu discursu & demonstratione." "Διανοητικὸς est, qui res intelligit cum discursu & ratiocinatione."

23 Goclenius, *Lexicon philosophicum, s.v. habitus*: "Intelligentia est habitus intellectus, quo principia cognoscimus. Principia sunt primæ propositiones necessariæ & immotæ, seu axiomata αναποδεικτα, id est, quibus per se fidem habemus." "Scientia est habitus intellectus apodicticus (qui comparatur per Demonstrationem) quo percipimus necessarias conclusiones è necessariis propositionibus."

of *intelligentia* and *scientia*, or *sapientia* (wisdom), the *summum bonum* of human understanding.[24] As already mentioned, the multiple definitions and interpretations of *habitus* came almost directly from scholasticism, especially Aquinas's interpretation—which itself was much influenced by Aristotle's.[25]

As will be discussed soon, Descartes criticized in his *Regulæ ad directionem ingenii* this conventional treatment of *habitus in anima* within the realm of *scientia*. He disapproved of the notion that the sciences had been connected to the uniqueness of *ars*—that is to say, specific knowledge demanded particular types of instruments and *habitus*, which led to a variety of training. Aristotle plainly expressed this condition when he wrote, for instance, that

> science is an instrument of the intelligence (for it is useful to the intelligence just as flutes are useful to the flute player), and many things in nature are instruments of the hands ... Now it is natural that where the instruments are prior, the faculties should also come into being in us first (for it is by using the instruments that we acquire a disposition); and the instrument of each faculty is related similarly to that faculty, and conversely, as the instruments are to one another, so are the faculties of which they are the instruments to one another.[26]

And as the Philosopher said elsewhere, "for the exercise of any faculty or art a previous training and habituation are required."[27] Hence every intellectual virtue or art necessitated its particular *organum* and *habitus*. Aquinas agreed

---

24 Goclenius, *Lexicon philosophicum, s.v. habitus*: "Habitus compositus ex Intelligentia Principiorum & scientia, dicitur sapientia." On the presentation as branching diagrams, Ann Blair, *Too Much to Know*, 144–52. For how influential this practice was, Ayelet Even-Ezra, *Lines of Thought: Branching Diagrams and the Medieval Mind* (Chicago: The University of Chicago Press, 2021).

25 See especially Thomas Aquinas, *The* Summa theologiae *of St. Thomas Aquinas*, 2nd ed. (1920), internet resource http://www.newadvent.org/summa/ (accessed 4 July 2019), prima secundæ partis, questions 49 to 58 (English version). For the original Latin, based on the 1891 edition, see http://www.corpusthomisticum.org/sth2049.html (accessed 4 July 2019). Some very good analyses exist, like René Arnou, S.J., *L'Homme a-t-il le pouvoir de connaître la vérité? Réponse de Saint Thomas: La connaissance par habitus* (Rome: Presses de l'Université Grégorienne, 1970); Yuji Nagamachi, S.J., *Selbstbezüglichkeit und Habitus. Die latente Idee der Geistmetaphysik bei Thomas von Aquin* (St. Ottilien: EOS Verlag Erzabtei, 1997); Rolf Darge, *Habitus per actus cognoscuntur. Die Erkenntnis des Habitus und die Funktion des moralischen Habitus im Aufbau des Handlung nach Thomas von Aquin* (Bonn: Bouvier Verlag, 1996).

26 Aristotle. *The Complete Works of Aristotle*, ed. Jonathan Barnes, 2 vols. (Princeton: Princeton University Press, 1984–85), *Problems**, book 30, §955b22–956a10, vol. 2, 1503.

27 Ibid., *Politics*, book 8, §1337a20–1337a32, vol. 2, 2121.

with that. Because a habit directed to an operation "is a perfection of a power," and since a power extends "to many things, in so far as they have something in common," a habit "extends to many things, in so far as they are related to one, for instance, to some specific objective aspect, or to one nature, or to one principle." He added that

> In any science, he who acquires, by demonstration, scientific knowledge of one conclusion, has the habit indeed, yet imperfectly. And when he obtains, by demonstration, the scientific knowledge of another conclusion, no additional habit is engendered in him: but the habit which was in him previously is perfected, forasmuch as it has increased in extent; because the conclusions and demonstrations *of one science* are coordinating, and one flows from another.[28]

Where Descartes saw a unity of *scientia*—the interconnectedness of knowledge corresponding to universal wisdom—Aristotle and the scholastics differentiated individual *scientia* from the uniqueness of the instrument and disposition required for acquiring and perfecting each of them, one at a time.[29] Later natural philosophers had a more nuanced understanding of habitus, which was closer to Goclenius's definition than what most historians of science might imagine. Before we turn to them, let's take a closer look at how Descartes understood *habitus* and endeavored to reframe it in a new natural philosophy contra Aristotle.

2 *Habitus* and Descartes's Logic of Practice

Artisans and Cartesian rational knowledge give the impression of an unusual pair. Yet, as it is well known, artisans were part of Descartes's material and intellectual world. The first rule of the *Regulæ ad directionem ingenii* questions the very role artisans play in Descartes's philosophy. Rule I opposes widening the established (and essentially artisanal) use of *habitus* to the realm of *scientia. Habitus, i.e.*, the 'bodily aptitude and practice' of artisans, first secures proficiency in individual arts; it is connected to the *uniqueness* of *ars*, each

28 Aquinas, *The* Summa theologiae *of St. Thomas Aquinas*, prima secundæ partis, Question 54, art. 4. (emphasis added)

29 On the interconnectedness of knowledge as one of the central components of Descartes's project, Daniel Garber, *Descartes's Metaphysical Physics* (Chicago: The University of Chicago Press, 1992).

and every art requiring a set of bodily deftness and movements (*gestes*) ordinarily distinct from one artisanal practice to another.[30] Aristotle developed this conception of art into a *habitus scientiarum*, a notion to which Aquinas, Suárez, and Eustachius a Sancto Paulo's philosophies adhered, fragmenting *scientia* into a multiplicity of independent knowledge components, each imposing their special intellectual skills and training. Akin to the various arts, every science—every knowledge—is thus created unique and is said to possess its own system of principles to ensure logical and coherent deductive links between objects of a same genus.[31] The Aristotelian model of an ideal science involved, by definition, a collection of untransferable principles, just as *habitus* was exclusive to either the farmer or the cithara player in Descartes well-known examples.[32]

Descartes argued forcefully in Rule I against the generalization of *habitus* because "the knowledge of one truth does not, like skill in one art, hinder us from discovering another; on the contrary it helps us." Descartes instead gave credence in Rule I to the *unity* of *scientia*, i.e., the idea mentioned above regarding an interconnectedness of knowledge commensurate with universal wisdom. Hence, to liken the artisan's *habitus* to *scientia* was sterile and vain "since what makes us stray from the correct way of seeking the truth is chiefly our ignoring the general end of universal wisdom and directing our studies

30 As Descartes explains, "for one man cannot turn his hand to both farming and harp-playing [cithara], or to several different tasks of this kind, as easily as he can to just one of them." Descartes, *Regulæ ad directionem ingenii*, AT X, 359–60; CSM i, 9. All quotes from Descartes are taken in *Œuvres de Descartes*, eds. Charles Adam and Paul Tannery, new edn., 11 vols (Paris: Vrin, 1996) [henceforth AT X, 359–60]. For the English translation I used *The Philosophical Writings of Descartes*, trans. John Cottingham, Robert Stoothoff, and Dugald Murdoch, 3 vols. (Cambridge: Cambridge University Press, 1984–1991) [henceforth CSM i, 9].

31 "Habituum autem varia sunt genera, alii enim sunt animi, alii vero corporis." Eustachius a Sancto Paulo, *Summa philosophica quadripartita*, 2 vols (Lyon, 1609), vol. 2, 121. See Etienne Gilson, *Index scolastico-cartésien* (PhD diss., Université de Paris, 1912), *s.v.* habitus. Descartes, *Règles utiles et claires pour la direction de l'esprit en la recherche de la vérité*, ed. and trans. Jean-Luc Marion (The Hague, Nijhoff, 1977), 90–91.

32 "Arithmetical demonstration and the other sciences likewise possess, each of them, their own genera; so that if the demonstration is to pass from one sphere to another, the genus must be either absolutely or to some extent the same. If this is not so, transference is clearly impossible, because the extreme and the middle terms must be drawn from the same genus: otherwise, as predicated, they will not be essential and will thus be accidents." Aristotle, *Posterior analytics* I.7. The Internet Classics Archives (classics.mit.edu, accessed 3 August 2005). A very good discussion is found in Peter Dear, *Discipline and Experience: The Mathematical Way in the Scientific Revolution* (Chicago: The University of Chicago Press, 1995), 36–46.

towards some particular ends."[33] Rather, it should be "acknowledged that all the sciences are so closely interconnected that it is much easier to learn them all together than to separate one from the other." To seek true knowledge, therefore, one must avoid the study of particular sciences and try instead to "increase the natural light of his reason," which will not only help solve this or that scholastic problem, but also show the will "what decision it ought to make in each of life's contingencies."[34] In contrast to the multiplicity of arts, only one science for Descartes existed, one universal knowledge guided by an all-encompassing wisdom. In other words, *habitus* was to the uniqueness of *ars* what wisdom was to the unity of *scientia*.[35]

Although the character of *habitus* is unequivocal in Rule 1, I would suggest it was not utterly useless or purposeless to the Cartesian method of knowledge production—and to control one's passions.[36] Looking beyond the uniqueness of *habitus*, to the higher ground of methodology, one witnesses how artisanal practices bring to light Descartes's epistemology of rational knowledge, namely the concept of order underpinning the Cartesian method.[37] The method is rarely described as a bona fide logic of practice. As such the method—not yet a metaphysics in the *Regulæ*—takes place before any theoretical or experimental activity of natural philosophy; it lies at the foundation of all Cartesian knowledge.[38] Within the method are embedded a series of well-defined logi-

33 Descartes, *Regulæ*, AT x, 359–61; CSM i, 9–10 for the quotes.

34 Descartes, *Regulæ*, AT x, 361; CSM i, 10.

35 On the uniqueness of *ars* and unity of *scientia*, Jean-Luc Marion, *Sur l'ontologie grise de Descartes. Science cartésienne et savoir aristotélicien dans les* Regulæ, 2nd edn (Paris: Vrin, 1981), 25–30.

36 Later, in *Les Passion de l'âme* (1649), Descartes explains that if we can dress animals devoid of reason to behave in certain ways through habit, it is obvious that even men of the weakest minds will be able to acquire a complete control (*un empire tres-absolu*) on their passions if we employ enough industry to train them as well. Descartes, AT xi, 368–70.

37 I take a different approach here than Matthew Jones, who carefully looked at Descartes's mathematical 'spiritual exercise' to cultivate the mind, eventually leading to the 'good life.' He writes: "Descartes's geometry [using algebra] was supposed to cultivate the mind, then, by habituating one to discerning the clear and distinct, by habituating one to the temporary use of powerful tools, and by accustoming one to the process of producing mechanisms capable of showing the interconnections among phenomena and among geometrical objects." Jones, *The Good Life in the Scientific Revolution: Descartes, Pascal, Leibniz, and the Cultivation of Virtue* (Chicago: The University of Chicago Press, 2006), chapter 1, quote on p. 34.

38 Jean-Luc Marion suggests that the *Regulæ* contain the seeds of the Cartesian metaphysics as found in the *Meditations*, but it does not then unfold because Descartes was unable to properly order the intellectual simple natures with the common simple natures. Marion, "Cartesian Metaphysics and the Role of the Simple Natures," in *The Cambridge Companion to Descartes*, ed. John Cottingham (Cambridge: Cambridge University Press, 1992), 115–39.

cal rules one must learn and—more importantly—learn to follow; being familiar with them does not guarantee success on the path to true knowledge. Consequently, one must practice, train in the method, because without practice it remains a mere *jeu de l'esprit.* Descartes wrote that the method consisted "more in practice than theory," and that the ultimate aim of the *Discours de la méthode* was to uncover "a practical philosophy which might replace the speculative philosophy taught in the schools."[39] Accordingly, the four-rule method unveiled in the *Discours*—summarizing the *Regulæ*—was not to be taught but rather continually exercised, Descartes himself being compelled to "practice [it] constantly ... in order to strengthen myself more and more in its use."[40]

This emphasis on practice was not merely rhetorical; through such a practical philosophy, the power and action of all things would be known "as distinctly as we know the various crafts of our artisans."[41] Although Descartes belittled *habitus* as a legitimate approach to knowledge production, he found an inherent logic beyond the observable gestures of artisans, an artisanal logic that could help natural philosophers acquire a universal method. Descartes explained in Rules IX and X of the *Regulæ* "how we can make our employment of *intuitus* and *deductio* more skillful," and likewise "how to cultivate two special mental faculties, *viz.* perspicacity [*perspicacitas*] in the distinct intuition of

39 Descartes to Marin Mersenne, March 1637?, AT i, 349. Descartes, *Discours de la méthode*, AT vi, 61; CSM i, 142. In a very insightful analysis of Descartes's famous anaclastic line, Daniel Garber shows how the programmatic statement of the method can be reconciled with a theory of practice. Following closely this example Garber explains that what the method gives is a "*workable procedure* for *discovering* an appropriate path" between the reductive steps the knower must take from a question asked to the actual *intuitus*, in this case, of a *potentia naturalis*. From there, the constructive steps (deductions) take us back to the question asked, for which we are now in possession of certain knowledge. Garber, *Descartes's Metaphysical Physics* (Chicago: The University of Chicago Press, 1992), 35–6 (emphasis original). For a very helpful diagram, see Garber, "Descartes and Experiment in the *Discourse* and *Essays*," in *Descartes Embodied: Reading Cartesian Philosophy Through Cartesian Science* (Cambridge: Cambridge University Press, 2001), 85–110, on p. 100.

40 Descartes, *Discours de la méthode*, AT vi, 22; CSM i, 122. Descartes emphasizes the same point a few pages later: "Moreover, I continued practising the method I had prescribed for myself. Besides taking care in general to conduct all my thoughts according to its rules, I set aside some hours now and again to apply it more particularly to mathematical problems." (AT vi, 29; CSM i, 125) See also Descartes, *Règles utiles et claires pour la direction de l'esprit*, 208. A more detailed and somewhat similar analysis is given in Peter A. Schouls, *Descartes and the Possibility of Science* (Ithaca: Cornell University Press, 2000), 63–91. It is interesting to note that Antoine Furetière says in his *Dictionnaire universel* (1690), *s.v.* Méthode, that Descartes's *Discours de la méthode* "est un ouvrage merveilleux," adding in a second entry that the notion of 'méthode' "est aussi la coustume, l'habitude, la maniere d'agir particuliere."

41 Descartes, *Discours de la méthode*, AT vi, 61–2; CSM i, 142–3.

particular things and discernment [*sagacitas*] in the methodical deduction of one thing from another."[42] *Perspicacitas* is linked to *intuitus*, the certain *experientia* of Rule III, as the natural mental ability to concentrate "upon the most insignificant and easiest of matters" and to focus intensively "to acquire the habit of intuiting the truth distinctly and clearly." Descartes contended that "[s]ome people of course are born with a much greater aptitude for this sort of insight than others; but our minds can become much better equipped for it through method and practice." Artisans, for instance, "who engage in delicate operations, and are used to fixing their eyes on a single point, acquire through practice the ability to make perfect distinctions between things, however minute and delicate."[43]

Perspicacitas was not a bodily disposition acquired through time and hands-on practice: it was rather a mental faculty innate to every thinking being. Artisans were not the only people endowed with such a power of perception.[44] However, according to Descartes, their *habitus*—the ability, for instance, to fix one's eyes on a single point—illustrated best how to attune one's own inherent *perspicacitas*. Echoing Renaissance humanists who urged natural philosophers to enter the craftsman's workshops and study the practice of their trade, Descartes suggested that the capacity to distinguish the most minute and delicate of things—to intuit the common simple natures—is strengthened by the action of artisanal practices.[45] From Descartes's point of view artisan's practical knowledge could not by itself produce ideas, yet it could exhibit a course of action guiding the mind toward *scientia*.

Regarding *sagacitas*, Descartes's other constitutive mental faculty, Rule X stipulates that in order to acquire it one should "methodically survey even the most insignificant products of human skill, especially those which display or presuppose order." In line with *perspicacitas*, *sagacitas*—the mental skill to exercise *deductio*—can be aptly brought to light from the arts and crafts:

42 Descartes, *Regulæ*, AT X, 400; CSM i, 33.

43 Descartes, *Regulæ*, AT X, 400–02; CSM i, 33–4.

44 Descartes, *Regulæ*, AT X, 371; CSM i, 16. Descartes maintains that "the power of judging well and of distinguishing the true from the false—which is what we properly call 'good sense' or 'reason'—is naturally equal in all men." Descartes, *Discours de la méthode*, AT vi, 2; CSM i, 111.

45 The classic reference remains Paolo Rossi, *Philosophy, Technology, and the Arts in the Early Modern Era*, trans. Salvator Attanasio (New York: Harper & Row, 1970). A more sophisticated analysis is found in Pamela Smith, *The Body of the Artisan: Art and Experience in the Scientific Revolution* (Chicago: The University of Chicago Press, 2004).

> Since not all minds have such a natural disposition to puzzle things out by their own exertions, the message of this Rule is that we must not take up the more difficult and arduous issues immediately, but must first tackle the simplest and least exalted arts, and especially those in which order prevails—such as weaving and carpet-making, or the more feminine arts of embroidery, in which threads are interwoven in an infinitely varied pattern. Number games and any games involving arithmetic, and the like, belong here. It is surprising how much all these activities exercise our minds, provided of course we discover them for ourselves and not from others. For, since nothing in these activities remains hidden and they are adapted to human cognitive capacities, they present us in the most distinct way with innumerable instances of order, each different from the other, yet all regular. Human discernment [*sagacitas*] consists almost entirely in the proper observance of such order.[46]

Sagacitas directly followed *perspicacitas:* it was the crucial mental faculty that put in series what *perspicacitas* had discovered. If *perspicacitas*, which aimed at "attentively noting in all things that which is absolute in the highest degree," is truly the "whole secret of the art," *sagacitas* was indispensable in figuring out the chain of inferences from the most absolute to the most relative of things.[47] And to create flawless chains of inferences Descartes knew no better way "than by accustoming ourselves to reflecting with some discernment [*cum quâdam sagacitate reflectere*] on the minute details of the things we have already perceived."[48] Referring once again to the most unassuming of artisans reinforced the epistemic connection between what lied beyond *habitus* and the way the mind ought to be disciplined.[49]

Descartes's artisanal knowledge-making model, or Cartesian empiricism one might argue, diverged from the scholastic weapon of choice, syllogism.[50] To rid philosophy of unproven truth-producing premises, he created a new

46 Descartes, *Regulæ*, AT x, 404; CSM i, 35.

47 Descartes, *Regulæ*, AT x, 382; CSM i, 22.

48 Descartes, *Regulæ*, AT x, 384; CSM i, 23.

49 According to Jean-Luc Marion, Descartes would have used in French the word '*adresse*'—and not '*sagacité*,' which is close to '*perspicacité*'—to designate this mental faculty, a word the natural philosopher happily applied to both mechanical and mental skills. For instance, "il faut de l'adresse et de l'habitude pour faire et pour ajuster les machines que j'ai décrites," and "savoir joindre l'adresse de la main à celle de l'esprit." Descartes, *Discours de la méthode*, AT vi, 77 and Descartes to Huygens, 1 November 1635, AT i, 330, respectively. Descartes, *Règles utiles et claires pour la direction de l'esprit*, 208–09.

50 Mihnea Dobre and Tammy Nyden, eds., *Cartesian Empiricisms* (Dordrecht: Springer, 2013).

method inspired by a general and inherent logic of practice found within *habitus*. Although the School's training in syllogism was better than nothing (given that it exercised the minds of the young, which without guidance "might head towards a precipice"), Descartes refocused the scholar's attention on artisanal practices, on the things that were "perfectly known and incapable of being doubted."[51] The method, Descartes explained,

> cannot go so far as to teach us how to perform the actual operations of *intuitus* and *deductio*, since these are the simplest of all and quite basic. If our intellect were not already able to perform them, it would not comprehend any of the rules of the method, however easy they might be.[52]

The Cartesian method did not teach how to use *intuitus* and *deductio*, both being natural abilities. The method was rather created to instruct how mastery of these two innate powers could be achieved. This meant the introduction, use, and continual exercise of matter-of-fact mental faculties such as *perspicacitas* and *sagacitas*, their proficiency drawing from basic mathematics and down-to-earth artisanal practices.

Artisanal practices, however, were not to be studied for their own sake, but as exemplary practices of the method. Take, for instance, blacksmithing. If first deprived of all the instruments of his trade the blacksmith—like any artisan exercising a self-supporting mechanical art—initially uses either a hard rock or an unformed mass of iron as an anvil, a stone as a hammer, pincers made of wood, and other rudimentary tools he might need to begin working. Of course, he will not start making swords, helmets, and metallic artifacts immediately, but rather will fabricate an anvil, all sorts of hammers, pincers, and metallic tools necessary to his trade. Only then will the blacksmith be able to undertake the production of commodities. This example taken from Rule VIII teaches that unmethodical and inconsistent operations should be eliminated to create merchandise—or to solve mathematical problems and settle philosophical disputes.[53] Here Descartes did not emphasize the blacksmith's bodily skills in making tools, swords or helmets—and he did not try to improve those skills per se. Rather, he focused on the blacksmith's working method and orderly approach toward the production of materials. The method, seen through the

51 Descartes, *Regulæ*, AT x, 362–4; CSM i, 10–11. In the *Discours de la méthode*, Descartes mentions that he did not "cease to value the exercises done in the Schools." Descartes, *Discours de la méthode*, AT vi, 5; CSM i, 113.

52 Descartes, *Regulæ*, AT x, 372; CSM i, 16.

53 Descartes, *Regulæ*, AT x, 397; CSM i, 31.

eyes of this mechanical art, becomes an unbroken 'mechanical' thinking process, one that is methodical and intended for a specific goal. Nothing is left to chance, random practices or accidental judgments.[54] In this well-known example, producing Cartesian knowledge and producing artisanal goods look as though they fundamentally came from the same logic of practice, the function of an ordered thinking process.

3 *Habitus* and the Concept of Knowledge Production

According to the lexicographer Goclenius, *philosophia*, as understood by the Peripatetics, is acquired by discipline and the habit of things (*pro disciplinis & habitibus rerum*). It is further divided into two separate concepts: *Vniversalis et particularis*. When universal, one can embrace all theoretical and practical habitus (*omnes Theoricos & Practicos habitus*); when particular, we speak of a science, which is said to contain a single habitus (*singulos habitus continent*), such as mathematics and natural philosophy.[55] Moreover, a science is characterized by an apodictic intellectual habit comparable to a demonstration—the latter being nothing shorter than a 'scientific instrument.'[56] Understanding how habit is defined and interpreted in early modern science provides an intellectual—one might say epistemic—strategy to correlate notions, attitudes, and perspectives of knowledge production formed by natural philosophers, which might at first seem incongruent with each other. In the following few paragraphs, I examine the concept of habit through the lens of

54 The "continuous and wholly uninterrupted sweep of thought" refers to Rule VII and is part of Descartes's theory of order. On the mechanical thinking process, this could explain some of Descartes's strange assertion like: "Ce qui cadre beaucoup avec ma manière de philosopher, et qui revient merveilleusement à toutes les expériences mécaniques que j'ai faites de la nature à ce sujet." Descartes to Villebressieu, summer 1631, AT i, 217. See also Descartes to Froidmont, 3 October 1637, AT i, 420–21. Descartes, *Règles utiles et claires pour la direction de l'esprit*, 204.

55 Goclenius, *Lexicon philosophicum, s.v. philosophia*. Contemporary adaptations are numerous, such as Daniel Sennert's, which explains that "*Phylosophy is an Habit of the Understanding* ... And it is such a kind of quality as is not born with us (for no man is borne a Phylosopher) but attained, being no other than an habitual ability, not of the body (for it is not the body of a man but his mind which wants perfection) but of the understanding. And therefore Phylosophy is defined to be a twofold Habit of the understanding, viz. Wisdom and Prudence, under which terms Intellection and science are comprehended." Sennert, *Thirteen Books of Natural Philosophy* ... (London, 1660), 2.

56 Goclenius, *Lexicon philosophicum, s.v. habitus*: "Scientia est habitus intellectus apodicticus (qui comparatur per Demonstrationem) quo percipimus necessarias conclusiones à necessariis propositionibus." And "Et Demonstratio Instrumentum Scientificum dicitur."

a few well-known English philosophers. It illustrates that although their philosophies might be radically different—as it is, incidentally, with the case of Descartes presented above—they all relied on a definition of habit in support of their claims to knowledge production.

In terms that would have been familiar to Goclenius, John Wilkins, an Anglican clergyman and one of the founders of the Royal Society, described in detail in *An Essay Towards a Real Character, and a Philosophical Language* (1668) what habit meant philosophically. We learn that habits—as in endowment, gift, talent—are "*superinduced Qualities*, whether infused or acquired, *whereby the natural Faculties are perfected*, and rend[e]red more readily and vigorous in the exercise of their several Acts, according to the *more* or *less* perfect Degrees of them."[57] He further divided habit into six 'general considerations,' the last of which is of particular concern to us: Acquired intellectual habits, a virtuous habit. Those, Wilkins explained, "may be gotten by Industry, and tend to the perfecting of the Mind or Understanding."[58]

Building on the scholastic interpretation mentioned above, acquired intellectual habits can be speculative, i.e., "furnishing the mind with due Notions and conceptions concerning the Nature of things, their Causes, Differences, Relations and Dependencies," such as *science* (knowledge, skill, theory, learning, insight). Their object can also be active, i.e., "denoting Skill in men and business, whereby we are inabled [sic] to judge what is fit and convenient, according to various cases and circumstances," such as *wisdom* (prudence, discretion, sapience, sage, politic) and *craft* (cunning, subtlety, shyness, policy, device, quirk, sleight, fetch, wile, trick, sly, shrewd, knave, shark). Finally, their object can be effective, i.e., "implying Skill in those several Operations and Works which concern Human life," such as *art* (skill, dexterity, craft, cunning, insight, knack, expert, well-seen in, good at, artificial, workman, and artist). The manner of acquiring these intellectual habits (science, wisdom, and art) was by observation and repeated trials, such as *experience*, described as "Practice, Exercise, Knowledge, conversant, versed, expert, Experiment, Empiric" and from the teaching of others, "either *vivâ voce*, or *ex scriptis*."[59] What Wilkins is describing in this section of the book, printed by the Royal Society, was basically the principles of Baconian natural philosophy: practice, experiment, empiric, and exercise were manners of acquiring intellectual habits (science,

57 John Wilkins, *An Essay Towards a Real Character, and a Philosophical Language* (London, 1668), 200 (italics original).

58 Ibid., 205.

59 Ibid., 205.

wisdom, art), where manners were "*customary*, and habitual *Actions* of men *considered as voluntary*."[60]

A habit was but a technique or process for perfecting a natural faculty or power. Once a habit was acquired, it was understood as fixed and immutable, compelling the body or mind to move in a particular way (old habits die hard, as we say).[61] Through the early modern concept of *habitus*, the epistemic nature of the mind and body was acknowledged. In his *Elements of Philosophy* (1656), Thomas Hobbes perhaps best explained how habit and actions (or motion) were linked together:

> *Habit* ... is a generation of motion, not of motion simply, but an easy conducting of the moved body in a certain and designed way. And seeing it is attained by the weakening of such endeavors as divert its motion, therefore such endeavors are to be weakened by little and little. But this cannot be done but by the long continuance of action, or by actions often repeated; and therefore custom begets that facility, which is commonly and rightly called *habit*; and it may be defined thus: HABIT *is motion made more easy and ready by custom; that is to say, by perpetual endeavour, or by iterated endeavors in a way differing from that in which the motion proceeded from the beginning, and opposing such endeavors as resist.*[62]

What Hobbes is describing here is evidently not limited to a *habitus in corpore*, namely to bodily health and disposition as understood since Greek antiquity. It was related to bodily gestures, motions, and actions, whether one was learning to play the clavichord or crafting an object. When Hobbes says that habit was the result of 'perpetual endeavor' or 'iterated endeavors,' he endowed habit with the notion of least effort. *Endeavor* indeed had a special meaning to Hobbes.

60 Ibid., 206.

61 Aquinas, quoting Aristotle's *Book of Predicaments* (Categor. vi), says something similar about habits: that "habit is a quality which is difficult to change." Aquinas, *The* Summa theologica *of St. Thomas Aquinas*, prima secundæ partis, question 49. A disposition—as in propensity, proclivity, proneness, inclination, readiness, aptitude—played a similar role as habit, but was usually perceived as less perfect. For example, Philippes Canaye in his *L'Organe, c'est à dire l'instrument du discours* (1589, 59–60) says that "l'Habitude est la perfection de la Disposition ... Car n'est Habitude, que quand elle est hors de toute alteration, & qu'elle a atteinct le souuerain degré de sa perfection, soit que ceste qualité soit bonne ou mauuaise. Et ceste perfection de l'Habitude s'acquiert par exercice, accoustumance, meditation, & estude."

62 Thomas Hobbes, *The English Works of Thomas Hobbes* [electronic resource], 16 vols. (Charlottesville, V.A.: InteLex Corporation [Past Masters Collection], 1995), vol. 4, 348. (emphasis original)

He defined it as a "Motion made in less Space and Time than can be given; that is, less than can be determined or assigned by Exposition or Number; that is, Motion made through the length of a Point."[63] A habit was more than an automatic movement: it was a force. Through the repeated action of bodily organs (or of the mind, as we saw with Descartes), habit became akin to a natural power—a second nature.[64] Hence, whether one dealt with logic as a *habitus organicus*, an instrumental habit (virtue) of the mind, or a mathematical instrument, both mind and body received a special training in the form of *habitus* in order to properly use these instruments. The *habitus*, moreover, was specific to the *organum* (logic, mathematical instrument) deemed suitable for the accomplishment of the desired task.

To Hobbes, and to most early modern savants, including Blaise Pascal in his *Pensées*, habits (or *coutume*, a similar notion but not completely identical to *habitude*[65]) were not about voluntary actions—or free will.[66] Pascal, according to Matthew Jones, "denied the artisans any theoretical or free agency in their habitual activity." In making new inventions, such as his arithmetical machine, Pascal held that artisans required the close supervision of savants "until practice has made the rules of theory so common that [the rules] have finally been reduced into art, and until continual exercise has given to the artisans the habit of following and practicing these rules with confidence." When such a course of action was contested, i.e., when artisans innovate without the guiding intellectual light of savants, they upset, Jones argues, "the epistemic, technical, and social order and, in so doing, produced only monsters or abortions

63 Quoted in Margaret J. Osler, "Whose Ends? Teleology in Early Modern Natural Philosophy," *Osiris*, 2nd series, 16 (2001): 151–68, pp. 166–7.

64 Hobbes, *The English Works*, vol. 4, 348, also mentioned that habit was not restricted to living organisms (contrary to the Renaissance understanding of *habitus*, see Canaye, *L'Organe*, p. 60): "Nor are habits to be observed in living creatures only, but also in bodies inanimate. For we find that when the lath of a cross-bow is strongly bent, and would if the impediment were removed return again with great force; if it remains a long time bent, it will get such a habit, that when it is loosed and left to its own freedom, it will not only not restore itself, but will require as much force for the bringing of it back to its first posture, as it did for the bending of it at the first."

65 According to Diderot, who wrote the article 'coûtume, habitude' for the *Encyclopédie*, the difference is one of degree: "La coûtume concerne l'objet, elle le rend familier; l'habitude a rapport à l'action, elle la rend facile." They are linked together, *habitude* always coming after *coûtume*. "La coûtume, ou plûtôt l'accoûtumance, naît de l'uniformité, & l'habitude, de la répétition."

66 Pascal, *Pensées*, ed. Gérard Ferreyrolles (Paris: Le Livre de Poche, 2000), S617: "La machine arithmétique fait des effets qui approchent plus de la pensée que tout ce que font les animaux. Mais elle ne fait rien qui puisse faire dire qu'elle a de la volonté, comme les animaux."

that were nonfunctional disunities."[67] As seen above, Hobbes contended that habits were "attained by the weakening of such endeavors as divert its motion." Habits were about constraining these 'endeavors' in order to compel the body or mind to carry on a specific undertaking. Whereas this constraining notion was made explicit in Pascal's writings on the arithmetical machine (more on this *infra*), Hobbes used musical instruments to explain the effect of habit. In the case of a "man that playeth on an instrument with his hand"

> it proveth only, that the habit maketh the motion of his hand more ready and quick; but it proveth not that it maketh it more voluntary, but rather less; because the rest of the motions follow the first by an easiness acquired from long custom; in which motion the will doth not accompany all the strokes of the hand, but gives a beginning to them only in the first.[68]

The will, or reason, provides the first 'motion' to the hands; the subsequent motions are not voluntary, but "acquired from long custom."[69] John Locke would later say that "[c]ustom settles habits of thinking in the understanding,

---

67 Matthew Jones, *Reckoning with Matter. Calculating Machines, Innovation, and Thinking about Thinking from Pascal to Babbage* (Chicago: The University of Chicago Press, 2016), quotes on p. 34.

68 Thomas Hobbes, *The English Works*, vol. 9, 354. Two decades earlier, Mersenne had a somewhat similar approach to Hobbes's when he designed a 27-key organ clavier, different from the customary ones and aimed at perfecting the harmony of the organ (see chapter 2). Because the new clavier was organized by the rules of reason, Mersenne argued, an organist will learn how to play on it "as easily as the others" once his "hands become accustomed" to it. This was a common trope. For instance, see Ralph Cudworth, *The True Intellectual System of the Universe, Wherein All the Reason and Philosophy of Atheism is Confuted and its Impossibility Demonstrated* (London, 1678), where he writes: "and first by that of *Habits*, particularly those Musical ones, of Singing, Playing upon Instruments, and Dancing. Which *Habits* direct every Motion of the Hand, Voice, and Body, and prompt them readily, without any *Deliberation* or *Studied Consideration,* what the next following *Note* or *Motion* should be." (p. 157)

69 Rebecca Cypess in a very interesting study looks at those musicians, 'instrumental artisans,' who possessed an impressive *dispositione di mano*, were able to play on the *affeti* of their listeners by their virtuosity and creativity. She shows, for instance, how the deployment of hand positions, repetitive figuration and open strings created sonic illusion (a violin imitating a *lira da braccio*, for example), demonstrating how the "virtuosic artisan, in total control over his instruments, actually changes its constitution." This, she calls, was the conquering *habitus* of the artisan's instrument. Rebecca Cypess, *Curious & Modern Inventions. Instrumental Music as Discovery in Galileo's Italy* (Chicago: The University of Chicago Press, 2016), 129. Even these very special 'sonic illusions' would enter into Hobbes's definition of musical habitus.

as well as of determining in the will, and of motions in the body; all which seems to be but trains of motion in the animal spirits, which once set a-going, continue in the same steps they have been used to, which by often treading, are worn into a smooth path, and the motion in it becomes easy, and as it were natural."[70]

In his *Essay Concerning Human Understanding* (1690), Locke discussed the "several ways wherein the Mind is possessed of Truth." First, *actual knowledge* is how the mind judged between the agreement or disagreement of ideas at the present moment. *Habitual Knowledge*, on the other hand, corresponds to a proposition known to be true but has since been lodged in one's memory. When reflected upon again at a later moment, "without doubt or hesitation" one "assents to, and is certain of the Truth of it." Locke further differentiated between two degrees of habitual knowledge: 1) when the mind, through 'intuitive Knowledge,' perceived, "by an immediate view," the relationship between ideas; 2) when the mind, "having been [once] convinced, it retains the memory of the Conviction, without the Proofs." The latter, called 'demonstrative Knowledge,' was more imperfect than 'intuitive Knowledge' and considered between opinion and knowledge, i.e., to "a sort of assurance that exceeds bare Belief, which relies on the Testimony of another, and yet comes short of perfect Knowledge."[71] To Locke, there existed only one way to acquiring good habitual knowledge: "As it is in the body, so it is in the mind; practice makes it what it is." As he wrote in a little-known text, "it is practice alone that brings the powers of the mind as well as those of the body to their perfection." The difference between people's abilities (from musicians and rope-dancers to geometricians) did not "arise so much from the natural faculties as [from] acquired habits."[72] A habit, defined as a "power or ability in Man, of doing anything, when it has been acquired by frequent doing the same thing," can become so perfected that they "often pass in us without our notice," escaping our observation.[73] This situation, obviously, turns into a huge problem if left unchecked:

> Many men firmly embrace falshood [sic] for truth not only because they never thought otherwise but also because thus blinded as they have been from the beginning they never could think otherwise, at least without a

70 John Locke, *An Essay Concerning Human Understanding* (London, 1690), book 2, chapter 33, §6.

71 Locke, *An Essay Concerning Human Understanding*, book 4, chapter 1, §8 and §9.

72 John Locke, "Practice and Habit," *The Chautauquann; A Weekly News Magazine* 3 (1883): 317–8.

73 Locke, *An Essay Concerning Human Understanding*, book 2, chapter 22, §10 and book 2, chapter 9, §10.

> vigor of minde able to contest the empire of habit, and looke into its own principles, a freedom which few men have the notion of in themselves and fewer are allowed the practice of by others.[74]

This 'empire of habit' was never far from Locke's mind when considering epistemology, politics, education, psychological, and sociological mechanisms.[75] It explained in several different ways how people socially behaved, performed their daily activities, and thought.

Habit, or custom, also played a fundamental role in David Hume's philosophy of knowledge.[76] Influenced by Locke, among others, the Scottish philosopher pointed out that "wherever the Repetition of any particular Act or Operation produces a Propensity to renew the same Act or Operation, without being impell'd by any Reasoning or Process of the Understanding; we always say, that this Propensity is the Effect of *Custom*."[77] Suppose that for the most common operations of nature, e.g., the descent of heavy bodies and the growth of plants, if one could "perceive the very Force and Energy of the Cause" and connect it to its effect, she would "acquire, by long Habit, such a Turn of the Mind, that, upon the Appearance of the Cause, [she would] immediately expect with Assurance its usual Attendant, and hardly conceive it possible, that any other Event could result from it." Only when 'extraordinary Phaenomena' such as earthquakes and 'Prodigies of any Kind' would emerge that she would find herself "at a Loss to assign a proper Cause, and to explain the Manner, in which the Effect is produc'd by it."[78] As Hume explained elsewhere, if "the nature of all our reasonings concerning matter of fact" is established on the relation of cause and effect, and all our reasonings concerning that relation is founded on

74 John Locke, *Of the Conduct of the Understanding*, ed. Paul Schuurman (PhD diss., University of Keele, 2000), par. 77, 229.

75 Ruth W. Grant, "John Locke on Custom's Power and Reason's Authority," *The Review of Politics* 74 (2012): 607–29. John Baltes, *The Empire of Habit. John Locke, Discipline, and the Origins of Liberalism* (Rochester: University of Rochester Press, 2016). Rita Koganzon, "'Contesting the Empire of Habit': Habituation and Liberty in Lockean Education," *American Political Science Association* 111 (2016): 547–58. Kathryn Tabb, "Locke on Habituation and the Association of Ideas," in *The Lockean Mind*, eds. Jessica Gordon-Roth and Shelley Weinberg (London: Routledge, 2022), 385–95.

76 John Christian Laursen, "David Hume on Custom and Habit and Living with Skepticism," *Daimon. Revista internacional de filosofia* 52 (2011): 87–99.

77 David Hume, *Philosophical Essays Concerning Human Understanding* (London, 1748), essay v, part I, 73.

78 Ibid., essay VII, part I, 113–4. This is when "Men, in such difficulties, have recourse to some invisible, intelligent Principle (*quasi Deus ex machina*) as the immediate cause of that Event".

experience, "What is the foundation of all conclusions from experience?" The principle of custom or habit: "All inferences from experience, therefore, are effects of custom, not of reasoning." The *customary conjunction* between, say, flame and heat, i.e., the repeated instances demonstrating that a flame burns, leads to a belief that it will always be the same. "It is that principle alone, which renders our experience useful to us, and makes us expect, for the future, a similar train of events with those which have appeared in the past." Without custom no reasoning about matter of fact exists; there is simply no knowledge of the world. "Custom, then, is the great guide of human life."[79]

Following the principle of *customary conjunction* one would indeed expect that the "connexion of the ideas is not habitual after one experiment." Hume, however, held that a single—likely non-preternatural—event could still turn into a custom in 'an oblique manner' by means of another habitual principle, which posited "*that like objects, plac'd in like circumstances, will always produce like effects*." This is how we can always "transfer our experience to instances, of which we have no experience, either *expressly* or *tacitly*, either *directly* or *indirectly*."[80] Such are the reasons why, as we mentioned above regarding the Scholastics, a habit was said to be a quality difficult to change. Once ingrained in one's mind (or body), it was not easy to abandon. Hume agreed, which is why the social and cognitive functions of custom (or habit) took such priority in his philosophy:

> All those opinions and notions of things, to which we have been accustom'd from our infancy, take such deep root, that 'tis impossible for us, by all the powers of reason and experience, to eradicate them; and this habit not only approaches in its influence, but even on many occasions prevails over that which arises from the constant and inseparable union of causes and effects.[81]

It is not so much of a stretch to use this early modern notion as an explanation to why early modern natural philosophers were said to have erred for so long, stuck as they were with the Aristotelian philosophy and its way of thinking. To Robert Boyle and his experimental followers, a new cognitive habit based on new philosophical principles could move them forward:

79 Hume, *Philosophical Essays Concerning Human Understanding*, essay v, part I, 73 and 75.

80 David Hume, *A Treatise of Human Nature: Being an Attempt to Introduce the Experimental Method of Reasoning into Moral Subjects*, 3 vols. (London, 1739–1740), vol. 1, 187. (italics original)

81 Ibid., 206.

> But a *Virtuoso*, that is wont in his Reasonings to attend to the Principles of Mathematicks, and sound Philosophy, and to the clear Testimonies of Sense, or well verifi'd Experiments, acquires a Habit of discerning the Cogency of an Argument, or way of Probation; and easily discerns, That Dialectical Subtilties [sic], and School Tricks, cannot shift off its force, but finds more Satisfaction in Embracing a Demonstrated Truth, than in the vain Glory of Disputing subtilly against it.[82]

Such a feat required a new frame of mind, a complete transformation of the way of thinking, in which habits played a significant role.

Though discussed, theorized, and formulated by many authors in the early modern period, no systematic work is written on the notion of habit—or custom, two terms as we saw above often used interchangeably, though not completely identical.[83] The goal of this section was not to be exhaustive, but rather to underline the fact that *habitus* was decidedly in the mind of natural philosophers when figuring out how to inspire and catalyze a novel way of thinking about nature—and doing science. Whether a force to eradicate or harness, habit left no natural philosopher indifferent: he or she had to somehow cope with it. I end this chapter with an example that will help further historicize this viewpoint.

4 Blaise Pascal, *coutume*, and the Arithmetical Machine

In the *Lettre dédicatoire* and *Avis nécessaire* written to describe and ask a privilege for his invention, Blaise Pascal emphasized several times that creating and making the arithmetical machine necessitated the alliance of theory and the mechanical arts. Taken separately, Pascal claimed, one was as deficient as the other; both were essential to the task, whether one liked it or not. Such a stance toward the nature of knowledge likely came from the time spent with his father at the Mersenne academy. I have shown in chapter 2 how the father Minim, especially in his *Traité de l'orgue*, underscored the role of theory, experiment, and musical instruments in achieving a complete understanding of the nature

82 Robert Boyle, *The Christian Virtuoso Shewing that by Being Addicted to Experimental Philosophy, a Man is Rather Assisted than Indisposed to be a Good Christian* (London, 1690), 46–7.

83 John P. Wright, "Ideas of Habit and Custom in Early Modern Philosophy," *Journal of the British Society for Phenomenology* 42 (2011): 18–32. In Diderot and d'Alembert's *Encyclopédie*, habit is still distinguished into Goclenius's two chief categories: "On distingue les *habitudes* en *habitudes* du corps & en *habitudes* de l'ame."

of sound. Pascal *père*, it would appear, fully endorsed this approach to natural philosophy. Mersenne dedicated the same *Traité de l'orgue* to the elder Pascal, praising Blaise's father not only for being an accomplished musician, but more importantly for having married his extensive knowledge of mathematics to the practical expression of mechanics.[84] To Pascal the conjunction of theory and practice must have felt natural, *allant de soi*, owing to the intellectual milieu he grew up in. Though Pascal firmly believed the artisan was subordinated to the *grandeur* of the natural philosopher's mind, the former's mastery of the mechanical arts was indispensable to the success of his enterprise—and the whole of natural philosophy.[85]

The pascaline, Pascal claimed, was designed to replace a faulty memory, and ipso facto the habit of calculating.[86] Mathematical practices such as logarithm and rabdology, both invented by the English mathematician John Napier, were intended to substitute the so-called weakness of memory with tools like printed logarithmic tables, paper or metallic rules (*reigles de proportion*) and Napier's bones.[87] The memory involved in making arithmetical operations like multiplication and subtraction, said differently, was supplanted by material devices created for specific mathematical practices. Abstract number concepts were materialized into these devices—they were blackboxed one would say today. They no longer required to be fully grasped or remembered by rote. A material device, to which was assigned a prescriptive technique, was replacing intangible memory. This trade-off points as well to another fundamental aspect regarding mathematical practices, and the pascaline in particular. A new material object (*organum*) necessitated learning new gestural habits, likely distinct from conventional ones. Using Napier's bones, for instance, bound a mathematician or 'the vnskilled and slothfull men' to bodily movements

---

84 Mersenne, "Traité de l'orgue," book VI, Epitre, in *Harmonie universelle, contenant la théorie et la pratique de la musique*, 3 vols. (Paris: Centre national de la recherche scientifique, 1963), vol.3, n.p. The Epitre is dated 1 November 1635.

85 On the grandeur of the natural philosopher vis-à-vis the artisan, see Jones, *Reckoning with Matter*, chapter 1.

86 Contrary to the pascaline, Descartes's proportional compass was not designed as a simple mechanical aid to *logistae*, or calculators. That said, it did have a role to play on the relationship between memory, intuition, and self-cultivation. As Matthew Jones explains: "The compass offered a crucial heuristic, a material propaedeutic, for Descartes's revised account of mathematics freed from the memory and subject to a criterion of graspable unity [in regard to a sequence of continued relationship]. A simple mathematical instrument offers a model and exemplar of the knowledge necessary for cultivating oneself." Jones, *The Good Life in the Scientific Revolution*, 29.

87 Napier's rabdology was described similarly by John Dansie, who gave the following subtitle to his 1627 *A Mathematicall manuel*: "*whereby* Any man that can but add and substract, may learne to multiply and divide in two houres by *Rabdologie, without any trouble* at all to the memorie."

drastically different from the pen and counters (*plume et jetons*) technique: the positioning and reading of rods had nothing to do with counters. Without the proper gestural habits, multiplication with Napier's bones would prove impossible. Pascal's arithmetical machine was no different than other mathematical instruments in this respect. To use it, one required a completely new *habitude*, specific bodily motions that ensured the machine's proper working. With time, a *habitude* is generalized from a *coutume*, which became part of Pascal's epistemology of knowledge.

Memory was not completely obsolete or useless. In fragment 536 of the *Pensées*, Pascal asserted that "memory is necessary for all the operations of reason" and in fragment 322 he described how central was genealogy (the memory of past generations) to the ancients' understanding of history.[88] Yet memory became an obvious problem when dealing with arithmetic. The pascaline's carry mechanism, the machine's most important mechanical feature, was meant to alleviate the mind from such a mental exertion:

> You [the reader] know as well that while using the *plume* one is compelled at any time to carry or borrow numbers, and how many errors slip by in these carries and borrowings, unless one has a very long habit [*habitude*] [in such things] and a deep concentration, which rapidly wears out the mind. This machine relieves anyone who uses it from this vexation. It suffices that one has judgment, whilst the machine takes over from the failing of memory. And, with no carrying nor borrowing [to execute], the machine does by itself what its user wants, without the latter even having to think about it.[89]

In a sonnet dedicated to Pascal's arithmetical machine, the poet Dalibray explicitly mentioned how wonderful was the 'artifice' of this 'amazing genius' in demanding no reason nor memory to help even the 'thickest minds' to calculate.[90] Pascal invented the arithmetical machine to curtail the superfluousness of the conventional reckoning technique. With this new machine, the "most ignorant [calculator] will find as much advantage as the most

88 Pascal, *Pensées*, ed. Gérard Ferreyrolles (Paris: Le Livre de Poche, 2000), S536, S322, and S618.

89 Pascal, *Avis nécessaire*, in *Oeuvres complètes*, ed. Jean Mesnard, 4 vols. (Paris: Desclée de Brouwer, 1964–1992), vol. 2, 337.

90 Dalibray, *Les Oeuvres poétiques*, in Pascal, *Oeuvres complètes*, vol 2, 692:
Après ton grand esprit, que sert-il d'en avoir?
Compter fut l'action d'un homme raisonnable,
Et voilà, maintenant ton art inimitable
Aux esprits les plus lourds en donne le pouvoir.

experienced one" because "the instrument makes up for the lack of knowledge or lack of *habitude*" of its user. Pascal added that through the machine's 'necessary movements,' the user of the pascaline could do with ease, without even thinking about it, every possible arithmetical operation.[91]

The swift mechanical actions of the machine (*mouvements nécessaires*) were made obvious when measured against the motions (*gestes*) required by the pen and counters technique. In fact, as Pascal argued to his reader, "if you want a more specific explanation of the speed of the machine's movements, I would tell you that it is equal to the quickness [*agilité*] of the hand working on it."[92] The gearing system of the machine did not impede at all the promptitude of the hand; owing to the machine's design the hand working on it could become as free and as fast as if it dealt with the *plume* and *jetons*. In contrasting with the swiftness of execution of both techniques, Pascal underscored one essential facet of his invention: the necessity of learning a new gestural *habitude*, one that was as efficient as the traditional method—perhaps more—albeit utterly dissimilar. This new *habitude* was confounding not because it required the training of the body per se, but because this bodily *accoutumance* suggested a corporeal 'memory' rather than an act of memory from the mind. Pascal's arithmetical machine, according to the rhetoric, stored mathematical knowledge within its gears and the practice of arithmetic. In the end, only movements—or *habitudes corporelles*—were left, which meant that no memorization nor mathematical practices were required in order to operate Pascal's machine. The pascaline, in a sense, demoted arithmetic from the realm of the mind to that of the body. It became a 'mere' addition to *la machine*, or Pascal's characterization of the human body. Arithmetic became a precise and rhythmic mechanical action generated by a body appendage.

Habitude and *coutume* were not exactly identical notions in the seventeenth century, yet they were related to each other. Usually, *coutume* encompassed both the general and the individual definitions (taking for example Mme de Sévigné's writings, who said about various subjects that *Je n'en ferai pas ma coutume*). Whereas *coutume* could stand in for *habitude*, the reverse could be considered an incorrect semantic use. Even though Pascal referred to both notions repeatedly, it was the more expansive concept of *coutume* that held his attention. In the *Pensées*, for instance, Pascal mentioned the substantive *habitude*

91 Pascal, *Avis nécessaire*, 337.

92 Ibid., 337: "Et, enfin, quant à la promptitude, elle paraît de même en la comparant avec celle des autres deux méthodes du jeton et de la plume; et si tu veux encore une plus particulière explication de sa vitesse, je te dirai qu'elle est pareille à l'agilité de la main de celui qui opère."

only three times compared to forty-nine times for *coutume*.[93] Therefore, to understand the *habitude corporelle* underlined in the *Avis nécessaire*, it is necessary to take a closer look at Pascal's concept of *coutume*.

In fragment 480 Pascal famously said that "*coutume* can do everything." It is all powerful because one cannot see it or feel it, because most of the time we are unaware of its presence and action. "The *coutume* is a second nature that destroys the first." (S159) *Coutume* is a major source of errors because it pulls a veil over the real nature of things. It influences every activity of life: intellectual, social, and religious. People, for instance, are so accustomed to seeing kings surrounded by royal guards, officers, and drummers, things that convey respect and fear, that they believe it all comes from a natural force instead of coutume.[94] In matters of faith, the lack of *coutume* made some individuals doubt Christ's resurrection: the *coutume* of creating something anew that never was (childbirth) largely supplanted the miraculous instance of the (re)creation of something that once was (resurrection).[95] As Gérard Ferreyrolles so aptly described, the notion of *coutume* created for Pascal social, ontological, and epistemological problems that needed to be addressed.[96] It was of great consequence, because once an individual was accustomed to using false reasoning to prove natural phenomena, he or she would not easily acknowledge the right ones when discovered. (S617)

Coutume, furthermore, constrained the human body; it bent *la machine* under the weight of habituation. (S59) As dangerous and problematic as *coutume* might be, it could nonetheless become a powerful epistemological instrument.[97] In matters of faith, Pascal determined three methods of believing: reason, *coutume*, and divine inspiration. A true Christian believer opened

93 Gérard Ferreyrolles, *Les Reines du monde. L'Imagination et la coutume chez Pascal* (Paris: Honoré Champion Editeur, 1995), 17–19. This is the best general analysis of Pascal's concept of *coutume*.

94 Pascal, *Pensées*, S59. Sometimes, however, it is the other way around: "C'est l'effet de la force, non de la coutume, car ceux qui sont capables d'inventer sont rares. Les plus forts en nombre ne veulent que suivre et refusent la gloire à ces inventeurs qui la cherchent par leurs inventions. Et s'ils s'obstinent à la vouloir obtenir et à mépriser ceux qui n'inventent pas, les autres leur donneront des noms ridicules, leur donneraient des coups de bâton." (S122)

95 Regarding atheists and *coutume*, Pascal writes: "Quelle raison ont-ils de dire qu'on ne peut ressusciter? Quel est plus difficile: de naître ou de ressusciter? Que ce qui n'a jamais été soit, ou que ce qui a été soit encore? Est-il plus difficile de venir en être que d'y revenir? La coutume nous rend l'un facile, le manque de coutume rend l'autre impossible. Populaire façon de juger!" (S444)

96 Ferreyrolles, *Les Reines du monde*, 17–37.

97 Ferreyrolles, *Les Reines du monde*, 65–119.

his/her mind to reasoned proofs, confirmed them by *coutume*, but most importantly let divine inspiration bring in the real and salutary effect of faith. (S655) Yet the influence of *coutume* here was perhaps more important than anything else. In fragment 661, Pascal referred again to matters of faith, but one can interpret the subtle rule of *coutume* beyond religion. Reasoned proofs, according to Pascal, convinced only the mind. It was *coutume* that brought the most powerful and raw evidence of faith and secular issues. Only *coutume* bent the body—the automaton—which would then drive the mind (like the gears of a machine) without the latter ever thinking about it. Because reasoned proofs were just too difficult to always remember, one had to acquire another, easier principle like *habitude*, which without violence, art, or argument "persuades all of our powers to this credence, in such a way that our soul falls in naturally." It was not enough to have the mind believes something. Both mind and body needed to be convinced: the former by reasoned proofs, which required to have been seen only once in life; the latter, the automaton, by the mechanical repetition of *coutume*. Hence fragment 661 outlined for matters of faith what the *Avis nécessaire* had already described several years earlier with regard to the arithmetical machine: first, visiting with Roberval, professor at the Collège Royal, to learn at least once the reasoned proofs of arithmetic; second, allowing the power of *habitude* to bend the body *qua* automaton into accepting the art of arithmetic as a *sentiment* rather than an act of the mind.[98]

"What are our natural principles if not our accustomed principles?" (S158) Such issues entered Pascal's *Pensées* not only from a cultural or literary perspective—Jean Nicot's definition of *accoutumé* is remarkably like Pascal's[99]—but from an epistemological one. To fix the instability of reason one must revert to the mechanical character of *coutume*. One must, to borrow Pascal's own expression, *s'abêtir* (to dumb down, or literally to lower oneself to the level of a beast), not in the sense of adopting pure animal instinct, but rather in embodying rational thinking during every human act. *S'abêtir* meant transforming the whole body into a thinking machine, each and every part

98 *Habitude* and *sentiment* act similarly, and Pascal is quite explicit about that fact: "La raison agit avec lenteur, et avec tant de vues, sur tant de principes, lesquels il faut qu'ils soient toujours présents, qu'à toute heure elle s'assoupit ou s'égare, manque d'avoir tous ses principes présents. Le sentiment n'agit pas ainsi; il agit en un instant, et toujours est prêt à agir. Il faut donc mettre notre foi dans le sentiment, autrement elle sera toujours vacillante." Pascal, *Pensées*, S661.

99 Jean Nicot, *Thresor de la langue françoyse tant ancienne que moderne* (1606), cf. *Accoutumé*: "Comme tu as accoutumé, c'est à dire suivant ton naturel." There are many other definitions of interest under that entry.

of it becoming what Pascal called a 'thinking limb,' or a *membre pensant*.[100] Though an appendage of *la machine*, the arithmetical machine had the same ontological value as any other thinking limb of the human body. Once one was accustomed to using the pascaline through *habitude*, this mechanical appendage became a 'natural' extension of his or her body. No ontological difference existed any more between the arithmetical machine and *la machine*'s thinking limbs. Through *coutume*, body, mind, and machine formed one epistemic unit dedicated to the production of knowledge.

As incomplete as this chapter might be, it indicates how significant *habitus*, and by extension *coutume*, was to natural philosophers who investigated and self-reflected on their own actions and thinking about nature. The habit Daston recognized and scrutinized in her ontology of observation enjoys a history and philosophy deeply rooted in the writings of most early modern natural philosophers. Goclenius's conventional lexicographical description of *habitus*, as a starting point, denotes a long tradition embedded in the epistemological Weltanschauung of Aristotle and Aquinas, among others. Later in the early modern period we see how frequently *habitus* is invoked as a positive force or nuisance on the road to true knowledge. The following chapter builds on this study by examining how *habitus* and its corollaries are deployed in the context of artisanal craftsmanship.

---

100 Pascal, *Pensées*, S403: "Qu'on s'imagine un corps plein de membres pensants!" See also S401. For a general discussion, Horia Lazar, "L'Habitude de penser: de la machine arithmétique à l'abêtissement'," *Chronique: Institut catholique de Toulouse* (1995): 57–70.

CHAPTER 4

From *idiotae* to *artistes*: Artisans, Instruments, and the Nature of Craftsmanship in Early Modern Europe

In 1671, the Capucin father Michel Lassere, known as Chérubin d'Orléans, published a study on the telescope entitled *La Dioptrique oculaire*, to which he added a remarkable and telling subtitle: *ou la théorique, la positive, et la méchanique, de l'oculaire dioptrique en toutes ses especes*.[1] The book's structure is analogous to René Descartes's *Dioptrique*—still very influential in France in the 1670s.[2] First comes the theory of light (how reflection and refraction work); second, the theory of the telescope (what d'Orléans called the *oculaire dioptrique*); third, the *positive* and *mechanique* of the telescope, respectively how to make and use it and how to cut, grind and polish the most perfect spherical (not Descartes's hyperbolic) lenses using specially designed devices.[3] Unlike his famous predecessor, however, d'Orléans emphasized a new way of thinking about machines, artisanal practices, and theory. Where Descartes endeavored to reform—rationalize—artisanal skills and mechanical practices by getting rid of the artisan altogether, *La Dioptrique oculaire* prioritized the creation of a novel kind of artisan, an *artiste curieux*, someone who would exist in the spatial and conceptual intersection of theory, practices, and mechanical devices. As D. Graham Burnett suggests, "The essential character of the [lens-grinding] machine could be discovered and demonstrated by mathematics, but the accidental character of the machine had to be shaped by experience and attention to the natural and real characteristics of materials, and the abilities of the human body."[4] Only the *artistes curieux* who embraced at once the theory of light, the *positive* of telescope making and use, and the necessary hands-on

1 Chérubin d'Orléans, *La Dioptrique oculaire, ou la théorique, la positive, et la méchanique, de l'oculaire dioptrique en toutes ses especes* (Paris, 1671).

2 The book review author of Isaac Barrow's *Lectiones opticæ et geometricæ* for the *Journal des sçavans* (Monday, 18 November 1675) mentioned upfront that "L'Optique est une des matieres sur lesquelles on a fait en ce siecle de plus belles & plus curieuses découvertes. On admire encore tous les jours la subtilité avec laquelle Mr. Descartes a découvert sur la Dioptrique plusieurs choses qui avoient été inconnuës aux Anciens ... Le principe de M. Descartes sur la Dioptrique est quasi universellement receu." (pp. 241–2)

3 A summary of the book in English is found in the *Philosophical Transactions* 6 (1671): 3045–50.

4 D. Graham Burnett has no doubt offered the best analysis of d'Orléans's work in the general context of early modern lens making machines. See his *Descartes and the Hyperbolic*

 | DOI:10.1163/9789004504615_008

skills provided by experience in the mechanical arts would gain complete knowledge of the science of optics. D'Orléans introduced one of the most remarkable and comprehensive epistemic schemes of the period—in immediate conversation one might say with Marin Mersenne's views about music and the organ a few decades earlier (see chapter 2).[5]

The term '*positif*' in early modern French meant something certain and real, regular and assured—think of Auguste Comte's *positivism* almost two centuries later.[6] The *positive* of telescope making was a specialized practice based on experiments and sometimes life-long experience. It had become, according to d'Orléans, as unquestionable as the 'practice' of theory. The final ambition was to achieve perfect spherical lenses, enabling an accurate method of telescope making. This stemmed from the domain of *mechanique*, essentially distinct from the *positive*.[7] Here, d'Orléans distinguished the simple artisan—the *ocularisté vulgaire*—and the *artiste curieux*, an enlightened craftsman combining reason and mechanical skills.[8] These skills, however, had to be used in conjunction with a lens-grinding machine. Not the compound sort, like Descartes's, which added two or more spinning movements and where the artisan was completely left out of the grinding process. The machine designed by d'Orléans was simpler: one motion mechanical and one regulated by the hands of the *artiste curieux*, artfully guided by the latter's reason. (Fig. 9) The lens-grinding machine and the *artiste curieux* thus created a new type of entity in d'Orléans rhetoric, the *artiste curieux* being seen as the soul of the device.[9] As Graham Burnett puts it: "The artisan was the maker, but he had to make through the machine, which had to be an extension of himself, just as he had to become a 'component' within it. The successful system for making had to

Quest: Lens Making Machines and Their Significance in the Seventeenth Century (Philadelphia: American Philosophical Society, 2005), 107–21, quote on p. 112.

5 I will come back to the description of *artiste* below, based on the pivotal book by Paola Bertucci, *Artisanal Enlightenment. Science and the Mechanical Arts in the Old Regime France* (New Haven: Yale University Press, 2017).

6 Antoine Furetière, *Dictionnaire universel* (1690), *s.v.* positif: "Qui est certain & effectif, qu'on met en fait comme une chose constante & assûrée. Cela n'est point imaginaire, mais est *positif.* Ce fait est *positif,* & il en offre la preuve. Ce ne sont pas des offres labiales, mais reelles & *positives.*"

7 D'Orléans, *La Dioptrique oculaire*, 337, "réellement, tres-differentes & distinctes."

8 D'Orléans, *La Dioptrique oculaire*, 393, for example, where he states: "La maniere, que tiennent les Ocularistes vulgaires, a concaver les verres, n'estant qu'une routine grossiere, & sans esprit: je ne l'ay pas jugée meriter, ny l'estime, ny la main, de nostre Artiste curieux, ny consequemment, la peine d'estre rectifiée, & corrigée des defauts tres-considerables, qu'elle admet dans la pratique."

9 D'Orléans, *La Dioptrique oculaire*, 380–81.

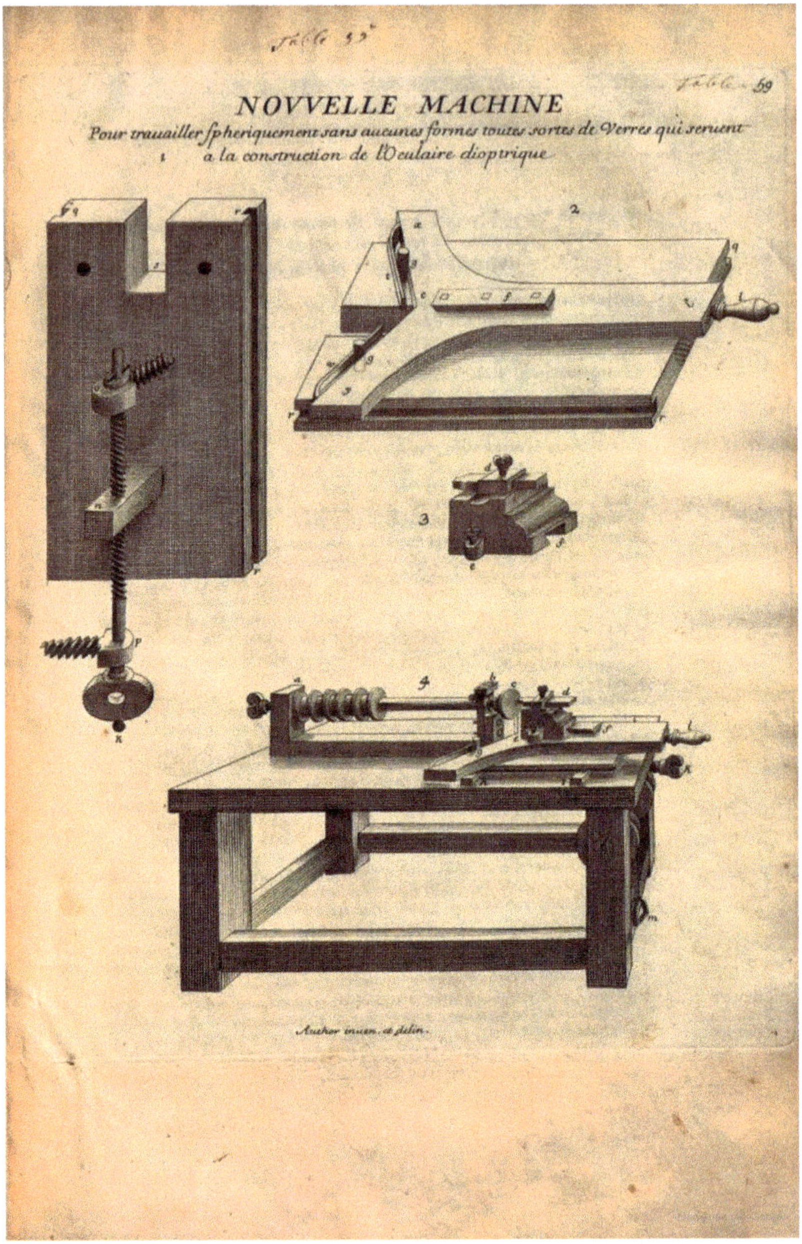

FIGURE 9 Chérubin d'Orléans's lens-grinding machine (1671), known as 'Nouvelle machine à travailler sphériquement sans aucunes formes toutes sortes de verres qui servent à la construction de l'Oculaire dioptrique.'
CHERUBIN D'ORLÉANS, *LA DIOPTRIQUE OCULAIRE, OU LA THÉORIQUE, LA POSITIVE, ET LA MÉCHANIQUE, DE L'OCULAIRE DIOPTRIQUE EN TOUTES SES ESPECES* (PARIS, 1671), 59. FROM GOOGLE BOOKS

bind the artisan, the mechanism, and the analytical 'essence' into a tight unit, a unit necessitating the mechanization of the artisan himself."[10] The *artiste curieux*, d'Orléans's intended audience, brought into play the theory of optics, experience, and experiments in the form of *positive*, and manual skills in order to manufacture an instrument such as the telescope, which was then employed to generate new knowledge about nature. Theory, practice, and instrumentation were no longer compartmentalized into separate activities.

Machines and their makers were in the second half of the seventeenth century important features of philosophizing. When Christiaan Huygens's visited Paris in the 1660s, he did not fail to mention in his *journal* what he saw in Henri Louis Habert de Montmor's cabinet. He called upon on many occasions the clockmaker Martinot, the optician Ménard, and the instrument maker Blondeau and was intensively involved in carriage design.[11] Samuel Sorbière, member and secretary of the Montmor academy, did not himself miss an opportunity to visit workshops and report on instruments and machines he came across while journeying across Europe. In personal letters and talks at the Montmor academy, he often mentioned the role of experiments and savoir-faire from cooks, architects, and other artisans in knowledge-production pursuits.[12] Balthasar de Monconys did likewise, especially during his stay in London, where he dropped by the workshops of Ralph Greatorex, Anthony Thompson, and Richard Reeve. According to Sorbière, who visited London at the same time, "I found [Monconys] in his element, immersed in the conversation of natural philosophers and breathing nothing but machines and new experiments." Huygens, who was also traveling to London, gave Monconys

10 Graham Burnett, *Descartes and the Hyperbolic Quest*, 120.

11 Henri-L. Brugmans, *Le séjour de Christian Huygens à Paris et ses relations avec les milieux scientifiques français, suivi de son* Journal de voyage à Paris et à Londres (Paris: Impressions Pierre André, 1935), 143: "Vu son cabinet. Tableaux, instruments de mathema[tique]. Venus d'Aleaume. Pierres d'aimant. Figures d'Albert Dürer. Jouet des petites planches liées avec des rubans. Esguille suspendue tournant a tous sens. Petites bouteilles dans l'eau qui montent et descendent, sans qu'on s'en aperçoive." Ménard is also mentioned several times in Huygens's *Correspondance*. On carriage design, Jean-François Gauvin, "Huygens's *Carriole*: Urban Commodity, Authorship, and the Intelligence of Carriage Design," *Nuncius* 38 (2023): 72–102.

12 Samuel Sorbiere, *Relations, lettres et discours de Mr de Sorbiere sur diverses matieres curieuses* (Paris, 1660) and Sorbiere, *Lettres et discours de Mr de Sorbiere sur diverses matieres curieuses* (Paris, 1660), in which he mentions during a talk at the Montmor academy that "Nous nous sommes dauantage preualus des experiences & de la pratique de nos Cuisiniers, de nos Architectes, & de tous nos autres Artisans; que de toutes les speculations des Physiciens qui ont pris l'essor, & que nous auons perdu de veuë incontinent ... Euitons le plus que nous pourrons ce langage, & fuyons les inutiles subtilités. Tendons au plustost à la pratique dans nos raisonnemens sur les causes naturelles que nous recherchons." (p. 200)

a drawing of his air pump, which Monconys reproduced in his *journal* and displayed to a few people afterward. There is even a useful table about machines and *artifices divers* at the end of Monconys's work.[13]

At the Paris Académie royale des sciences, as discussed in the introduction, instruments and machines were omnipresent, as the well-known engravings from Sébastien Le Clerc illustrate—especially the one depicting the king's apocryphal 1671 visit to the Academy. (Louis XVI did come to the Academy a decade later.) When William III, king of England, graced the Academy of his company in 1690, he was shown all sorts of astronomical instruments made for the Observatory, and even the *salle des machines*, where the English monarch admired Ole Rømer's two planetaria (built by the clockmaker Thuret) and discoursed on various mechanical devices. Experiments were performed, machines were seen and operated, and theoretical claims discussed in the presence of His Royal Highness, all under one roof.[14] By fully reintegrating the notion of material culture (*organum*) into the domain of early modern knowledge production, I show how it might structure the manner we think about scientific practices.[15] The latter, interpreted in chapter 3 as *habitus*, is examined

---

13 Balthasar de Monconys, *Journal des voyages de Monsieur Monconys, Conseiller du Roy en ses Conseils d'Estat & Privé, & Lieutenant Criminel au Siege Presidial de Lyon. Où les Sçavants trouveront un nombre infini de nouveautez, en Machines de Mathematique, Experiences Physiques, Raisonnemens de la belle Philosophie, curiositez de Chymie, & conversations des Illustres de ce Siecle ...* 3 vols. (Lyon, 1665–6). On Monconys's collection, Anthony Turner, "Grollier de Servière, the Brothers Monconys: Curiosity and Collecting in Seventeenth-Century Lyon," *Journal of the History of Collections* 20 (2008): 205–15. For Huygens's air pump, see vol. 2, p. 73. For Sorbière's quote, see James A. Bennett, "Shopping for Instruments in Paris and London," in *Merchants & Marvels: Commerce, Science and Art in Early Modern Europe*, eds. Pamela H. Smith and Paula Findlen (New York: Routledge, 2002), 370–95, on p. 372. Huygens had an unfavorable bias against Monconys's work, mentioning in one of his correspondence that "Ie n'ay pas leu encore les voiages de Monsieur de Monconis, ma curiositè ayant estè diminuee par le raport de quelques uns qui m'ont dit que c'est un ouurage fort mal digerè, et ou il y a quantitè de choses de peu d'importance, l'autheur ayant eu de curiositez vaines en plusieurs choses comme en l'astrologie, alchimie &c. Mais puis que vous dites y avoir pris de la satisfaction j'ay envie de veoir ce que c'est." Huygens to Philips Doublet, 6 August 1666, in *Oeuvres complètes*, 22 vols. (The Hague: M. Nijhoff, 1888–1950), vol. 6, no. 1555, 73. [Hereafter cited as *OC*(6), no. 1555, 73.]

14 E. C. Watson, "The Early Days of the Académie des sciences as Portrayed in the Engravings of Sébastien Le Clerc," *Osiris* 7 (1939): 556–87. On William III's visit, see Bernard Le Bovier de Fontenelle's *Histoire de l'Académie royale des sciences*, 2 vols (Paris, 1733), vol. 2, 94–104, esp. pp. 102–4.

15 In a statistical study based solely on the *Dictionary of Scientific Biography*, Richard Westfall came to the conclusion that 62.3% of the scientific community during the Scientific Revolution was involved in technological projects. (If one adds medicine, the percentage goes up to about 75%.) Of that number, the largest community of scientists was engaged

here from the perspective of the 'artisan.' We discover how the 'artisan' emerged as a plurality of concepts and images, how it mutated and grew into various forms over the course of centuries. From the heuristic idea and inspiration brought forth by an artisanal activity to the actual work and output of craftsmen, this chapter traces some of the disparate locations and theoretical trajectories taken by artisans and natural philosophers in producing the instruments and machines that ushered in the so-called era of modern science.

1 Descartes, Artisans, and *âmes réglées*

Let us begin with René Descartes, again. Descartes's whole "project of self-instruction," we learn in Part VI of the *Discours de la méthode*, "is suffering because of the need for innumerable observations which I cannot possibly make without the help of others [*sans l'aide d'autrui*]."[16] Who are these 'others' fit to be of service to Descartes in his quest for truth?[17] Not just any individual was right for such an important task:

> True, as regards observations which may help in [natural philosophy], one man could not possibly make them all. But also he could not usefully employ other hands than his own, except those of artisans, or such people as he could pay, who would be led by the hope of gain (a most

in what he called 'instrumentation,' 22.7%, a number higher than military engineering, hydraulic engineering, navigation, and cartography. Westfall, "Science and Technology during the Scientific Revolution: An Empirical Approach," in *Renaissance and Revolution: Humanists, Scholars, Craftsmen and Natural Philosophers in Early Modern Europe*, eds. J. V. Field and Frank A. J. L. James (Cambridge: Cambridge University Press, 1993), 63–72.

16 Descartes, *Discours de la méthode*, *Œuvres de Descartes*, eds. Charles Adam and Paul Tannery, new edn., 11 vols (Paris: J. Vrin, 1996), vol. 6, 75 [hereafter cited as AT vi, 75]. For the English translation I used *The Philosophical Writings of Descartes*, trans. John Cottingham, Robert Stoothoff, and Dugald Murdoch, 3 vols (Cambridge: Cambridge University Press, 1984–1991), vol. 1, 149 [hereafter cited as CSM i, 149].

17 On the role of observations and experiments for Descartes, see Daniel Garber, "Descartes and Experiment in the *Discourse* and *Essays*," in *Descartes Embodied: Reading Cartesian Philosophy Through Cartesian Science* (Cambridge: Cambridge University Press, 2001), 85–110; Ralph M. Blake, "The Rôle of Experience in Descartes' Theory of Method (I)," *The Philosophical Review* 38 (1929): 125–43; Blake, "The Rôle of Experience in Descartes' Theory of Method (II)," *The Philosophical Review* 38 (1929): 201–18; Alan Gewirtz, "Experience and the Non-Mathematical in the Cartesian Method," *Journal of the History of Ideas* 2 (1941): 183–210; Desmond M. Clarke, *Descartes's Philosophy of Science* (Manchester: Manchester University Press, c1982); Spyros Sakellariadis, "Descartes's Use of Empirical Data to Test Hypotheses," *Isis* 78 (1982): 68–76.

> effective motive) to do precisely what he ordered them to do. For voluntary helpers [*volontaires*], who might offer to help him from curiosity or a desire to learn, usually promise more than they achieve and make fine proposals which never come to anything. In addition, they would inevitably wish to be rewarded by having certain difficulties explained to them, or at any rate by compliments and useless conversation, which could not but waste a lot of time.[18]

Taken literally, this quotation from the *Discours de la méthode* establishes the status of artisans as a docile workforce, whose mechanical skills should help uncover, under the tutelage of natural philosophers, nature's deepest secrets.[19] *Volontaires*, or *honnêtes curieux*, were given even less credit here, perceived as a nuisance rather than the source and authority of knowledge—in contrast to Robert Boyle's conception of gentlemanly science.[20] Although Descartes despised the *curieux-volontaires* in his search for truth, he judged them allies in his fight against scholasticism. In addressing the honnêtes curieux outside the Republic of Letters, Descartes believed the acceptance of his natural philosophy hinged on the yeas and nays of the emerging early modern public.[21] In this sense Descartes was akin to a Cornelian hero, who was nothing "[w]ithout the recognition accorded to his name." Descartes, as the protagonist to his own natural philosophy, knew to the same extent as Corneille's Cid, that he was "as dependent on his admirers as they [were] dependent on him."[22]

18 Descartes, *Discours de la méthode*, AT vi, 72–3; CSM i, 148.

19 Money sometimes was not a good enough incentive to hold artisans in check. When the time came, for instance, to engrave the plates for the *Discours* and *Essais*, Descartes and his printer made sure the engraver would not leave without address or procrastinate for too long. The only way to enforce their wish was to keep this engraver (Franz Schooten the younger) 'under house arrest': "Celui qui les taille [the plates] me contente assez, et le libraire le tient en son logis, de peur qu'il ne lui échappe." Descartes to Constantijn Huygens, 30 October 1636, AT I, 614.

20 Steven Shapin, *A Social History of Truth: Civility and Science in Seventeenth-Century England* (Chicago: The University of Chicago Press, 1994).

21 On this, see Alain Viala, *Naissance de l'écrivain: sociologie de la littérature à l'âge classique* (Paris: Editions de Minuit, 1985).

22 Jean-Pierre Cavaillé, "Descartes stratège de la destination," *XVIIe siècle* 177 (1992): 551–9, on p. 555; Cavaillé, "'Le plus éloquent philosophe des derniers temps': les stratégies d'auteur de René Descartes," *Annales: histoire, sciences sociales* 49 (1994): 349–67. Lionel Gossman, *Men and Masks: A Study of Molière* (Baltimore: The Johns Hopkins University Press, 1963), 188–9. Gossman explains that "Corneille presents the awed admiration of the world as the consequence of his heroes' absolute independence and superiority, whereas in fact it is the *condition* of it." (p. 181, emphasis original) According to Furetière's dictionary under *gloire*: "la louange se donne par les particuliers, et la gloire par le général du monde."

While *volontaires* were Descartes's vectors of knowledge dissemination, artisans, alternatively, became his archetypal model of rational discipline and ordered thought.[23]

In *La Recherche de la vérité* Descartes points out that "enough truth can be known in each subject to satisfy amply the curiosity of orderly souls [*âmes réglées*]."[24] The pursuit of knowledge must be guided by an orderly soul, which ought not search for "those simple [and textual] forms of knowledge which can be acquired without any process of reasoning, such as languages, history, geography and in general any subject which rests on experience alone" but should rather inquire into "ordinary facts about which everyone has heard"—and which artisans are thoroughly accustomed to.[25] A closer look at Descartes's early writing shows the artisan emerging under a more nuanced and fundamental light. In fact, I claim the artisan depicted in the early 1620s *Regulæ ad directionem ingenii* should not be seen as an amenable aide but rather as a heuristic model for the production of rational knowledge (see chapter 3). Remarkably enough, some of the most unassuming and mechanical artisans are converted into archetypal examples of rational discipline and orderly thoughts. However, it is the *concept* of artisan and artisanal practices that characterizes the Cartesian method. The artisan as genuine *homo faber* is transformed in Descartes's writings into an idealization, a disembodied epistemic metaphor compelled by order. The body techniques, the gestural knowledge of practices were completely dropped and superseded by an abstract rationalization, an *âme réglée*, which communicated a new theoretical understanding of the practice of natural philosophy.[26] The Cartesian artisan (mostly male) became not so much 'invisible' as 'idealized.'

Descartes doesn't seem to make this distinction. In *Les Passions de l'âme*, he explains that "Ce que j'appelle icy du nom de Gloire, est une espece de Ioye, fondée sur l'Amour qu'on a pour soy mesme, & qui vient de l'opinion ou de l'esperance qu'on a d'estre loüé par quelques autres. Ainsi, elle est differente de la Satisfaction interieure, qui vient de l'opinion qu'on a d'avoir fait quelque bonne action." AT xi, 482.

23 Jean-François Gauvin, "Artisans, Machines, and Descartes's *Organon*," *History of Science* 44 (2006): 187–216, from which this section was taken.

24 Descartes, *La Recherche de la verité*, AT x, 500; CSM ii, 402.

25 Descartes, *La Recherche de la verité*, AT x, 502–3; CSM ii, 403–4.

26 The literature on this topic is rich. See, for instance, Marcel Mauss, "Body Techniques," in *Sociology and Psychology: Essays*, ed. Ben Brewster (London: Routledge, 1979), 97–135 and Otto Sibum, "Reworking the Mechanical Value of Heat: Instruments of Precision and Gestures of Accuracy in Early Victorian England," *Studies in History and Philosophy of Science* 26 (1995): 73–106.

The chief characteristic of Cartesian artisans is unquestionably their condition of being 'properly mechanic' individuals.[27] Descartes (so far as I know) always used in French, in published and manuscript texts, the word 'artisan' rather than 'artiste' in identifying craftsmen. This distinction is clearly not arbitrary and denotes another epistemic dichotomy between forms of learning and, ultimately, the nature of knowledge ascertained (as Chérubin d'Orléans defended a few decades later). Socially and etymologically speaking, artisan and artiste meant essentially the same thing in the sixteenth and most of the seventeenth centuries (both derived from the Latin *artifex*).[28] To Descartes, however, an 'artiste' was and remained someone who dealt with the Grand Art, or alchemy, one of these 'false sciences' for which he knew enough about "to be liable not to be deceived" by their promises.[29] Marin Mersenne believed that such artistes could assist natural philosophers, although his faith they would do so was low.[30] Nonetheless, behind the substantive 'artisan'

27 Charles Loyseau, *A Treatise of Orders and Plain Dignities*, ed. and trans. Howell A. Lloyd (Cambridge: Cambridge University Press, 1994 [1610]), 179–81. Tradesmen such as "apothecaries, goldsmiths, jewellers, haberdashers, wholesalers, drapers, hosiers, and others like them," gained some prominence because their crafts involved commerce. The latter artisans, who called themselves 'honourable men' and 'bourgeois,' were morally superior to other tradesmen whose métiers "consist[ed] rather in physical labour than in commercial activity or in shrewdness of mind." Mere manual laborers were almost by definition the basest artisans of them all since "there is no worse occupation than to have no occupation." To qualify as a *honnête homme* an artisan had to leave the manual labor almost entirely to others, thus transforming himself into a merchant. See also Giulia Belgioioso, "Descartes e gli artigiani," in *La Biografia intellettuale di René Descartes attraverso la* Correspondance, eds. Jean-Robert Armogathe, Giulia Belgioioso, and Carlo Vinti (Naples: Vivarium, 1999), 113–65.

28 See Jean Nicot, *Thresor de la langue française* (1606) in which one reads under artisan: "*Artisan, ou Artiste*, Artifex, Opifex." L'Académie française made the distinction we are accustomed to nowadays in 1762: "*artiste*, celui qui travaille dans un art où le génie et la main doivent concourir (un peintre, un architecte sont des artistes); *l'artisan* est un ouvrier dans un art mécanique, un homme de métier." *Le Grand Robert de la langue française*, ed. Alain Rey, *s.v.* artisan. For a historical analysis of this significant shift, Larry Shiner, *The Invention of Art: A Cultural History* (Chicago: The University of Chicago Press, 2001), 99–120. Jean de La Fontaine, for instance, in one of his fables—*Le Lion abattu par l'homme*—used 'artisan' to describe a painter.

29 Descartes, *Discours de la méthode*, AT vi, 9; CSM i, 115. In the Furetière and Académie française dictionaries, 'artiste' is used to portray alchemists. In the Middle Ages, it became common to name 'artistes' (or sometimes *artiens*) those who studied the liberal arts—scholars *en devenir*—and 'artifex' those who practiced the mechanical arts. Shiner, *The Invention of Art*, 30.

30 "Ce que l'on pourroit desirer d'eux [Peripateticiens] (au cas qu'ils voulussent ayder à establir la vraye Philosophie) consiste seulement à dresser des memoires fidelles de leurs obseruations, & de leurs experiences: ce qu'il ne faut pas esperer iusqu'à ce que les

Descartes recommended we look specifically at the mechanical arts and their practitioners—not at alchemists and Paracelsians—to acquire an orderly soul.

The Cartesian artisan, therefore, was not guided by an 'epistemology of the hunt.' Unconstrained by a pure perception of order the 'venatic methodology,' as described by William Eamon, was "a kind of practical intelligence based upon acquired skill, experience, subtle wit, and quick judgment: in short, cunning."[31] The venatic methodology was needed, it was believed, to chase the signs and clues, the signatures of substances and materials to pick up the scent and as a result uncover these secrets—the modern facts—"tucked snugly in under the blanket of *scientia*." Cunning, or *mètis* as known to the Greeks, supplanted orderly thoughts.[32] In Descartes's judgment, artistes rather than artisans possessed and exercised such a *mètis*; one could say they were endowed with an *âme* d*éréglée*. When Descartes contemplated artisans in the early 1620s, he saw something completely different: no *venatio*, no conjectural knowledge, no cunning, only a pure and uncompromising order in their logic of practice.

For Descartes, nature could not act as a model leading to *mathesis*. The *mathesis* was instead at the origins of the 'grande mécanique de la nature.' Rediscovering *mathesis* (suppressed "with a kind of pernicious cunning" from the writings of the Ancients, as artisans customarily do with their inventions[33]) is to uncover how the universe was built and set in motion—the great Cartesian fables of *Le Monde* and *L'Homme*. Art did not imitate nature; art and nature were rather guided by a more general *mathesis*. According to Descartes, anyone endowed with this science of order would not have to physically struggle with matter and *ars*. Such a struggle with reality was indicative of a disorder in knowledge-making practices. Artisans and artistes laboring, toiling in the workshops, manipulating and transforming matter through sweat, burns, grease, and heat "practiced knowing … that constituted a bodily engagement with nature."[34] To Descartes, it merely gave them the illusion of acquiring a first-hand understanding of reality—that said, later in the *Principes de la*

honnestes hommes s'employent à cet art, & iusques à ce que les Artistes & Operateurs ayent quitté l'imagination de la poudre de projection, de la Magnesie des sages, & de la pierre Philosophique." Mersenne, *Qvestions inovyes, ov recreation des sçavans* (Paris, 1634; facsimile Stuttgart: F. Frommann, 1972), question xxviii, 126–7.

31 William Eamon, *Science and the Secrets of Nature: Books of Secrets in Medieval and Early Modern Culture* (Princeton: Princeton University Press, 1994), 281.

32 Eamon, *Science and the Secrets of Nature*, 284. On the concept of *mêtis* more generally, Marcel Detienne and Jean-Pierre Vernant, *Les Ruses de l'intelligence. La mètis des Grecs* (Paris: Flammarion, 1974).

33 Descartes, *Regulæ*, AT x, 375–7; CSM i, 18–9.

34 Pamela Smith, *The Body of the Artisan. Art and Experience in the Scientific Revolution* (Chicago: The University of Chicago Press, 2004), 142.

philosophie, Descartes explained that all arts contained some truth (*quelque chose de vray*), even though they were crude and imperfect in the beginning, ever improving with skill.[35] This 'artisanal epistemology,' epitomized by the art of alchemy, and especially by the writings of Paracelsus, went against the approach of an orderly soul.[36] The Cartesian *âme réglée* was only concerned with strategies inculcated by the *mathématicité* of the *mathesis*, which was not derived from the knowledge of artisanal practices but from the inbuilt structured discipline of *habitus* (see chapter 3).

Take one of Descartes's paradigmatic examples from the *Regulæ*, weaving. The choice of this craft is historically charged and not taken randomly. In France, after the devastation caused by the Wars of Religion, the Bourbon economic restoration instigated by Henry IV was wholly felt within the textile industry. In Beauvais alone, 700 to 800 looms were continuously in operation, employing roughly half of the city's population in the first half of the seventeenth century. Dijon was even more important as a *ville drapante*, surpassed only by the new manufactures opening at that time in the Parisian region.[37] Interestingly enough, the technological design of the horizontal loom had not changed significantly since its introduction in the twelfth century. This technical revolution was important for at least one medieval philosopher. Five centuries before Descartes's birth, weaving became one of Hugh of St. Victor's archetypes of the mechanical arts, inspired by the *trivium* and *quadrivium* of the liberal arts. Hugh argued in his *Didascalicon* for a division between *ratio* (wisdom, order) and *administratio* (actual practices) in the mechanical arts, maintaining that a *ratio* of mechanical origins should also be regarded as an integral part of philosophy.[38] Descartes similarly saw *ratio*—not just the *gestes* of *administratio*—behind the mechanical arts, what we associated with the structured discipline of *habitus*.

Weavers on their looms did not physically struggle with the machine and the fabric, as Descartes may have himself noticed. In fact, such bodily struggles would have resulted in a poor standard of the manufactured goods since everything about mechanical weaving was governed by a strict order of

35 Descartes, AT ix, 18.

36 Smith, *The Body of the Artisan*, esp. chapters 4 and 5.

37 Pierre Goubert, *Beauvais et le Beauvaisis de 1600 à 1730: contribution à l'histoire sociale de la France du XVIIe siècle* (Paris: S.E.V.P.E.N., 1960), 281–2, and 585. Pierre Deyon, "Variations de la production textile aux XVIe et XVIIe siècles," *Annales: ESC* 18 (1963): 921–55. James R. Farr, *Hand of Honor: Artisans and Their World in Dijon, 1550–1650* (Ithaca: Cornell University Press, 1988). Henri Heller, *Labour, Science and Technology in France, 1500–1620* (Cambridge: Cambridge University Press, 1996), chapter 6.

38 Roger Baron, *Science et sagesse chez Hugues de Saint-Victor* (Paris: Lethielleux, 1957), 60–87.

procedure and the smooth, continuous, and regular movements of the couple man/woman machine (mostly women until the fifteenth century). (Notice here two fundamental concepts of the *Regulæ:* order [Rule IV] and regular and uninterrupted motion [Rule VII].) To achieve the best quality of draperies, weavers had to become one with the machine, as if they were just another link in the great chain (*tela*) driving the looms. (This body-machine symbiosis was so manifest that weavers were often called *telier* in old French and *teler* in old English.) The body and hands-on experience of weavers did not by themselves guarantee excellence; one needed to look instead at the orderly and uninterrupted movements of the body-machine entity taken as a whole. Weavers in Descartes's observation and contemplation became an abstraction of order owing to their symbiosis with a mechanical device. Their orderly souls thus emanated naturally from the technology of weaving.[39] (Fig. 10)

Although Descartes skillfully used the artisan as a rhetorical trope, there is no indication he himself ever dabbled with crafts or worked with artisans before his optical days in Paris in the mid 1620s.[40] We know that when he left the Collège La Flèche to travel across Europe, to find knowledge "in the great book of the world," he mixed "with people of diverse temperaments and ranks [*diuerses humeurs & conditions*]," which no doubt suggests some acquaintances with artisans and instrument makers.[41] In Holland, particularly, the

39 Dominique Cardon, *La Draperie au Moyen Âge. Essor d'une grande industrie européenne* (Paris: CNRS, 1999), 416–7 and 539–63. See also Giorgio Israel, "Des *Regulæ* à la *Géométrie*," *Revue d'histoire des sciences* 51 (1998): 183–236, in which he discusses the role of weaving in Descartes's thinking. The social anthropologist Tim Ingold makes a fascinating case about the conceptual role of weaving as regards the notion of making, where he writes: "Weaving the world, then, turns out to be a matter of 'making culture,' of submitting the disorder of nature to the guidelines of traditional design." Ingold, "Making Culture and Weaving the World," in *Museum Objects. Experiencing the Properties of Things*, ed. Sandra H. Dudley (New York: Routledge, 2012), 368–84, quote on p. 381. On weaving and mathematics, Michael Friedman, "Joachim Jungius and the Transfer of Knowledge from Weaving to Mathematics in the 17th Century," *Cahiers François Viète* III-13 (2022): 43–82.

40 According to Adrien Baillet, if Descartes had been raised in a condition allowing him to become an artisan, he would have been a skillful one because, we learn, he had in his youth a particular inclination for the arts. Like so many other claims made by Baillet this one could be utterly wrong, or at best a misinterpretation. Adrien Baillet, *La Vie de Monsieur Des-Cartes*, 2 vols (Paris, 1691), vol. 1, 35. Geneviève Rodis-Lewis believes Baillet has his chronology wrong here. This remark should be associated to a much later phase in Descartes's life. Rodis-Lewis, "Descartes' Life and the Development of his Philosophy," in *The Cambridge Companion to Descartes*, ed. John Cottingham (Cambridge, Cambridge University Press, 1992), 21–57, on p. 26. Descartes himself often contradicts Baillet's assertion. He said, for instance, that he was born without any manual abilities: "pour moy … i'estois venu au monde sans mains." Descartes to ***, [Nov.–Dec. 1638?], AT ii, 452.

41 Descartes, *Discours de la méthode*, AT vi, 9; CSM I, 115. On the role of travel during the early modern period, see the remarkable opus by Daniel Roche, *Humeurs vagabondes. De la*

FIGURE 10 Rational and geometric weaving patterns. Ellen Harlizius-Klück, a resident scholar at the Deutsches Museum in 2006 when I met her, was working on an early eighteenth-century German book on textile patterns (based on a previous book dated 1677) that demonstrates precisely the orderly fashion (and not the randomness) with which these "threads are interwoven in an infinitely varied pattern," to quote Descartes again. Her work (and similar analyses done by computer scientists, see: www.handweaving.net) shows that these textile patterns were rationalized in such a way that simple mathematical equations can be derived from the lines and dots printed in the book and destined to guide the weavers' hands on the mechanical looms. Textile patterns, therefore, were an artisanal form of reason, one that appears to have been noticed by Descartes. I would like to thank Dr. Harlizius-Klück for sharing her research with me. NATHANIEL LUMSCHER, *NEU EINGERICHTETES WEBER KUNST UND BILD BUCH* … (CULMBACH, 1709). FROM HATHI TRUST DIGITAL LIBRARY (ORIGINAL SOURCE: STERLING AND FRANCINE CLARK ART INSTITUTE), ONLINE HTTPS://BABEL.HATHITRUST.ORG/CGI/PT?ID=CAIA.ARK:/13960/T9M34G171&VIEW=1UP&SEQ=91, IMAGE 91, FOLIO 36. PUBLIC DOMAIN

collective embarrassment of riches and the Baconian style of natural philosophy caught his eyes.[42] There, during his well-known stay with Isaac Beeckman in 1618, he realized that *scientia* and *ars* should not be subordinated to one another; he recognized that both were needed in concert to reveal the true nature of the world.[43] As the first so-called physico-mathematici in Europe, Beeckman and Descartes claimed, they tried—in the hydrostatic manuscript for example—to unify the mathematical study of nature with true ontological, corpuscular-mechanical causes. Trained in the candle making and water conduit laying trades, Beeckman could theorize about hands-on knowledge of the mechanical arts in order to raise to the realm of concepts the operations of artisans and machines. "What Beeckman was demanding in natural philosophy was the application of the criteria of meaningful communication between mechanical artisans—the appeal to a pictorial or imaginable structure of parts whose motions are controlled within a putative theory of mechanics."[44] Beeckman gave the initial impetus to Descartes's later contention that the mechanical arts could function as a methodological guideline for the purpose

circulation des hommes et de l'utilité des voyages (Paris: Fayard, 2003). On the early years of Descartes, Harold C. Cook, *Young Descartes* (Chicago: The University of Chicago Press, 2018). It is worth noting here the evidential criticisms in the review written by Erik-Jan Bos in *Isis* 110 (2019): 172–3.

42 Simon Schama, *The Embarrassment of Riches: An Interpretation of Dutch Culture in the Golden Age* (New York: Vintage, 1997). On Dutch Baconianism, Svetlana Alpers, *The Art of Describing: Dutch Art in the Seventeenth Century* (Chicago: The University of Chicago Press, 1983), esp. chapter 1. Smith, *The Body of the Artisan*, esp. chapters 5 and 6.

43 This is the first thing Beeckman wrote down in his *Journal*: "Quæritur cur artes inter se non sint subordinatæ" On the importance of both *scientia* and *ars*, he also penned on the first page of his *Journal*: "Ad excitandum artium studium illud maximè faceret, si immunitates alicujus vectigalis etc. ijs qui Euclidis *Elementa* intelligerent, promitterentur. Quibus bene intellectis, pauci cætera studia negligerent, etiam in medijs occupationibus mechanicis." Isaac Beeckman, *Journal tenu par Isaac Beeckman de 1604 à 1634*, ed. Cornélis de Waard, 4 vols (The Hague: M. Nijhoff, 1939–53), vol. 1, 1 for both quotes.

44 Stephen Gaukroger and John Schuster, "The Hydrodynamic Paradox and the Origins of Cartesian Dynamics," *Studies in the History and Philosophy of Science* 23 (2002): 535–72, quote on p. 552. For a general appraisal of the Beeckman-Descartes relationship, Gaukroger, *Descartes: An Intellectual Biography* (Oxford: Clarendon Press, 1995), chapter 3; Klaas van Berkel, "Descartes' Debt to Beeckman: Inspiration, Cooperation, Conflict," in *Descartes' Natural Philosophy*, eds. Stephen Gaukroger, John Schuster, and John Sutton (London: Routledge, 2000), 46–59. The classic work on Beeckman remains Van Berkel, *Isaac Beeckman (1588–1637) en de mechanisering van het wereldbeeld* (Amsterdam: Rodopi, 1983), 217–35 for the role of technology in Beeckman's thinking.

of discovering the laws of nature.[45] Descartes employed to his advantage the trope of artisans (whether weavers or blacksmiths) because their *ratio* appeared universal, reaching the *mathématicité* of mathematics. Descartes's few years spent in Paris in the mid 1620s, however, would significantly modify this methodological point of view. He soon shifted his portrayal of artisans; he began to see them as individuals possessing unreliable physical dispositions, in serious need of a rigorous rational training in the logic of practice. This change of heart was due to Descartes's direct dealings (at long last) with artisans. It led, I would also claim, together with more profound epistemological problems, to the abandonment around 1628 of the *Regulae* and its methodological program.[46]

2 Who Assists Whom: the Structural Dynamic of Artisan and Savant Interactions

Descartes's initial, amenable artisanal outlook was, one could argue, an echo of the past. In the early sixteenth century, in the circle of the University of Paris's professor Jacques Lefèvre d'Étaples, the theme of the 'untutored mind' symbolized a new and unusual way of looking at the fabric of nature. Seen through the library of one of Lefèvre's students, Beatus Rhenanus, Richard Oosterhoff explains how the persona of such an untutored mind, the *idiota*, or someone—an artisan—who lacked mastery of Latin, represented for the Fabrist circle a matter-type epistemology, where artisans got "knowledge from things themselves." Offering the *idiota* as a *sui generis* authority, three classic authors valued by the Fabrist circle exemplified this stance in diverse ways: Ramon Lull, Ramond de Sebonde, and especially Nicholas of Cusa, a famous fifteenth-century ecclesiastical reformer, administrator, and cardinal. For the latter, according to Oosterhoff, "the *idiota* emerges as a special authority in measuring and governing rough material, since by performing such activities in daily life he more deeply understands the basis of the Divine Artisan's

---

45 The French philosopher Georges Canguilhem is among the few to see a century ago how Descartes's musings about technology during the course of his life was central to the development of the Cartesian system. Canguilhem, "Descartes et la technique [1937]," in *Œuvres complètes tome 1: Écrits philosophiques et politiques (1926–1939)*, ed. Jean-François Braunstein and Yves Schwartz (Paris: Vrin, 2011), 490–98. On the role of Descartes in Canguilhem's thought, Vincent Guillin, "Les études cartésiennes de G. Canguilhem," *Études philosophiques* 114 (2008): 65–84.

46 John Schuster, *Descartes-Agonistes. Physico-mathematics, Methods & Corpuscular-Mechanism 1618–1633* (London: Springer, 2013), chapter 7.

creation in number, weight, and measure."[47] *Idiotae* did not, and could not, fully replace conventional savants and scholarship: though the *idiota*'s 'naïve experience' was considered a key starting point for a more applied and material natural philosophy, he himself lacked the skills—i.e., proficiency in Latin—to properly communicate his acquired knowledge. *Idiotae* and scholars thus required a shared community in order to circulate knowledge. Cusanus's attitude, ascertained and disseminated within the Fabrist circle, "fed a growing desire to simplify reasoning, to emphasize 'naïve invention' in 'method'. ... Put another way, the trope of the *idiota* and the push toward immediate knowledge was as much an attempt to naturalize intellection as it was anti-intellectual."[48]

The *idiota*'s 'untutored mind' is what Pamela Smith comparably refers to as the "bodily experience of the particulars of matter." Taking the potter Bernard Palissy as a paradigmatic example, we discover how the sixteenth-century alchemical teachings of Paracelsus, inter alia, inspired and influenced the production of natural philosophical knowledge—and, congruently, the creation of works of art. In his famous *Discours Admirables* (1580) written in a dialog form between *Theorique* and *Practique*, Palissy (personified by *Practique*) unfolds in excruciating detail his physical struggle with matter, to the final, utter disgust of *Theorique*. Genuine, hard manual labor was at the core of Palissy's project. Yet not everything was based on close observations, sensibility, and mechanical habits, contrary to what *Theorique* believed—and what *Practique* insinuated *ad nauseam*. Some operations required intelligence (*gentil esprit*) and geometry:

> You [Theorique] need to know that to successfully carry out a batch of work, whether or not it is enamelled, one has to govern the fire with a thorough philosophy [*vne philosophie si soingneuse*] such that even a carefully-formed and fair mind [*gentil esprit*] is often disappointed. Regarding the proper way of using the kiln one needs a specialized geometry.[49]

---

47 Richard J. Oosterhoff, "*Idiotae*, Mathematics, and Artisans: The Untutored Mind and the Discovery of Nature in the Fabrist Circle," *Intellectual History Review* 24 (2014): 301–19, on p. 304.

48 Ibid., 312.

49 Bernard Palissy, *Discours admirables de la nature des eaux et fontaines, etc.* in *Œuvres complètes de Bernard Palissy, édition conforme aux textes originaux imprimés du vivant de l'auteur*, ed. Paul-Antoine Cap (Paris: J. J. Dubochet et Cie, Éditeurs, 1844), 322–3. On Palissy, Pamela H. Smith, *The Body of the Artisan*, 100–06 and Hanna Rose Shell, "Ceramic Nature," in *Materials and Expertise in Early Modern Europe*, eds. Ursula Klein and Emma C. Spary (Chicago: The University of Chicago Press, 2010), 50–70.

To govern fire in a kiln, using a specially designed loading technique (hence the necessity of geometry), required a thorough philosophy not easily attainable.[50] Palissy was not alone in arguing for such a philosophy: other artisans untiringly working with materials such as metals, ink, blood, mineral water, dyestuffs, and milk were all secured in "the conviction that knowledge is active and knowing is doing."[51] Even doing it wrong, as Sven Dupré suggests, translated into practical knowledge-producing actions; i.e., codifying the sources of error in print was equally informative as the 'true and successful' recipes.[52] From Palissy to Antoine Lavoisier, the study of materials created an epistemic and ontologically productive transmission channel between artisans and savants, ultimately creating during the Enlightenment "a whole range of slightly nuanced figures:" the artisanal-scientific expert. This truly hybrid individual, "[l]iving in both an academic and an industrial world, and further participating in the world of state bureaucracy, … personally spurred the circulation of knowledge and objects between these worlds."[53]

Mathematics played a similar role in commercial centers of Europe. Oronce Fine, one of Paris's finest mathematicians and instrument makers of the sixteenth century, recognized along with members of the Fabrist circle that mathematics could be defined as naïve knowledge; not only that, but it rested at the foundation of all perfect knowledge ("*clefz de tout perfet sçavoir*") and the mirror of all certainty. At the center of Fine's assertion was the (allegedly) straightforward use of his specially designed mathematical instruments, some of which he made available through early *fabrica et usus* type of printed books—such as his *Composition et usage du quarré geometrique* (1556). Fine's intention was to connect practitioners with mathematics and, through his teachings at the newly founded Collège Royal, to expand practical geometry to educated Paris

50 The workshop built in Paris by Palissy in ca 1567 was rediscovered during the archeological digs of the Grand Louvre in the mid 1980s. It is described in Bruno Dufay *et al.*, "L'atelier parisien de Bernard Palissy," *Revue de l'Art* 78 (1987): 33–60.

51 Smith, *The Body of the Artisan*, 149.

52 Sven Dupré, "Doing it Wrong: The Translation of Artisanal Knowledge and the Codification of Error," in *The Structures of Practical Knowledge*, ed. Matteo Valleriani (Cham: Springer International Publishing, 2017), 167–88.

53 Ursula Klein, "Artisanal-Scientific Experts in Eighteenth-Century France and Germany," *Annals of Science* 69 (2012): 303–6 (introduction to a special issue on the topic). On the same theme, Klein, "Hybrid Experts," in *The Structures of Practical Knowledge*, 287–306 and Ursula Klein and Wolfgang Lefèvre, *Materials in Eighteenth-Century Science: A Historical Ontology* (Cambridge, MA: MIT Press, 2007). On early modern expertise in general, Eric H. Ash, ed., *Expertise. Practical Knowledge and the Early Modern State* (Chicago: The University of Chicago Press [*Osiris*, vol. 25], 2010).

residents, soldiers, and courtiers.[54] Knowledge of basic mathematics was, according to Anthony Turner, what chiefly distinguished early modern instrument makers from other artisans such as locksmiths, cabinet makers, gilders, and book binders. The outcome of this intellectual aptitude for the instrument makers' social status was considerable: "it led them into relation with *savants*." In sixteenth-century France, for example, the type-founder and cutter Philippe Danfrie exemplify this trend better than anyone else. His work on paper instruments, and especially the development in the middle 1580s of a *radio latino* (a surveying instrument) with the mathematics teacher Milles Denorry, had a lasting influence on the later French instrument-making trade.[55]

Mathematical instrument makers came from various craft activities, dedicated workshops only arising late in the sixteenth century in large European capitals—especially in London, showcased by the output of advertising.[56] Specialized shops like the ones established in Nuremberg in the early cinquecento (Georg Hartmann) or in Louvain in the middle of the same century (Gemma Frisius) were exceptional. When, in 1547, natural philosopher and court magician John Dee returned to England from the Low Countries with accurate globes drawn by Gerard Mercator and fine lined-engraved astronomical instruments from Frisius's shop, nothing comparable in workmanship existed in England. An analytical look at the famous artwork *The Ambassadors*, painted by Hans Holbein the Younger in 1533, confirms this assertion. It depicts an array of instruments such as a pair of globes, a torquetum, and sundials, all 'state-of-the-art' instruments, all German in their manufacture—though the work was painted in London.[57] Instruments are key here. As the industry and

54 Oosterhoff, "*Idiotae*, Mathematics, and Artisans," 311. On Fine, Alexander Marr, ed., *The Worlds of Oronce Fine. Mathematics, Instruments and Print in Renaissance France* (Donington: Shaun Tyas, 2009). On the role of mathematics in the Fabrist circle, Richard J. Oosterhoff, *Making Mathematical Culture. University and Print in the Circle of Lefèvre d'Étaples* (Oxford: Oxford University Press, 2018).

55 Anthony Turner, "Mathematical Instrument-Making in Early Modern Paris," in *Luxury Trades and Consumerism in ancien régime Paris: Studies in the History of the Skilled Workforce*, eds. Robert Fox and Anthony Turner (Aldershot: Ashgate, 1998), 63–96. Anthony Turner, "Paper, Print, and Mathematics. Philippe Danfrie and the Making of Mathematical Instruments in late 16th-Century Paris," in *Studies in the History of Scientific Instruments*, eds. Christine Blondel, *et al.* (London: Rogers Turner Books Ltd, 1989), 23–41.

56 D. J. Bryden, "Evidence from Advertising for Mathematical Instrument Making in London, 1556–1714," *Annals of Science* 49 (1992): 301–36.

57 Anthony Turner, "John Dee, Louvain and the Origins of English Instrument-Making," in *Musa musei: Studies on Scientific Instruments and Collections in Honour of Mara Miniati*, eds. Marco Beretta, Paolo Galluzzi, and Carlo Triarico (Firenze: L. S. Olschki, 2003), 63–78. Elly Dekker and Kristen Lippincott, "The Scientific Instruments in Holbein's Ambassadors: A Re-Examination," *Journal of the Warburg and Courtauld Institutes* 62 (1999): 93–125.

commerce of these commodities grew in size, a new culture of practitioners rose simultaneously in London: 'the mathematicalls.' Combining practical know-how with mathematical knowledge in fields such as astronomy, gunnery, surveying, architecture, and navigation, they championed a novel approach to knowledge and, through their array of instruments, endeavored to "convince Elizabethan audiences of the dignity, utility and pleasure" of their work. They were not mere 'doers of tricks,' or 'Juglers' as some liked to claim. By the end of the sixteenth century, Londoners fully embraced this new instrumental education and mathematical literacy. It brought together theory and matters of practical concerns, as well as "students and scholars, artisans and apprentices, into fruitful exchange and collaboration." It likewise helped renegotiate the relationship between 'the mathematicalls' and natural philosophers. As Jim Bennett convincingly argued decades ago, this relationship was not about 'high science, low science,' or how artisans helped shape the new science, i.e., mechanical philosophy. Rather, one needs to understand that the perfection of the "mechanical philosophy legitimized the use of instruments—previously the domain of the mathematical sciences—in natural philosophy." One of the many steps leading toward the 'instrumentality' of natural philosophy.[58]

The negotiation, exchange, and trade-off between artisanal know-how and natural philosophical knowledge took place in a wide array of spaces and locations. As Gottfried Wilhelm Leibniz discovered while visiting Paris in the early 1670s, the city had plenty to offer to anyone in search of specialized skills:

> They [artisans] are marvelous in the art of varnishing in the Chinese style; ... The art of molding or casting every kind of material in sand-forms is in its greatest perfection, as is the casting of medals. ... They work well in iron and they have people with the secret of casting iron into forms, and of filing it in ways not commonly possible. ... In sum, there is an

---

Gerard L'E. Turner, *Elizabethan Instrument Makers. The Origins of the London Trade in Precision Instrument Making* (Oxford: Oxford University Press, 2000).

58 Stephen Johnston, "Mathematical Practitioners and Instruments in Elizabethan England," *Annals of Science* 48 (1991): 319–44, quote on p. 320. Katherine Hill, "'Juglers or Schollers?': Negotiating the Role of a Mathematical Practitioner," *The British Journal for the History of Science* 31 (1998): 253–74. Deborah Harkness, *The Jewel House: Elizabethan London and the Scientific Revolution* (New Haven: Yale University Press, 2007), chapter 3, quote on p. 141. Jim Bennett, "The Mechanic's Philosophy and the Mechanical Philosophy," *History of Science* 24 (1986), 1–28, quote on p. 3. Jim Bennett, "Presidential Address: Knowing and Doing in the Sixteenth Century: What Were Instruments For?," *The British Journal for the History of Science* 36 (2003): 129–50. Peter Dear, *The Intelligibility of Nature. How Science Makes Sense of the World* (Chicago: The University of Chicago Press, 2006).

> infinity of curiosities, in gold making, enameling, glassmaking, clock making, leather working, [and] pewter pottery.[59]

Early modern cities supplied a plethora of such sites to investigate and discourse about natural knowledge in all its intricacies.[60] Marketplaces and piazzas, for instance, were "a repository of empirical knowledge" where butcher shops, fishmongers, apothecaries, and distillers provided, besides common commodities, rare items and collectible objects from local villages and faraway places—worldly goods, fast becoming significant matters of exchange. Coffeehouses and print shops both exploited a thick, black liquid that facilitated and prompted conversations between what might appear strange bedfellows. As Adrian Johns reminds us, coffeehouses and print shops were "culturally volatile, and their mixture proved both potent and dangerous." They "became locations for both the building and destruction of belief." Someone like Robert Hooke easily transitioned from Gresham College (another urban site, and early setting of the Royal Society) to an instrument maker's shop, and finally Garraway's, where one day he would humiliate the Astronomer Royal in the company of savants and gentlemen—around a bitter hot coffee. On a much larger scale, the Venetian Arsenal, the high-tech center of the Republic of Venice, as labeled by Matteo Valleriani, would train and help Galileo focus his attention on the science of the resistance of materials. Observing mastmakers and shipbuilders at work, asking admirals and coxswains questions, Galileo developed his "first new science" by exercising a technical modus operandi specific to the Arsenal's artisans, and by being challenged in his core theoretical assumptions by shipwrights, who questioned parts of the Aristotelian doctrine supported by Galileo. This, in fact, is not entirely unlike how Galileo acquired an understanding of the optics of the telescope, by combining his mathematical knowledge of sixteenth-century optics to practices found in optical workshops.[61]

---

59 Quoted in Matthew L. Jones, *Reckoning with Matter. Calculating Machines, Innovation, and Thinking about Thinking from Pascal to Babbage* (Chicago: The University of Chicago Press, 2016), 66. On Leibniz in Paris, Maria Rosa Antognazza, *Leibniz. An Intellectual Biography* (Cambridge: Cambridge University Press, 2009), chapter 3.

60 Bert de Munck and Antonella Romano, eds., *Knowledge and the Early Modern City* (New York: Routledge, 2020).

61 William Eamon, "Markets, Piazzas, and Villages," in *The Cambridge History of Science, Volume 3: Early Modern Science*, eds. Katharine Park and Lorraine Daston (Cambridge: Cambridge University Press, 2008), chapter 8. Harold J. Cook, *Matters of Exchange. Commerce, Medicine, and Science in the Dutch Golden Age* (New Haven: Yale University Press, 2007). Adrian Johns, "Coffeehouses and Print Houses," in *The Cambridge History of*

Private settings also offered plenty of opportunities to enrich negotiations and exchange between artisans and savants. It is well known how John Dee's household, under the careful supervision of his wife Jane, became a site visited by a high volume of individuals—besides assistants and foreign natural philosophers *de passage*: "clients seeking astrological advice, messengers, potential patrons from the court of Elizabeth, and students of navigation and natural philosophy." Despite her good will and managerial skills, Jane's domestic sphere was overtaken by her husband's work in natural philosophy and the occult sciences. A century later, Lady Ranelagh, Robert Boyle's sister, famously opened her London house to his brother so he could continue performing his experiments. Beyond the fact that part of her house was transformed into a chemical laboratory (tolerating "the sporadic explosions and the stench of luminescent rotten fish"), she fed Boyle's friends and assistants and restored him to health in time of dire need. Her role was so significant in organizing the domestic workflow and natural philosophical traffic in the house that she was, in Steven Shapin's words, "Boyle's greatest [invisible] technician." In such a house of experiments, a specialized laboratory was not always available or even necessary. The kitchen, for instance, provided all the tools, wares, and utensils one required to cook, heat, purge, and conduct distillations. Thrifty science, or making use of what was at hand, was common praxis in the early modern household oeconomy, straightforwardly customized for a vast array of operations in experimental natural philosophy. Other more refined and lavish rooms were transformed into cabinets of curiosities, or Wunderkammern, where a cornucopia of items from the art and the sciences was assembled to the benefit of a few distinguished guests. Occupying and maintaining an open space clearly designed as an intellectual social boundary, individuals such as Ulisse Aldrovandi in Italy and Hans Sloane in England demonstrated how "the subject of science, natural philosophy, could contain its object, nature." In these early 'museum' settings (Sloane's collection seeded the British Museum) "[k]nowledge, formerly embedded in texts, was created by a community of collectors, experimenters, and visitors whose viewing of nature established its authoritative image." Though a *boudoir* likewise showcased a woman's connoisseurship and ambition as a collector in natural history, it was the

Science, chapter 15. Matteo Valleriani, *Galileo Engineer* (Dordrecht: Springer, 2010), chapter 4. Sven Dupré, "Ausonio's Mirrors and Galileo's Lenses: The Telescope and Sixteenth-Century Practical Optical Knowledge," *Galilaeana: Journal of Galilean Studies* 2 (2005): 145–80.

FIGURE 11 Jacques de Lajoüe, *Le cabinet de physique de Bonnier de La Mosson*, 1734 (detail). BEIT COLLECTION, BLESSINGTON, IRELAND. COURTESY OF WIKIMEDIA COMMONS: HTTPS://COMMONS.WIKIMEDIA.ORG/WIKI/FILE:LE_CABINET_DE_PHYSIQUE_DE_BONNIER_DE_LA_MOSSON.JPG

larger and more extravagant cabinets, such as Joseph Bonnier de la Mosson's in Paris in the 1730s (Fig. 11), which inspired conversations, created novel ideas, and challenged conventions about physics, mechanics, chemistry, pharmacy, and natural history.[62]

62 Deborah E. Harkness, "Managing an Experimental Household: The Dees of Mortlake and the Practice of Natural Philosophy," *Isis* 88 (1997): 247–62. Steven Shapin, *A Social History of Truth*, 371. On Lady Ranelagh, offering a somewhat divergent version, Michelle DiMeo, *Lady Ranelagh: The Incomparable Life of Robert Boyle's Sister* (Chicago: The University of Chicago Press, 2021). Simon Werrett, *Thrifty Science. Making the Most of Materials in the History of Experiment* (Chicago: The University of Chicago Press, 2019). Paula Findlen, *Possessing Nature. Museums, Collecting, and Scientific Culture in Early Modern Italy* (Berkeley: University of California Press, 1994), 199. James Delbourgo, *Collecting the World.*

2.1 *A Structural Approach to Natural Philosophical Knowledge?*

The experimental spaces mentioned above have not exhausted the variety and diversity of such sites in early modern Europe. From private homes and societies to state-sponsored institutions, from simple workshops and laboratories to grand anatomical theaters, the mutable nature of exchange routinely displayed between scholars and craftsmen powered the early modern knowledge economy.[63] To fully embrace the robust and dynamic interplay between practical (artisan) and theoretical (savant) knowledge in early modern Europe, a broad and encompassing conceptualization is required, one that looks at the structures of practical knowledge. It is not enough to recognize "the epistemic value of practical knowledge," contends Matteo Valleriani, we need to "investigate how and through which steps the structure of knowledge used in the frame of practical activities was reflected, conceptualized and externalized." The dichotomy between practical and theoretical knowledge, as useful a heuristic scheme it might have been in the past for historians of science, needs a redefinition, a new perspective reframing "the relationship between different structures of knowledge, where the qualitative difference between the different structures is given by the degree of abstraction and by the range of connections between different fields of knowledge."[64] Valleriani suggests three analytical layers of inquiry to achieve this goal: "These are: (1) the knowledge structure of practical activities; (2) the social structuring of practical knowledge; and (3) the conceptual structures of knowledge." Theory, practice, and society. These layers, though self-consistent and structurally sound in themselves, do not work separately, but are fully interconnected to one another; they "trade with one another and develop dynamics that lead to an association of different aspects of knowledge that, finally, can also explain how, in a later epoch, scientific knowledge had re-impacted practical activities."[65]

Hans Sloane and the Origins of the British Museum (Cambridge, MA: Harvard University Press, 2017). Krzysztof Pomian, *Collectionneurs, amateurs et curieux. Paris, Venise: XVIe–XVIIIe siècle* (Paris: Gallimard, 1987). Margaret Carlyle, "Collecting the World in her Boudoir: Women and Scientific Amateurism in Eighteenth-Century Paris," *Early Modern Women: An Interdisciplinary Journal* 11 (2016): 149–61. C. R. Hill, "The Cabinet of Bonnier de la Mosson (1702–1744)," *Annals of Science* 43 (1986): 147–74. For a broad view, Jim Bennett and Sofia Talas, eds., *Cabinet of Experimental Philosophy in Eighteenth-Century Europe* (Leiden: Brill, 2013).

63 Larry Stewart, "Experimental Spaces and the Knowledge Economy," *History of Science* 45 (2007): 155–77.

64 Matteo Valleriani, ed., *The Structures of Practical Knowledge* (Cham: Springer, 2017), 2.

65 Ibid., 2–3.

My interest here lies not with the numerous historical examples provided by Valleriani supporting his structural concept of early modern knowledge—especially the role of codification, in many ways interlocking the three analytical layers of knowledge structures mentioned above. I want instead to go one step further by harnessing what has been known for the past several decades as structural anthropology. In a lecture given at Barnard College in 1972 entitled *Structuralism and Ecology*, Claude Lévi-Strauss spelled out his position vis-à-vis the nature/culture dichotomy debate. He established that ethnographical observation did not act as a demarcator between one position or the other. In his own words, one didn't have to choose "a plastic mind passively reflecting the ecology outside," or, as he was himself often inaccurately criticized, to entertain a set of "universal psychic laws unfolding everywhere the same inborn properties, mindless of the ongoing history of each group and of the concrete features of its natural and social surroundings." For Lévi-Strauss, nature and culture are in constant flux, reinforcing one another until "human history and natural ecology become articulated so as to make up a meaningful whole."[66] Philippe Descola, his student and commentator, explains that Lévi-Strauss fully rejected philosophical dualism à la Descartes, while at the same time carrying out a dualistic methodology.[67] Based on linguistics studies, Lévi-Strauss forms the notion of binary oppositions to navigate "through sensory perception and through the processing of sensory data which takes place in the brain." Recovering these underlying cultural and natural oppositions enables one to find the structural rules that make the thought processes of a culture unique. In fact, Lévi-Strauss posits that "structuralism recovers and brings up to awareness deeper truths that are already latent in the body itself. By reconciling soul and body, mind and ecology, thought and the world, structuralism tends toward the only kind of materialism consistent with the ways in which science is developing."[68]

---

66 Claude Lévi-Strauss, "Structuralism and Ecology," *Social Science Information* 12 (1973): 7–23, quotes on p. 10. To understand the capital role the pair nature/culture plays in Lévi-Strauss's structural anthropology, see Philippe Descola, "The Two Natures of Lévi-Strauss," in *The Cambridge Companion to Lévi-Strauss*, ed. Boris Wiseman (Cambridge: Cambridge University Press, 2009), 103–17.

67 Descola, quoting Lévi-Strauss, emphasizes that the opposition between nature and culture is "neither a primeval fact, nor a concrete aspect of universal order. Rather it should be seen as *an artificial creation of culture*." Descola, "The Two Natures of Lévi-Strauss," 109.

68 Lévi-Strauss, "Structuralism and Ecology," 23. On the use of linguistics, Lévi-Strauss, "Structural Analysis in Linguistics and in Anthropology," in *Structural Anthropology* (New York: Basic Books, 1963), 31–54.

I use the nature/culture apparent disjuncture at the core of Lévi-Strauss's structuralism to frame the antinomy of theory/practice scrutinized by historians, sociologists, and philosophers of science. I would suggest that underneath the three structural layers suggested by Valleriani are found a number of binary oppositions that create a theoretical architecture explaining and defining the nature (bodily gestures) and culture (conjectural interpretations) of the theory/practice alleged divide. A non-exhaustive list of binary oppositions, separately all well studied, reads thusly: *episteme/technē*, art/science, matter/ether, body/mind, oral/print, client/patron, openness/secrecy, tacit/explicit, public/private, university/academy, observation/experiment, and artisan/savant. Though more dyads could be found, the argument I want to defend here is that these binary oppositions bring about the knowledge structures found in Valleriani's subsets of theory, practice, and society. History is important here, and though he was often criticized for denigrating it, Lévi-Strauss valued the discipline, for "by examining institutions in the process of transformation (that is, historically), scholars learn to appreciate how structures underlie the succession of event."[69] Some proponents in the field of New Archaeology have likewise argued it was no longer sufficient to come up with better descriptions of culture history, but that we should rather aim at making and expanding "*explanatory* understanding or the underlying structure and dynamics of cultural systems—the cultural processes—that are responsible for the forms of life and trajectories of development documented by culture historians." Structural archaeology considers the material culture of civilizations 'well-formed cultural constructs' offering 'models of intelligibility' or 'innate logics of classification.' As Edmund Leach suggested, strangely echoing Gaston Bachelard's notion of instruments as reified theories, "archaeologists must appreciate that the material objects revealed by their excavations are not things in themselves, nor are they just artifacts—things made by men—they are representations of ideas." A structuralist approach to archaeology, in other words, "challenges archaeologists to come to terms with the distinctively cultural aspects of the material they study."[70]

I believe it shouldn't be different for the material culture of science. In fact, as the prohibition of incest became the 'primal base' of Lévi-Strauss's structural analysis of marital institutions, epitomized in *The Elementary Structures of Kinship* (1949), I propose that instrument—*organum*—emerges

69 Elizabeth A. Clark, *History, Theory, Text. Historians and the Linguistic Turn* (Cambridge, MA: Harvard University Press, 2004), 51.

70 Alison Wylie, *Thinking from Things. Essays in the Philosophy of Archaeology* (Berkeley: University of California Press, 2002), 3 (italics original), 129 (quote by) and 135.

as the logical foundation to a structuralist examination of the theory/practice antinomy.[71] As demonstrated in chapter 1, whether material (body organ, machine), immaterial (faith, geometry), cultural (musical instrument), social (legal document), religious (Erasmus's *Novum instrumentum*) or theoretical (Bacon's *Novum organum*), an instrument usually constitutes the bedrock of practical, theoretical, and social practices in early modern Europe. Natural philosophical knowledge structures can be understood with and built upon instruments as the prohibition of incest helps us conceptualize marital institutions—or, to take another core example from Lévi-Strauss's structural anthropology, how myths are assimilated to and founded upon language. Resting on the keystone of *organum*, I would contend, a structural analysis of early modern philosophical knowledge is made available.

3 From *artiste* to Toyware Manufacturing

A famous *cause célèbre* opposing the instrument makers to the Parisian glassmaker guild provides a helpful snapshot to the fact that specialized instrument making in eighteenth-century Paris required more than elementary—some would say witless—artisanal skills. In 1767, the guild decided unexpectedly to enforce their legal rights[72] on a few eminent Parisian thermometer makers (or *ingénieurs en instrumens de physique*), namely Toussaint Cappi (or Cappy), Assier Périca, and Bourdon. We learn that

> The guild's jurors, jealously keeping the privileges of their profession [Corps], went to the sieur Cappy's workshop, where they found several barometers, thermometers, a large quantity of glass tubes, etc. and all the instruments necessary to his art. They put everything into boxes, seized and transported them to their headquarters.[73]

71 Descola, "The Two Natures of Lévi-Strauss," 108.

72 According to the 1566 guild statutes, glassmakers became the sole and rightful manufacturers and sellers of glass products, such as vases and drinking glasses. They held that everything involving the transformation of sand and ashes by the fire of a furnace into blown glass of various shapes was the sole prerogative of glassmakers. This included fashioning barometers and thermometers, even though such instruments did not yet exist when their statutes were ratified.

73 M. Des Essarts, *Choix de nouvelles causes célebres, avec les jugemens qui les ont décidées, extraites du Journal des Causes célebres, depuis son origine jusques & compris l'année 1782* (Paris, 1785), vol. 4, "Les Fabricateurs de barometres, contre les Faïenciers & Emailleurs," 354–72, on p. 356. Such *saisi* happened regularly between guilds. See, for instance, a *saisi* in 1680 by the drapers against the merchant guild regarding cloth made in Languedoc.

Cappi fought back *sur le champ* and could retrieve one box of materials. Over the next five years the instrument makers' legal defense, led by Mr. Linguet, attempted to establish that they were making something more than everyday glass commodities. Their products went beyond the art of *verrerie*. It was simply not enough to throw pieces of glass together and sell them as barometers or thermometers:

> The enamelers [Boutonniers en émail] saw for a long time these makers, without jealousy, exercise an art to which they had no rivals. But slowly the members of this guild, seeing ever more barometers and thermometers, became accustomed to these instruments. They thought it was easy to put together glass tubes and balls, fill them with mercury and spirit of wine then fix everything on a board equipped with a random graduation scale, a golden border, and selling this object under a familiar name.[74]

There was, the defense argued, an original, legitimate and distinct practice behind the art of instrument making.

Mathematician, astronomer, and mayor of Paris at the dawn of the French Revolution, Jean Sylvain Bailly used this *cause célèbre* to demonstrate that not only the *ingénieurs physiciens*, but opticians and mathematical instrument makers as well deserved a special status. They could perhaps, he suggested, be sanctioned by the Academy of Sciences through a privilege—alike the one given by the king to artisans housed in the Galerie du Louvre. They required, Bailly argued, a greater, non-corporate freedom to produce high-quality instruments, which often involved an amalgam of glass, brass, wood and other materials. Such freedom of practice was one way to ensure that French instrument making would ultimately challenge London's supremacy—where guilds were slowly losing most of their authority over the work of instrument makers.[75]

Registre des délibérations et ordonnances des marchands merciers de Paris (1596–1696) (Paris: Léon Willem, Éditeur, 1878), 145–54.

74 Ibid., 362.

75 About the ingénieurs mathématiciens, Bailly wrote: "S'il faut qu'il ait recours à un serrurier et à un fondeur pour les métaux, à un menuisier-ébéniste pour les bois, à un miroitier-lunetier pour les verres, à un faïencier pour les tubes, il ne pourra jamais faire un bon instrument de ces pièces de rapport exécutées par des mains différentes; il serait absurde d'exiger qu'il fût de vingt communautés, et si chacune de ces communautés a le droit de le saisir, ses travaux seront sans cesse troublés, il se découragera; et pour faire son métier tranquillement, il prendra le parti d'employer les ouvriers de chaque genre, et il fera de mauvais instrumens." Jean Sylvain Bailly, "Exposé de ce que l'Académie des Sciences a fait, et de ce qui lui reste à faire pour la description complète des Arts," in *Recueil de pièces intéressantes sur les arts, les sciences et la littérature, ouvrage posthume de Sylvain*

Jacques-Dominique Cassini, director of the Paris Observatory from 1784 to 1794, jailed during the Terror, similarly commented that the *ingenieurs en instrumens de mathématiques* were repeatedly embroiled in such controversies with the tyrannical pretensions of guilds. The reason, he thought, was simple: "of all the arts [instrument making] is the one requiring the most skills, knowledge and *génie* from the *artiste*. To succeed the latter must gather in himself the proficiency of ten different *artistes*."[76] The *Faïenciers* and *Emailleurs* lost the *cause célèbre* through an act of the Parlement de Paris, dated 31 August 1772 (reiterated on 19 March 1774). The guild was ordered to return the material taken illegally from Cappi's house, and "the enamelers were forbidden to trouble again in their art the makers of barometers."[77] The act did not decree stopping barometer and thermometer fabrication. Though no longer a threat to instrument makers, glassmakers could still flood the marketplace with instruments of dubious quality, far from any philosophical standards. *Artistes*, or instrument makers of *génie*, however, were rewarded with renewed freedom in creating and manufacturing their goods.

The *artiste* as a distinct category of highly skilled individuals fully emerged in France, according to Paola Bertucci, in the early part of the eighteenth century. Though the appellation already stood for those who knew how to conduct chemical experiments, and Chérubin d'Orléans envisioned the *artiste curieux* as someone with an unusual mental acuity, the lexicographer Pierre Richelet extended around 1700 the term *artiste* to mean something much more, i.e., an artisan with *esprit*, one who could combine ingenuity—*génie*—with finely honed practical skills. The individuals who best represented—and exploited to its fullest—the broader meaning of *artiste* were found in the Société des arts, a specialized group of skilled artisans (up to 200 in the 1730s) who

Bailly ... (Paris, 1810), 126–46, on p. 140. Larry Stewart, "Science, Instruments, and Guilds in Early Modern Britain," *Early Science and Medicine* 10 (2005): 392–410. A good survey on the role of guilds, and especially metalworkers (*fondeurs*) crafting scientific instruments in eighteenth-century Paris is found in Jean-Dominique Augarde, "La fabrication des instruments scientifiques du XVIIIe siècle et la corporation des fondeurs," in *Studies in the History of Scientific Instruments*, ed. Christine Blondel, *et al.* (London: Rogers Turner Books Ltd, 1989), 53–72.

76 Jacques-Dominique Cassini, *Mémoires pour servir à l'histoire des sciences et à celle de l'Observatoire royal de Paris suivie de la vie de J. D. Cassini écrite par lui-même—et des éloges de plusieurs académiciens morts pendant la Révolution* (Paris, 1834), 219. See Turner, "Mathematical Instrument-Making in Early Modern Paris," 84.

77 Des Essarts, *Choix de nouvelles causes célebres*, 372. More than that, Assier Périca was apparently taking the liberty to teach glass-blowing techniques to enthusiastic amateurs. Abbé Rozier, *Observations sur la physique, sur l'histoire naturelle et sur les arts* ... (Paris, 1775), vol. 6, 89.

strongly believed that only useful knowledge produced by the mechanical arts could "effectively contribute to the economic and imperial advancement of the French state."[78] Openly going against members of the Académie des sciences, these *artistes* aspired and actively worked to position themselves as better decision makers in the global match of political economy. As Bertucci keenly argues, *artistes* from the Société des arts were not busily trying to "readjust the social perception or widespread poor consideration of artisans," as often reported by savants and academicians alike. They endeavored to elevate themselves *above* mere artisans, to claim for themselves, as we just saw with the *cause célèbre*, an intermediary technical niche more useful to the public good than the artisan's 'mindless' commodities or the savant's theoretical knowledge.[79] By focusing on the mechanical arts as their exclusive domain of knowledge and practice, the enlightened *artistes* used their newfound *génie* to call into being a technical authority aimed at advancing France's commercial, military, colonial, and imperial power. Their success was mitigated.

From inspiring *idiotae* to self-governing and politically acute *artistes*, the role and image of the artisan were reshaped several times throughout the early modern period. Taking yet another form during the industrial Revolution, the artisan remained at the center of the scientific and economic growth of Europe. Analyzing Joel Mokyr's *The Gifts of Athena* (2002), an influential study on the historical function of technology and useful knowledge to the developing knowledge economy of the modern West, Liliane Hilaire-Pérez argues that Mokyr's emphasis on 'pure' science over the mobility, public culture, and networks of arts and trades is upside down. She contends rather that "the crucial factor in the industrial Enlightenment was the part played by artisans"—or, said otherwise, the part played by markets, not by institutions.[80] Her essay, among other things, draws attention to the healthy, creative, and resourceful artisanal culture, demonstrating, for instance, that technology and engineering practices were not set up by scientists but "were rooted in the practices and in the cultural itineraries of artisans. … They were the main workforce [of

78 Bertucci, *Artisanal Enlightenment*, 23.

79 Ibid., 25–6.

80 Liliane Hilaire-Pérez, "Technology as a Public Culture in the Eighteenth Century: The Artisan's Legacy," *History of Science* 45 (2007): 135–53, quote on p. 137. Recently, see the collection of essays on artisanal mobility in David Garrioch, ed., *The Republic of Skill. Artisan Mobility, Innovation, and the Circulation of Knowledge in Premodern Europe* (Leiden: Brill, 2022).

the late Enlightenment] and highly innovative."[81] Investigating a specific and popular type of consumer objects, the so-called toywares (tableware, watches, optical instruments, cooking apparatus, etc.), Hilaire-Pérez's extensive study into the London manufacturing communities and networks of jewelers, watchmakers, instrument makers, ironmongers, and other specialist workers uncovers an 'operative knowledge' (*savoir opératoire*) based on the breaking down of objects into several parts (*pièces*), requiring nearly as many gestural operations (*gestes*) for their production. From the study of Enlightenment toywares, regarded as a genuine consumerist explosion of luxury products, one recognizes "an operative understanding of objects" at work that wields a 'ratiocination of action' (*rationalité de l'action*), combining the object to the gestural act of production. Together, Hilaire-Pérez argues, they form the "cornerstone to the exchange value of products."[82] It is in the powerful hands of the market, not in learned institutions, that the late eighteenth-century *génie* of artisans is best identified. One might even say the concept of *génie* shifted during that time from an individual appropriation (the *artiste* from the Société des arts) to a collective hive (artisans working on specific parts in manufacturing toyware products). From the sixteenth to the nineteenth centuries, the 'artisan' skills, productivity, and material output were scrutinized, analyzed, and deconstructed to determine and better understand their value to the production of scientific knowledge. If the *organum* is considered the keystone to a structural analysis of early modern philosophical knowledge, *habitus* (as defined in chapter 3) ought to be assumed as the main source of an artisanal worldview. *Organum* and *habitus* together form a dyad that, unfortunately, is too often forgotten in *museums*.

---

81 Hilaire-Pérez, "Technology as a Public Culture," 138–9. A very good accompanying survey on a related topic is Liliane Hilaire-Pérez and Catherine Verna, "Dissemination of Technical Knowledge in the Middle Ages and the Early Modern Era. New Approaches and Methodological Issues," *Technology and Culture* 47 (2006): 536–65.

82 Liliane Hilaire-Pérez, *La pièce et le geste. Artisans, marchands et savoir technique à Londres au XVIIIe siècle* (Paris: Albin Michel, 2013). On the London luxury market, Maxine Berg, *Luxury & Pleasure in Eighteenth-Century Britain* (Oxford: Oxford University Press, 2010).

PART 3

Museum

Introduction to Part 3

Making things and knowledge is unpredictable, contingent, fluid, and messy. In the early modern period it is carried out—performed would be a better verb—through body and mental instruments (our description of *organum*) and practices (our description of *habitus*). The first two parts of this book explore and define at length both concepts in order to comprehend how knowledge production was construed by early modern natural philosophers. *Organum* and *habitus* are models, insofar as I understand both heuristic types, finding strong resonance in our contemporary ways of thinking about the production of knowledge. They offer, I believe, new possibilities of investigating the early modern world using categories often too loosely specified, not anchored enough in the epistemology of the past.

In the third part of the book, I want to focus on *museum*, an institution mostly designed to display (one's power) and safeguard the (ostentatious) objects and the (lost and neglected) customs of past and present societies.[1] My goal is not to provide yet another interpretation of what a museum was, is, could be or should have been, but rather to explore how we can expand the notions of *organum* and *habitus* to *museum*. It is safe to say that as sociocultural and political institutions, museums have developed and customized over the last few centuries instruments and practices to fulfill their unique and distinct goals—beyond the technical instruments and practices put in place to care for the material and now digital collections. Before moving into the realm of instrument collections, I want to elaborate on the notion of 'instrument' as regards museum practices, concentrating upon the most visual, public, and appealing activity generated by museums all around the world: exhibitions.

For Tristan Garcia and Vincent Normand, museum exhibitions boast a split identity, ontologically defined by two dominating ideas. Primo, exhibitions have developed over time a *form* of their own allowing—especially in modern art—"a critical reflection on the conditions of aesthetic experience" and, perhaps more significantly, their "recomposition in a medium accommodating various kinds of representations." The 'exhibition as a critical form' has among other things transformed the medium of exhibition into a genuine 'prime material of art.' Secondo, instead of form, exhibitions have been marshaled (particularly in the fields of history, museum studies, anthropology, and art theory) into an *apparatus*, a Foucauldian *dispositif* enabling the museum to

1 For a *longue durée* history of museums, Krzysztof Pomian, *Le musée, une histoire mondiale*, 3 vols (Paris: Gallimard, 2020–2022).

 | DOI:10.1163/9789004504615_009

become "a privileged vehicle of subjectivation ... a site of materialization of the modern episteme, the concrete stage on which a new experience of knowledge emerged." Such apparatuses of modernity are "*spatial structures determining the social inscription of objects*."[2]

The conceptual dyad composed of form and apparatus, making up the 'exhibitionary complex,' is based on Tony Bennett's seminal article, in which he convincingly makes the *rapprochement* between Foucault's analysis of the carceral archipelago and the opening-up of museums by means of exhibitions. He contends that from the end of the Enlightenment to the mid-nineteenth century, the *apparatus* developed around prisons and museums was commensurable, the latter institutions emerging as "vehicles for inscribing and broadcasting the messages of power ... throughout society."[3] The spectacle created within the exhibitionary complex offered, according to this Foucauldian logic, a "*permanent* display of power/knowledge. ... [A] power made manifest not in its ability to inflict pain but by its ability to organize and co-ordinate an order of things and to produce a place for the people in relation to that order."[4]

For philosopher Giorgio Agamben, Foucault's *apparatus* has an interesting genealogy that stretches back to Antiquity. Though Jean Hyppolite's notion of *positivité* (1948) probably heralds that of *apparatus/dispositif*, Agamben finds much further in time, in the Greek *oikonomia*, or the "administration of the *oikos* (the home)," i.e., management, how "the Trinitarian dogma and the idea of a divine providential governance of the world were introduced into the Christian faith." This very idea of a "divine providential governance of the world" is of interest since the Greek *oikonomia* was translated into Latin by the Church Fathers as *dispositio*. An apparatus, therefore, is for Agamben "anything that has in some way the capacity to capture, orient, determine, intercept, model, control, or secure the gestures, behaviors, opinions, or discourses of living beings." To him, an apparatus "is first of all a machine that produces subjectifications, and only as such it is also a machine of governance."[5] Museums do not

2 Tristan Garcia and Vincent Normand, eds., *Theater, Garden, Bestiary. A Materialist History of Exhibitions* (Berlin: Sternberg Press, 2019), 11–23, quotes on pp. 13–4 (emphasis original). Chapters by Kim West and Vincent Normand are particularly relevant here. An interesting review from the historian's point of view remains Randolph Starn, "A Historian's Brief Guide to New Museum Studies," *American Historical Review* 110 (2005): 68–98.

3 Tony Bennett, "The Exhibitionary Complex," *New Formations* 4 (1988): 73–102, quote on p. 74.

4 Ibid., 79–80 (emphasis original). There is a fascinating connection to make here with the classic analysis of spectacle by Guy Debord, *La société du spectacle* (Paris: Gallimard, 1992), originally published in 1967. The spectacle is fundamentally a concealed means of individual and social control.

5 Giorgio Agamben, *What is an Apparatus? And Other Essays*, trans. David Kishik and Stefan Pedatella (Stanford: Stanford University Press, 2009), 10–11, 14, and 20.

escape this diagnosis and it is why the overarching and expansive abstraction that is an *apparatus* is such a relevant and influential tool of analysis. Yet the missing link here is what composes the DNA of a museum: its objects, its collections, and how they are studied and cared for. Without them, there is no discourse, no 'machine of governance' to fine-tune and exploit. Sans objects there is only partial knowledge of the world—and no museum.

Through a variety of critical and analytical tools, we can deconstruct all types of museum exhibitions until they reveal to us their innermost complexities, contested nature, and inherent power/knowledge relationships to the world. Yet exhibits are made by people, well-meaning people, usually grappling, managing, and taking on a myriad of possibilities and limitations.[6] *And they are created with objects*. The latter's built-in agency brings to the fore poignant feelings and emotions, which people greet and respond to differently depending on their level of affinity with the material, history, craftsmanship, and culture of the object they see and discover. When Sandra Dudley, for instance, a well-known scholar in museum studies, visited the art gallery at Compton Verney (Warwickshire) for the first time, she experienced a strong emotion while standing in front of a Chinese artifact, a bronze horse dating to the Han Dynasty:

> I was utterly spellbound by its majestic form, its power, and, as I began to look at it closely, its material details: its greenish colour, its textures surface, the small areas of damage. I wanted to touch it, though of course I could not—but that did not stop me imagining how it would feel to stroke it, or how it would sound if I could tap the metal, or how heavy it would be if I could try to pick it up. I was, in other words, sensorially exploring the object, even though I had to intuit and imagine rather than directly experience most of the encounter. … Nonetheless its three-dimensionality, tactility and sheer power had literally moved me to tears.[7]

6 A very informative and still relevant behind-the-scene analysis is provided by Sharon Macdonald, *Behind the Scenes at the Science Museum* (Oxford: Berg, 2002). A more recent account is Rebekah Higgitt, "Challenging Tropes: Genius, Heroic Invention, and the Longitude Problem in the Museum," *Isis* 108 (2017): 371–80. See also Simon Schaffer, "Chronometers, Charts, Charisma: On Histories of Longitude," *Science Museum Group Journal* vol. 2 (2014), online http://journal.sciencemuseum.ac.uk/browse/issue-02/chronometers-charts-charisma/ (accessed 27 July 2021).

7 Sandra Dudley, "Encountering a Chinese Horse: Engaging with the Thingness of Things," in *Museum Objects: Experiencing the Properties of Things*, ed. Sandra Dudley (New York: Routledge, 2012), 1–15, quote on p. 1. Dava Sobel, who wrote the bestselling (and somewhat controversial) book on the history of longitude, burst into tears the first time she saw John Harrison's chronometers. Higgitt, "Challenging Tropes," 378.

Is Dudley's reaction unusual? Absolutely not. We all experience at one moment or another such powerful sensations at the sight of objects.[8] What Dudley is attempting to convey and emphasize here is how significant the *material* qualities of objects are in producing these emotional reactions. The latter are not spawned by a descriptive label on a wall, but by the object itself—by the *thingness* of the thing. In the same edited volume, Simon Knell calls this inherent state of objects the 'intangibility of things.' He argues that any object, ipso facto a museum artifact, is composed of two things: the material and tangible (real) thing itself—not always reachable—and an intangible one, "the product of experience and negotiation ... [It] exists in our world but is made in our thoughts; it is ever present and inescapable." Though this immaterial creation "*appeals* to the truth of the material object" it remains detached and fluid. "The object in our thoughts may seem material, definite and fixed but it is in fact intangible, contingent and transient."[9] In other words, what Dudley experienced so strongly in front of the Chinese artifact is completely dependent on her experience and personal being—we could say on her *habitus*. Without the possibility of experiencing the real and tangible museum object with the help of more than one sense, in this case sight, there is no experiential common ground possible. Each visitor is left with her own and distinct contingent 'intangible object' based on her idiosyncratic knowledge and familiarity with matter.[10]

Though this sensation might be true for all objects stuck inside a museum case, some of them might be harder to figure out, to get an emotional reaction. They lack 'presence.' On the opposite side of the emotional spectrum, Thomas Söderqvist, Adam Bencard, and Camilla Mordhorst look at the 'presence effect' of contemporary biomedical objects and the challenge they pose to medical museums. Using Hans Ulrich Gumbrecht's distinction between 'presence culture' and 'meaning culture,' they argue that a *GeneChip*, for example, designed

---

8 There is a good amount of literature on the topic. Recently, see the special issue edited by Gaëlle Crenn and Jean-Christophe Vilatte, "L'émotion dans les expositions," *Culture et Musées* 36 (2020): 1–196. Daniel Fabre, ed., *Émotions patrimoniales* (Paris: Éditions de la Maison des sciences de l'homme, 2013).

9 Simon Knell, "The Intangibility of Things," in *Museum Objects*, 324–35, quotes on pp. 326 and 328 (emphasis original).

10 A very interesting sociological analysis of the *amateur's* taste is Antoine Hennion, "Those Things That Hold Us Together: Taste and Sociology," *Cultural Sociology* 1 (2007): 97–114, where he writes that "Through comparison, repetition and so on, things that are less inert than they appear are made more present. They must be made to appear in and through contact: to taste is to *make feel*, and *to make oneself feel*, and also, by the sensations of the body, exactly like the climber [on a rock wall], *to feel oneself doing*." (p. 101, emphasis original)

to analyze the genome, "lacks the qualities of a traditional evocative museum object"—such as a surgical saw or a trepan. What the biomedical artifacts do not have is the 'personal bodily experiences': they are less emotionally evocative than traditional medical instruments and body parts. According to the authors, it is the very "presence-producing features of medical artefacts that lures the public into medical museums." Twenty-first-century biomedical artifacts are, conversely, 'bad' museum artifacts since they do not stimulate memories or force visitors to pause in front of them with a sense of horror, curiosity or wonder.[11] How, one might ask, can such artifacts redeem themselves in a science museum?

The Medical Museion in Copenhagen decided to take on the challenge head-on. It created, under the leadership of Söderqvist, an art installation using fiber-optic cables to hang over 600 Illumina Genotyping BeadChips utilized in an ongoing research project with 17,000 Danish individuals, developed to identify genotypes that might lead to increased risk of obesity, type 2 diabetes, and hypertension. The art installation is one answer to the 'bad' museum objects discussed above: it is visually evocative, it elicits wonder and curiosity, and it contextualizes the real artifacts of present-day biomedical research.[12] Another answer is to fully combine the aesthetic and scientific methods into one

---

11 Thomas Söderqvist, Adam Bencard, and Camilla Mordhorst, "Between Meaning Culture and Presence Effects: Contemporary Biomedical Objects as a Challenge to Museums," *Studies in History and Philosophy of Science* 40 (2009): 431–38, quotes on pp. 434–5. Hans Ulrich Gumbrecht, *Production of Presence. What Meaning Cannot Convey* (Stanford: Stanford University Press, 2003) is a reaction to the so-called *linguistic turn* that finds meaning in language (and reason) and no longer in things themselves: "Therefore, 'production of presence' points to all kinds of events and processes in which the impact that 'present' objects have on human bodies is being initiated or intensified. All objects available in 'presence' will be called the 'things of the world'" (p. xiii). For a theoretical analysis of 'presence' v. 'meaning,' see also Eelco Runia, "Presence," *History and Theory* 45 (2006): 1–29, where the author argues that "Presence—being in touch with reality—is … just as basic as meaning." (p. 5). He explores the concept of *metonymy* and believes that with it "a discourse of presence can be established that does not explain discontinuity away in some sort of 'meaning,' but gives it its due" (p. 6). Another example of the use of emotion in a medical museum is provided by Cecilia Rodéhn, "Emotions in the Museum of Medical History. An Investigation of How Museum Educators Employ Emotions and What These Emotions Do," *International Journal of Heritage Studies* 26 (2020): 201–13.

12 The art piece, entitled 'Genomic Enlightenment,' is explained on the Medical Museion website: http://www.museion.ku.dk/whats-on/exhibitions/genomic-enlightenment/ (accessed on 27 July 2021). On evocative objects, Sherry Turkle, ed., *Evocative Objects: Things We Think With* (Cambridge, MA: MIT Press, 2011). More recent examples of such projects are found in Edward Juler and Alistair Robinson, eds., *Post-Specimen Encounters Between Art, Science and Curating. Rethinking Art Practice and Objecthood Through Scientific Collections* (Bristol and Chicago: Intellect, 2020).

project/product in order to embrace 'nature in its complexity,' to bring together the 'sensual and analytical' and become "comfortable with uncertainty and [be] able to frame a problem." This way of doing, thinking, and knowing is what David Edwards named *artscience*. Projects challenging the boundary of art and science are developed in places called 'the lab.' Neither a science center nor an art gallery, this space is designed to facilitate the exchange between various partners—social, research and educational, industrial and cultural—including the public. The latter is the "validating partner ... who review[s] whatever comes out of a lab" before it can be considered seriously as a final product. This concept has been taken up and adapted by the popular and successful Science Gallery Network, which to date features nine such galleries across the world.[13]

Though much more could be said here, this short analysis demonstrates how meaningful it is to pay attention to museums as complex knowledge-producing institutions. Now focusing fully on museum collections, part 3 of this book, based on my experience and praxis within the field, endeavors to examine and question a series of museum activities and undertakings relevant to the way we study and approach scientific instrument collections. The fundamental query underlying chapters five and six is simple, yet remains open-ended (no one can claim a final solution): what happens to an instrument (*organum*) and its built-in practice (*habitus*) when it is acquired by a *museum*? Ontologically speaking an instrument appears to retain its inherent features when introduced into a museum collection. But since and *organum* is also defined by its interdependent *habitus*, what happens to the former when the latter is entirely changed, obliterated, forgotten? Such a crucial interrogation allows us to delve further into the materiality of objects and discover some of the challenges and opportunities connected to today's collection management.

13 David Edwards, *Artscience. Creativity in the Post-Google Generation* (Cambridge, MA: Harvard University Press, 2008), quotes on pp. 7 and 182. David Edwards, *The Lab. Creativity and Culture* (Cambridge, MA: Harvard University Press, 2010). Information on the Science Gallery International Network is found here: https://sciencegallery.org/about-network (accessed on 6 August 2021). The evolution of the concept is described by the founder of the original Science Gallery in Trinity College Dublin, Michael John Gorman, *Idea Colliders: The Future of Science Museums* (Cambridge, MA: MIT Press, 2020). We must note here that the Science Gallery in Dublin closed recently.

CHAPTER 5

Repair, Restoration, Exhibition: Instruments and the Epistemic Value of Brokenness

During the 15 April 1755 meeting of the President & Fellows of Harvard College, Andrew Oliver, a Boston merchant (and generous Harvard donor) received an official acknowledgment for "his gift of a very large Pair of fine Globes." The gift came with restrictions. Oliver required that the globes always be kept in the Library, unless "the Mathematical Professor shall need them, for the Instruction of the Pupills that attend his Lectures." In such cases, said professor (and tutors teaching astronomy and geography) may carry them into the Apparatus Chamber; once the instruction was over, the globes "shall be immediately return'd to the Library." Moreover, the globes may not leave the campus's vicinity, unless the President himself has "Occasion for them."[1]

Likely lost in 1764 to the fire of Harvard Hall, the globes were replaced in 1765 with a pair manufactured by Benjamin Martin circa 1757. (Fig. 12L) Wear and tear over the years, however, were not kind to these new globes. In a letter written in 1804 to the Harvard Corporation, Parker Cleaveland (1750–1858), Tutor of Natural Philosophy and Mathematics, lamented that not a globe was in good enough condition for teaching purposes:

> Gentlemen, the subscriber begs leave to state to you, that there are belonging to the department of Nat. Philosophy &c, in which he is employed, three globes, one terrestrial & two celestials; all of which are entirely unfit for use, & can afford the pupil no assistance, excepting what may arise from simple inspection. One of the celestial globes is very old … [t]he surface of the other celestial is a little soiled … [and the] terrestrial globe is entirely worn out. Its surface is very much soiled, in many parts torn off; and some of its circles appear in furrows, made by pins & nails … The subscriber believes that one of the celestial globes may be repaired and rendered fit for use. The others are not worth repair.[2]

1 Harvard University. Corporation. College Books, 1636–1827. College Book 7, 1750–1778. UAI 5.5 Box 5, Harvard University Archives, seq. 95–96 (pp. 37–8). *Colonial North American Project at Harvard*, http://nrs.harvard.edu/urn-3:HUL.ARCH:10880062 (accessed 7 November 2022). For other donations by Oliver, see Ethan Lasser, ed., *The Philosophy Chamber: Art and Science in Harvard's Teaching Cabinet, 1766–1820* (New Haven: Yale University Press, 2017).

2 Quoted in David P. Wheatland, *The Apparatus of Science at Harvard, 1765–1800* (Cambridge, MA: Harvard University Press, 1968), 63–4. On the 1764 fire and the cabinet reconstruction,

 | DOI:10.1163/9789004504615_010

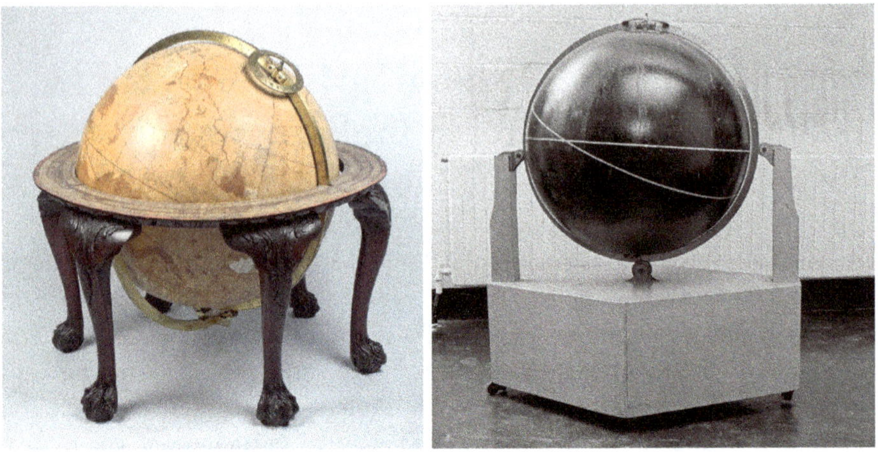

FIGURE 12.L–R Two globes dated ca 1757 from Benjamin Martin. The celestial globe is still in good original condition and on permanent display at the CHSI (left); the terrestrial globe was badly damaged and transformed into a teaching device in the 1930s (right).
© COLLECTION OF HISTORICAL SCIENTIFIC INSTRUMENTS, HARVARD UNIVERSITY, 0071A AND 0071B

Nine years later, Harvard's President authorized the purchase of a pair of globes "for the use of the students, if they find it necessary."[3] The Martin terrestrial globe, which in 1804 was already "entirely worn out," was repurposed in the early 1930s at the request of physicist Edwin Kemble into a spherical blackboard to teach positional astronomy and spherical geometry. Though it still holds its original graduated brass meridian and hour circle, it received an unassuming gray-painted wooden rolling stand. (Fig. 12R) Interestingly, this 'repair' is contemporaneous to the establishment of a conservation laboratory at the Harvard Fogg Museum, becoming since then a world leader in scientific museum restoration.[4] Whereas the Martin celestial globe is on permanent dis-

see Lasser, *The Philosophy Chamber*. Regarding the instruments' use at Harvard, see Jean-François Gauvin, "The Next Level of Play: How Cartesian Devils, Volcanoes, and Quantum Toys Taught General Science at Harvard, 1730–1970," *The Journal of the History of Collection* 31 (2019): 531–46.

3 Harvard University. Corporation. College Books, 1636–1827. College Book 10, 1810–1827. UAI 5.5 Box 8, Harvard University Archives, seq. 228 (p. 161). *Colonial North American Project at Harvard*, http://nrs.harvard.edu/urn-3:HUL.ARCH:10647834 (accessed 7 November 2022).

4 Francesca G. Bewer, *A Laboratory for the Art. Harvard's Fogg Museum and the Emergence of Conservation in America, 1900–1950* (New Haven: Yale University Press, 2010). This laboratory is known today as the Straus Center for Conservation and Technical Studies.

play at Harvard in the exhibition gallery of the Collection of Historical Scientific Instruments (CHSI), the other remains in storage, where no one can see it; it appears to be in such 'disrepair' it is no longer considered beautiful—despite its fascinating history of recovery and transformation.[5]

Rarely does one see an incomplete—or broken—scientific instrument in the context of a science and technology museum exhibition. Though common in history, art, and archaeology museums (one thinks of Greek or Roman pottery fragments and statues, or vestiges of a Spanish galleon's wreckage), the practice of displaying damaged and ruined objects is far less familiar within science museums. A rusted and twisted astrolabe, a dislocated and chipped sextant or a broken-down aircraft are usually put in storage and considered objects of study and research only, unworthy of public display. Historian of science Simon Schaffer believes this is a mistake. In a short article focused on the circulation of astronomical instruments in the eighteenth and nineteenth centuries, Schaffer suggests that instruments 'in a state of disrepair' are exhibited next to others considered more prestigious and dazzling. The former, he contends, have as much to say—perhaps even more to teach—about the circulation of knowledge and the epistemic and sociological value of repair, use, and maintenance as the latter, when delving into the material culture of science.[6] The state in which an object is found at a time *t* remains in and of itself a convincing indication of its historical and material trajectory. This is one of the main features connected to the field of conservation-restoration.

The damaged and dislocated object, although it provides important factual answers to its 'biography,' is itself often in jeopardy. Museums take advantage of the forthcoming public display of objects to restore them and thus improve their long-term condition in storage—an essential component of the 'object-love' felt by both curators and conservators regarding the objects they care for in storerooms.[7] Is it, however, always necessary to do so, insofar as the main goal is to make it more 'beautiful'? Must it be ipso facto integrated

5 On 'beautiful' instruments, Richard Dunn, "'More Artistic than Scientific': Exhibiting Instruments as Decorative Arts in the Victoria and Albert Museum," in *Scientific Instruments on Display*, eds. Silke Ackermann, Richard Kremer, and Mara Miniati (Leiden: Brill, 2014), 61–76. On the concept of art in science museums, Catia Rodrigues Barbosa, "L'esthétique dans les musées des sciences et des techniques," *La Lettre de l'OCIM* 95 (2004): 27–30.

6 Simon Schaffer, "Easily Cracked: Scientific Instruments in States of Disrepair," *Isis* 102 (2011): 706–17. See also John Durant, "The Challenge and the Opportunity of Presenting 'Unfinished Science'," in *Creating Connections. Museums and the Public Understanding of Current Research*, eds. David Chittenden, Graham Farmelo, and Bruce Lewenstein (Walnut Creek: AltaMira Press, 2004), 47–60.

7 Hilary Geoghegan and Alison Hess, "Object-Love at the Science Museum: Cultural Geographies of Museum Storerooms," *Cultural Geographies* 22 (2015): 445–65.

into a context of 'artistic capitalism' (*capitalisme artiste*), resulting in a mere increase in the number of visitors? Is the care put into the object's integral estheticization (because, concretely, it is as much about the aesthetic value of the exhibition as it is about the object's material preservation) going beyond the historical, pedagogical, and epistemic relevance it can provide—especially from a so-called object in a state of disrepair?[8]

By exploring the field of conservation-restoration applied to scientific instruments, this chapter delves into the question of an object's materiality (its ontology) and what we can learn from its study. A fundamental (and controversial) question is asked: if a broken instrument, in a state of disrepair, requires conservation-restoration to ensure its preservation, how best to proceed? What are the practical, philosophical, and deontological divergences between conservation-restoration and repair treatments? In a series of examples taken from the eighteenth to the twentieth centuries, the chapter argues that the careful analysis of materials provides a deeper understanding of an instrument (*organum*) and its uses (*habitus*) in a *museum* context. In support of my analysis, I find the philosophy of conservation-restoration by Jean-Pierre Cometti especially useful, and largely based on the paradigm of action, i.e., examining an object's operation and its operating conditions (*le fonctionnement et les conditions de fonctionnement*).

1 Restorer v. Conservator: How to 'Repair' Damaged Instruments?

To video-document a musical automaton clock designed and made by the Soho clockmaker William Carpenter (fl. 1770–1805), the Victoria and Albert Museum (V&A) in London decided in the early 2000s to examine how it worked. (Fig. 13) The clock, finely chiseled and executed, embodying exoticism, was intended for the Asian market.[9] The object's initial material conditions analysis, carried out by a clock *restorer* and a metalwork *conservator*, showed that despite some corrosion and a thin layer of dust on the external structure, a conservation-restoration was viable following art museum standards. Regarding the internal mechanism, largely covered with a sticky black paste made up of dust and residues of degraded oil lubricants, a superficial

8 Gilles Lipovetsky and Jean Serroy, *L'Esthétisation du monde. Vivre à l'âge du capitalisme artiste* (Paris: Gallimard, 2013).

9 On the significance of the Asian market for English merchants, Simon Schaffer, "Instruments as Cargo in the China Trade," *History of Science* 44 (2006): 217–46. For a broader context, Schaffer, "L'inventaire de l'astronome. Le commerce d'instruments scientifiques au XVIIIe siècle (Angleterre-Chine-Pacifique)," *Annales. Histoire, Sciences Sociales* 60 (2005): 791–815.

FIGURE 13 Musical automaton clock manufactured in London around 1780 by the watchmaker William Carpenter. This highly sophisticated clock was intended for the Asian market. Made of brass and enameled glass, it depicts exotic scenes and incorporates moving figurines that ring quarter hours and strike the hour.
© VICTORIA AND ALBERT MUSEUM, LONDON, M. 1108–1926

and sometimes vigorous brushing (though nothing as strong as a commercial cleaning with an ultrasonic bath, as is often the case when refurbishing a dirty clockwork mechanism) was performed to ensure free movement of the cogs and cogwheels. A missing part is replaced and marked 'VA' with the date, not to confuse it with an original part. After cleaning the surface, the cogs are lubricated with commercial oil. The video reveals the internal mechanism in action and the movement of the small figurines producing the ringtones.[10]

The virtual mechanical display, showcasing simultaneously the object's integral beauty and technological prowess, compels a dialogue between the practice of the expert clockmaker, *restorer*, "whose aim is to restore the clock to reliable day-to-day running," and the specialist in museum conservation, *conservator*, whose work testifies to an approach, "where the object is stabilised and brought to a condition where it is visually understandable."[11] This dichotomy in know-how and professional practices are justified since beyond its decorative scope the object, we learn, was sought for the perfection of its mechanical design. By keeping the conservation-restoration of the clock to a strict minimum to produce a single video, the compromise reached between a straightforward 'repair' and a 'standard conservation procedure' makes it possible to limit the risks incurred by the object, for a better overall understanding of its original function.[12]

This decorative/scientific/technological object is not strange to the V&A Museum since the famous South Kensington institution was originally, in the middle of the nineteenth century, an integral component of the London Science Museum. Branching out at the beginning of the twentieth century, compartmentalizing art and science, their current vocation attests to a fundamentally, and undoubtedly irreversible, epistemological and disciplinary

10 Christopher Powell and Sophy Wills, "Conservation or Restoration? Treatment of an 18th-Century Clock," *V & A Conservation Journal* 39 (2001): 21–2. The video is online: http://www.vam.ac.uk/content/articles/v/video-a-musical-automaton-clock/ (accessed 6 February 2019).

11 Powell and Wills, "Conservation or Restoration?," 22.

12 The authors do note, however, that such an intervention is not the usual modus operandi on museum objects: "The Carpenter clock, as a museum object, would not normally be subject to the risks of physical damage posed by the repeated assembling, disassembling and running of the object." Powell et Wills, "Conservation or Restoration?," 22. The decisions taken here, however, contrasts greatly with how a Marie-Antoine's armchair was fully restored by Zoe Allen, Senior Gilded Furniture and Frames Conservator. See her blog entry at the V&A: https://www.vam.ac.uk/blog/caring-for-our-collections/the-conservation-of-marie-antoinettes-chair (accessed 18 May 2020).

transformation.[13] Early in the history of the Science Museum, not unlike what happened at the Conservatoire national des arts et métiers (CNAM) in Paris, it was decided to 'destroy' objects either by revealing their internal mechanisms (sectioning) or by switching them on (operating). This apparently controversial destruction of historical clues and traces, almost impossible to imagine in museums today, is nonetheless one of the Science Museum's signature identities. It also illustrates the central paradox of science and technology museums: the need to preserve objects *and* demonstrate their operation. As we will see in detail below, this paradox is closely linked to how we think about the conservation and restoration of scientific objects.[14]

On a larger scale, the International Council of Museums (ICOM), with close to 45,000 members worldwide, adopted a resolution in 2008 advocating for a conservation-restoration terminology that would facilitate communication between various actors, including the public. ICOM defines conservation as "all measures and actions aimed at safeguarding tangible cultural heritage while ensuring its accessibility to present and future generations." Under this general definition, ICOM describes three subtypes of conservation: 1) *Preventive conservation*, or "all measures and actions aimed at avoiding and minimizing future

13 Richard Dunn, "'More Artistic than Scientific'." Alison Boyle, "'Not for Their Beauty': Instruments and Narratives at the Science Museum, London," in *Scientific Instruments on Display*, 37–60. Dominique Poulot, *L'Art d'aimer les objets* (Québec: Presses de l'Université Laval, 2016), 113–129. D. E. Cosgrove, "Should We Take It All so Seriously? Culture, Conservation, and Meaning in the Contemporary World," in *Durability and Change: The Science, Responsibility, and Cost of Sustaining Cultural Heritage*, eds. W. E. Krumbein, *et al.* (London: Wiley, 1994), 259–66.

14 Peter Robert Mann, "Working Evidence and the Destruction of Evidence in the Science Museum," *The International Journal of Museum Management and Curatorship* 8 (1989): 369–87. The author argues that this paradox between conservation and operation keeps museum curators in a state of 'schizophrenia': "Because they [curators] have been labouring under an inappropriate ethic derived from other types of museum, they have been living in a schizophrenic state in which their actions contradict their beliefs. They are not dishonest, thoughtless and uncaring, but merely confused." (p. 384) Idem, "The Restoration of Vehicles for Use in Research, Exhibition, and Demonstration," in *Restoration: Is It Acceptable?* (London: British Museum, Department of Conservation, 1994), 131–8. For conservation work on computers at the Science Museum, W. S. Elliott and A. E. Sale, "The Computer Conservation Society—Machines of the 1950's Restored to Life in the Science Museum, London," *Fifth National Conference on Engineering Heritage 1990: Interpreting Engineering Heritage; Preprints of Papers*, 39–44. For the CNAM, Louis André, "Scientific Instrument Restoration: Problems and Practice at the Musée des arts et métiers and in France," in *Proceedings of the Workshop Held in Florence, December 14–15, 1998*, ed. Laura Manetti (Florence: Le Lettere, 2000), 23–30. Bruno Jacomy, "Du cabinet au conservatoire: Les instruments scientifiques du Conservatoire des arts et métiers à Paris," *Journal of the History of Collections* 7, no. 2 (1995): 227–33.

deterioration or loss"; 2) *Remedial conservation*, or "all actions directly applied to an item or a group of items aimed at arresting current damaging processes or reinforcing their structure"; 3) *Restoration*, or "all actions directly applied to a single and stable item aimed at facilitating its appreciation, understanding and use."[15] The latter subtype, furthermore, is explicit about the end result:

> These actions are only carried out when the item has lost part of its significance or function through past alteration or deterioration. They are based on respect for the original material. Most often such actions modify the appearance of the item.[16]

The latter definition helps us understand how fluid and debatable is the dissimilitude introduced by the V&A between *restorer* and *conservator*. As discussed below, one key issue is the concept of *authenticity* and how it relates to museum objects.[17]

One relevant example is found in the memoir written by Emmanuel Aguila, a graduate from the Institut national du patrimoine (INP) in Paris, where he took a degree to become a heritage conservator (*restaurateur du patrimoine*), specializing in metalwork. The memoir's sophisticated analytical study of an early nineteenth-century tellurium (a mechanical system illustrating the combined Earth-Moon motion around the Sun) from the Musée des arts et métiers (CNAM) is resolutely grounded on the interplay between repair and restoration (as defined by ICOM). For Aguila, repairing the instrument's mechanism to restore its original operation is integral to its conservation. Though the conservator in training explains that his work on the instrument was done "with the minimum interventions necessary," one still notices a clear distinction between the latter and those carried out on the automaton clock from the V&A described above. For example, following the conventional method of clock repair, Aguila favored metal-cleaning baths (*bains de nettoyage*) to remove corrosion, before giving every piece of mechanism a careful brushing with a special type of flexible brass brush. The varnish on each brass part was removed by mechanical and chemical action then revarnished (contrary to Paolo Brenni's advice, see *infra*[18]). Finally, 'sacrificial parts' were inserted where needed to

---

15 ICOM, "Terminology to Characterize the Conservation of Tangible Cultural Heritage," http://www.icom-cc.org/242/about/terminology-for-conservation/#.XpcuBFNKjOQ (accessed 10 April 2020).

16 Ibid.

17 François Mairesse and Renata F. Peters, eds., *What is the Essence of Conservation? Materials for a Discussion* (Paris: ICOFOM, 2019).

18 Another memoir from the INP looks at the varnish found on a theodolite from the CNAM. Following Paolo Brenni's recommendation, Anne-Marie Geffroy made the effort

protect other components of the mechanism.[19] The primary intention behind this restoration was to make the instrument operational and functional again.

This repair philosophy commonly used in clockmaking, transferred to a prestigious conservation training program, was likely applied to the restoration and reconditioning (i.e., putting in working order) of a celebrated pendulum: *The Creation of the World*, a masterpiece engineered by Claude-Siméon Passemant. (Fig. 14) This task, performed by the Atelier Chronos, where Aguila works, aimed at "showcasing the exceptional complexity and remarkable precision" of the various mechanisms designed for the pendulum. The movement of the earth, the phases of the moon and the dials indicating the hour, day and month were made to run *de novo*.[20] Like the V&A automaton clock, combining both technological prowess and aesthetic splendor, this pendulum was originally intended to enrich the cabinet of an oriental prince. Disfavor falling on its sponsor, however, aborted the project. The pendulum was thereafter presented to Louis XV at Versailles, and, during the Revolution, it was integrated into the national collection. Unlike the restoration carried out by the V&A, the Louvre appears to have encouraged an approach more in line with science and technology museums than art museums. The contrast between these two exquisite automaton clocks, genuine works of art, science, and technology, demonstrates the wide deontological spectrum corresponding to the principles of conservation-restoration.[21]

---

to preserve and clean as much of the original varnish as possible—which, it turns out, was only doable on 15% of the total surface of the instrument. A special type of wax replaced the lost varnish. Anne-Marie Geffroy, *Conservation-restauration d'un instrument scientifique: un théodolite de la fin du XIXe siècle du musée des arts et métiers. Étude comparative du comportement au vieillissement artificiel de trois vernis sur laiton* (Paris: Institut national du patrimoine, mémoire de fin d'études, 2004).

19 Emmanuel Aguila, *Conservation-restauration du "Tellurium ou loxocosme" (no 07255-0000) du Conservatoire national des arts et métiers. Étude de l'usure des pivots dans un mécanisme d'horlogerie classique* (Paris: Institut national du patrimoine, mémoire de fin d'études, 2015), 90, 153–4, 162–4 et 177. A summary of this work is found in Aguila, "La mécanique céleste. La restauration du tellurium du CNAM," *Patrimoines. Revue de l'Institut national du patrimoine* 12 (2016): 104–9. This type of study follows in large part the methodology described in Mette Tang Simpson and Michael Huntley, eds., *La restauration des objets anciens* (Paris: Armand Colin, 1992), 81–93.

20 Press release dated 20 July 2017, category 'Restauration,' Musée du Louvre. It is interesting to note that the masterpiece was restored with the financial assistance of the Manufacture horlogère Vacheron Constantin.

21 The field of clockmaking is particularly sensitive to this question. René Beguin, for example, mentions that the "restoration of clocks ought to be conducted in the same spirit as any other work of art restoration." The final destination of a clock (whether it lands in a private collection or a public museum) requires a serious study in order to determine the best possible restoration options. That said, according to Beguin, any type of action going

FIGURE 14 Pendulum clock known as *La Création du monde*, 1754
We owe to the *ingénieur-mécanicien du roi* Claude-Siméon Passemant the design of the clockwork mechanism. It was executed by the clockmaker Joseph-Léonard Roque. The artistic décor was created by the goldsmith François-Thomas Germain. The pedestal on which the pendulum stands dates from 1834. On loan from the Musée national des châteaux de Versailles et de Trianon to the Musée du Louvre. The pendulum is exhibited with the scientific instruments collection integrated to the artworks Department. Musée du Louvre, VMB1036-1; VMB1036-2
PHOTO © RMN-GRAND PALAIS (CHÂTEAU DE VERSAILLES) / FRANCK RAUX

The repair/restoration of *The Creation of the World* pendulum largely conforms to the principles developed and practiced beginning in the early 1980s by the late instrument expert Paolo Brenni. For over thirty-five years, the scientific instrument historian and restorer successfully brought back to life the physics cabinet of the Istituto tecnico toscano (opened in 1857), now under the auspices of the Fundazione scienza e tecnica in Florence.[22] Of the Fundazione's circa 2,000 physics devices from the latter half of the nineteenth century, about 50% required only cleaning and minimal intervention to exhibit them. Another quarter necessitated a more thorough conservation-restoration, while the rest was heavily damaged. Several of these were 'dislocated' (*smembrati*), with parts scattered all over the building. In a laboratory specifically dedicated to the restoration of the Fundazione's instruments, Brenni and Anna Giatti have over the past decades systematically disassembled, cleaned, repaired, and reassembled all the instruments of the collection. In most cases it was feasible and meaningful to make them operational again (as seen on the Fundazione's YouTube channel showcasing corresponding lecture demonstrations).[23] Brenni and Giatti acknowledged that bringing the instruments back to working order was carried out only when the restoration remained non-invasive, since the chief goal was to preserve these objects of historical value—not to repair them: "un restauro inteso soprattuto a preservare (e non a riparare!) degli oggetti di interesse storico." Between 2004 and 2007, the room where the instruments were stored and displayed originally was refurbished and opened to the public.[24]

Contrary to some who would prefer that human intervention on science and technological objects remained parsimonious and kept to a strict minimum,[25] Brenni opts for a conservation-restoration (and repair when necessary) that allows an instrument to regain its luster lost through time—too often by pure negligence. When, for example, he tackled the restoration of a major

against a recognized practice is nothing short of vandalism. René Beguin, *Restauration des horloges, montres et pendules* (Fribourg: Office du livre S. A., 1979).

22 The Fundazione website: http://www.fstfirenze.it/gabinetto-di-fisica/ (accessed 3 April 2020).

23 YouTube channel: https://www.youtube.com/user/florencefst?feature=watch (accessed 3 April 2020). See also Paolo Brenni, "Filming Nineteenth-Century Physics Demonstrations with Historical Instruments," in *Historical Scientific Instruments in Contemporary Education*, eds. Elizabeth Cavicchi and Peter Heering (Leiden: Brill, 2022), 34–49.

24 Paolo Brenni, *Il gabinetto di fisica dell'Istituto tecnico toscano. Guida alla visita* (Florence: Edizioni Polistampa, 2009), 28–9 (for the instruments' restoration).

25 Hermann Kühn, "The Restoration of Historic Technological Artefacts, Scientific Instruments and Tools," *The International Journal of Museum Management and Curatorship* 8 (1989): 389–405, where he writes that "The maxim of restoration should be to reduce all measures to the absolute minimum necessary." (p. 405)

meteorological device—Father Angelo Secchi's meteorograph—several lost parts were reconstructed not only to bring back the device's original configuration, but also to facilitate the performance of lecture demonstrations. Brenni clarified, however, that any urge to 'invention or interpretation' was avoided. Missing parts were recreated only when a solid documentation existed, allowing complete knowledge of their exact shape, size, and characteristics. Sometimes old recipes or nineteenth-century craft methods (the preparation of varnish, for instance) were put to good use during the conservation-restoration of an instrument. To Brenni, it is not an attempt at replicating old technique, but simply a fact that they have proven to be effective in the past.[26] More recently, working on a refractor built by Joseph von Fraunhofer and delivered to the Observatory of Tartu (Estonia) in 1824 (Fig. 15), Brenni and his team deployed a similar know-how in restoring the instrument: the old and chipping varnish on the large wooden beams supporting the telescope was reconditioned; some missing mahogany fragments were replaced to fill the gaps; the parts made in brass and bronze, much oxidized, were cleaned and revarnished (unfortunately, the original nineteenth-century lacquer was removed a decade before); only the telescope optics, properly cleaned in 1993, did not require any treatment.[27]

The school of thought supported by Brenni is similar to the conservation and restoration practices adopted for musical instruments, whose function as sound producers also grants them aesthetic and epistemic values when their operation is brought back to life.[28] What Brenni proposes is in the middle of a spectrum advocating, at one end, repair at any cost restoring the operation of a device and, at the other end, a circumscribed and parsimonious conservation

---

26 Paolo Brenni, "Strumenti scientifici: il restauro del meteorografo Secchi," *OPD Restauro* 5 (1993): 99–107. On the history and use of this instrument, Brenni, "Il meteorografo di padre Angelo Secchi," *Nuncius* 8 (1993): 197–247. When historical and technical information is lacking, it is better not to attempt bringing the object back to working order, a decision Brenni and his colleagues made restoring an air pump. Laura Serra, Anna Giatti and Paolo Brenni, "The Pneumatic Pump of the Mariano College in Bergamo," in *Atti del XXXVI convegno annuale SISFA* (Naples, 2006), ed. Salvatore Esposito (Pavia: Pavia University Press, 2017), 371–9. Brenni, "Better than New? Scientific Instrument Restoration in Italy," in *Proceedings of the Workshop Held in Florence, December 14–15, 1998*, ed. Laura Manetti (Florence: Le Lettere, 2000), 51–6. In French, see Brenni, "Quelques considérations sur la restauration des instruments scientifiques," *CoRé* 17 (2006): 8–11.

27 Paolo Brenni, "The Fraunhofer's Refractor of Tartu (Dorpstat) and Its Restoration," *Bulletin of the Scientific Instrument Society* 113 (2012): 2–7.

28 Robert Barclay, "Instruments, Conservation, Restoration, Copying of." Grove Music Online, 2001 https://www-oxfordmusiconline-com.ezp-prod1.hul.harvard.edu/grovemusic/view/10.1093/gmo/9781561592630.001.0001/omo-9781561592630-e-0000046526 (accessed 12 September 2022). Stewart Pollens, *The Manual of Musical Instrument Conservation* (Cambridge: Cambridge University Press, 2015).

FIGURE 15
Fraunhofer's great refractor as seen after its final restoration. It is on exhibit at the Tartu Old Observatory, which was completely renovated as well. University of Tartu Museum, AM_11701:1 AjAM
PHOTO: ANDRES TENNUS / UNIVERSITY OF TARTU

that fetishizes the dust of time and other dubious historical traces (often simple dirt, resulting from conservation neglect). This middle-of-the-road stance is widely accepted and recognized today by those who are determined to safeguard the more recent scientific and technological objects, particularly those of the nineteenth and twentieth centuries. We find it, for example, in the conservation-restoration of the Gautier meridian circle of the Rio de Janeiro Observatory and in the public display project of the Cockcroft-Walton particle generator formerly built for the Institut de physique nucléaire de Lyon.[29] In

29 Marcus Granato and Françoise Le Guet Tully, "Les principes de la restauration d'instruments scientifiques: le cas du cercle méridien Gautier de l'Observatoire de Rio de Janeiro," *In Situ* [online], 10 (2009), http://journals.openedition.org/insitu/3865 (accessed 19 February 2019). See also Françoise Le Guet Tully, Joël Françoise, William Mourey, Jacques Rebière and Jean Davoigneau, "L'Observatoire de Nice Côte d'Azur, site de Nice. Projets et réalisations," in *Métal. Actes de la conférence internationale sur la conservation des métaux*, eds. William Mourey and Luc Robbiola (London: James & James, 1998), 22–5. Rémy Geindreau, "Étude préalable à la restauration d'un objet technico-scientifique contemporain. Le générateur

this last case, the hulking size, the composite materials employed to manufacture it, the incomplete and dispersed documentation, and the complexity of its operation compels one to rethink the overall value of such a scientific instrument, including its heritage and public display. Currently stored in fragments, it will be viewable to the public once the conservation-restoration project is completed.[30]

2 Reconstruction and Restoration: Abbé Nollet's Scientific Instruments

To broaden our exploration of conservation-restoration work on scientific instruments, one could ask the following question: what should be the conservation-restoration intellectual and experiential investment given to a reconstructed scientific instrument?[31] For one thing, it is not usually considered 'authentic' (we will explain what this term means in the next section). It is recreated according to past knowledge with techniques often long forgotten (but not always); the goal of a reconstruction is to recover this historical savoir-faire. But what if the reconstructed instrument originated in the late Enlightenment? Does it change anything in our conservation-restoration approach? Should a historical replica be treated as a genuine and documented instrument? To investigate these issues, let me turn to a collection I know well. In the early 2000s, the Stewart Museum in Montréal sent seven instruments from its Abbé Nollet collection to the Centre de conservation du Québec (CCQ), a world-class institution for conservation treatment. Some items of this collection, bought in 1983 by David M. Stewart from the founder of the Harvard Collection of Historical Scientific Instruments (CHSI), David P. Wheatland, were manufactured in the late eighteenth century for the Collège des Godrans, in Dijon (23 instruments).[32] Abbé Jean-Antoine Nollet (1700–1770) was an

Cockcroft-Walton de l'Institut de Physique Nucléaire de Lyon," *CeROArt* [online], EGG 5 (2016), http://journals.openedition.org/ceroart/4933 (accessed 19 February 2019).

30 Séverine Derolez, "La patrimonialisation des objets scientifiques contemporains et leurs contextes de valorisation: Cas de l'accélérateur de particules Cockroft-Walton," *Culture & Musées* 29 (2017): 147–50.

31 My definition of a reconstructed instrument is based on Klaus Staubermann, ed., *Reconstructions: Recreating Science and Technology of the Past* (Edinburgh: National Museums of Scotland, 2011). For some specific context, Deborah Warner, "Replicating Historic Scientific Instruments," *Bulletin of the Scientific Instrument Society* 151 (2021): 10–13.

32 Jacques Dubois, "An Eighteenth-Century Collection of Scientific Instruments at Dijon," in *The Art of Teaching Physics: The Eighteenth-Century Demonstration Apparatus of Jean-Antoine Nollet*, eds. Lewis Pyenson and Jean-François Gauvin (Sillery, Qc: Les Éditions

illustrious French Enlightenment savant.[33] A protégé of both Charles-François de Cisternai Dufay and René-Antoine Ferchault de Réaumur, Nollet was a prototypical 'artiste' early in life—an artisan with 'esprit,' as described by Paola Bertucci—elected to the Société des Arts in 1728.[34] A singular mind, gifted for mechanics, he built for Réaumur and manufactured for the wider public hundreds of so-called Réaumur thermometers beginning in 1730.[35] Nollet's election in 1739 to the Académie royale des sciences in Paris, which secured his status as a respected natural philosopher, did not interrupt a lifelong and fruitful occupation as a popular lecture demonstrator and a highly regarded instrument maker—providing sumptuous physics cabinets to enlightened and wealthy patrons, colleges and military schools, and the monarchy.[36] The artisanal entrepreneurship Nollet established was financially profitable[37] and so meaningful he transformed it into a lasting heritage: a three-volume work meticulously describing the manufacture of his uniquely designed instruments—such as the ones made for the Collège des Godrans mentioned above.

Dedicated to the Dauphin, heir to the throne of France, and the future Louis XVI, *L'Art des expériences* (1770) is an instructional work based on the principle that experimental physics cannot improve without proper instrumentation. As imparted to the Dauphin, Nollet believed learning to build

du Septentrion, 2002), 99–118. Dubois found a total of 39 instruments from Dijon, two smaller collections now preserved in Munich and Tours. On David P. Wheatland and the founding of the Collection of Historical Scientific Instruments (CHSI), William J. H. Andrewes, "The Life and Work of David Pingree Wheatland (1898–1993)," *Journal of the History of Collections* 7 (1995): 261–8 and Andrewes, "The Legacy of David Wheatland," *Nuncius* 16 (2001): 687–701.

33 One of the best books describing Nollet's scientific contributions remains John Heilbron, *Electricity in the 17th and 18th Centuries: A Study of Early Modern Physics* (Berkeley: University of California Press, 1979).

34 Paola Bertucci, *Artisanal Enlightenment: Science and the Mechanical Arts in Old Regime France* (New Haven: Yale University Press, 2017), 104–42 for a description of the Société des arts; pp. 181–93 about Nollet.

35 Jean-François Gauvin, "The Instrument That Never Was: Inventing, Manufacturing, and Branding Réaumur's Thermometer During the Enlightenment," *Annals of Science* 69 (2012): 515–49.

36 Regarding the purchase of Nollet's instruments by individuals, see Jean-François Gauvin, "Eighteenth-Century Entrepreneur: Excerpts from the Correspondence Between Jean-Antoine Nollet, Étienne-François Dutour and Jean-Jallabert, 1739–1768," in *The Art of Teaching Physics*, 47–69.

37 According to Nollet's after death inventory (*inventaire après-décès*), everything he owned was worth more than 20,000 livres tournois. Nollet died in his house, rue Hautefeuille, in Paris, on 25 April 1770. Archives nationales, Minutier central, Minutes et répertoires du notaire Charles Le Brun, 7 mars 1761–15 juin 1787, cote MC/ET/XVII/935 (14 mai 1770).

scientific instruments was as valuable as knowing how to conduct experiments with them:

> It is the description of all these instruments that I have displayed in front of you [the Dauphin] over the past ten years, *and that you have enjoyed disassembling and rebuilding to understand their mechanism*, and with which you have seen me make all these experiments that led you to understand natural phenomena and their causes.[38]

It was after noticing the difficulty "new schools of physics, spawning in universities all over the kingdom," had in acquiring proper instrumentation to deliver lecture demonstrations that Nollet committed to write this work. He vowed that

> anyone who has watched or had made [by someone else] everything I included in my *Avis* will be able to execute himself, or have made by capable craftsmen [*ouvriers un peu intelligents & passablement adroits*], most machines described in the *Mémoires* of the Academy of Science, in s'Gravesande's and Désaguilliers's [sic] physics, etc. There are few experiments he won't be able to perform successfully.[39]

L'Art des expériences forms, in a way, the rhetorical culmination of the artisan with 'esprit,' the *artiste*, at the center of Nollet's educational approach and scientific practice since his introduction to the Société des arts in the late 1720s. Already in 1735, in his first published work, Nollet developed a dissemination strategy, which positioned him advantageously vis-à-vis ordinary skilled workers. In an *Avertissement*, a preamble announcing a *catalogue raisonné* for the

---

38 Jean-Antoine Nollet, *L'Art des expériences, ou Avis aux amateurs de la physique sur le choix, la construction et l'usage des instrumens* ..., 3 vols. (Paris: Durand neveu, 1770), vol. 1, iv (traduction and emphasis are mine). This type of court teaching through experimental physics was not unique to France. At about the same time, the young English king, George III, received a similar education thanks to the acquisition of a large physics cabinet provided by the instrument maker George Adams. Florence Grant, "Mechanical Experiments as Moral Exercise in the Education of George III," *The British Journal for the History of Science* 48 (2015): 195–212. This collection is thoroughly documented in Alan Q. Morton and Jane A. Wess, *Public & Private Science: The King George III Collection* (Oxford: Oxford University Press, 1993). This type of scientific education contrasts heavily with the humanist one provided during the Renaissance to the English crown. Aysha Pollnitz, *Princely Education in Early Modern Britain* (Cambridge: Cambridge University Press, 2015).

39 Nollet, *L'Art des expériences*, xiij.

345 instruments he could manufacture, Nollet explained that the care taken in crafting precision instruments must necessarily go through the hands of "excellent *artistes*, though this sort of individual is rare." It is even more consequential when you consider that

> any machine is subject to breakage [*sujet à se déranger*]; use modifies parts that then require repairs; the latter necessitate various degrees of attention and abilities, according to the instrument's complexity; and since intellectual sagacity and hands-on skills are two independent endowments [*qualités*], the most intelligent man could be stopped in his study if his instruments required too frequent or difficult repairs.[40]

Nollet was the ideal type *artiste*, one who combined intellectual sagacity with hands-on skills. The manufacture and repair of scientific instruments remained throughout Nollet's life an essential component—and a constant preoccupation—of his scholarly and popular program of experimental physics.

The Stewart Museum's abbé Nollet collection is likely composed of instruments made 'according to' Nollet, i.e., they were 'reconstructed' following instructions from *L'Art des expériences*. Some instruments in the collection, however, were not designed and described by Nollet, but by one of his successors, Joseph-Aignan Sigaud de Lafond (1730–1810).[41] From a payment he received in 1776, we know that Lafond's contributions, a long-time scientific advisor to the Collège des Godrans, were much valued: "one hundred bottles of fine, aged wine shall be sent to Mr. Sigaud de la Fond to thank him for his help in providing the physics cabinet with the necessary machines."[42] What interest me is how these various 'according to' Nollet instruments were reconstructed: did they follow Nollet's prescriptions *à la lettre*? A thorough examination of three instruments provides an opportunity to do just that. In this selection three types of instruments are examined: An instrument that did not come from the Collège des Godrans (Fig. 16); an instrument from the Collège des Godrans described by Nollet (Fig. 17); finally, an instrument from the Collège

40 Jean-Antoine Nollet, *Programme, ou Idée générale d'un cours de physique expérimentale, avec un Catalogue raisonné des instrumens qui servent aux expériences* (Paris: P. G. Le Mercier, 1738), 114. Bertucci, *Artisanal Enlightenment*, 168.

41 The complete inventory of Nollet's apparatuses known in the world is listed in Pyenson and Gauvin, eds., *The Art of Teaching Physics*, 119–89. Sigaud de Lafond, *Description et usage d'un cabinet de physique expérimentale*, 2 vols. (Paris: P. Fr. Gueffier, 1775).

42 Cited in Dubois, "An Eighteenth-Century Collection of Scientific Instruments at Dijon," 104.

des Godrans inspired by Sigaud de Lafond. (Fig. 18) From the conservation analyses provided by the Centre de conservation du Québec (CCQ), it is fairly safe to claim these three apparatuses were reconstructed following Nollet's carefully written instructions. They fulfilled his wish that *L'Art des expériences* be used to teach specialized craftsmen on making and fixing instruments, who would then equip schools in provincial towns with an appropriate physics cabinet.

We note that the CCQ did not 'repair' the Stewart Museum's instruments, i.e., it construed these instruments as historical artifacts and not reconstructed replicas or facsimiles.[43] It has abided by the conventional conservation-restoration protocols (deontological code) defined and refined by the profession over the last 60 years. One can ask, though: why not take this opportunity to learn more about the (re)construction of these historical objects? How well did they follow Nollet's exceptional written descriptions to learn more about late eighteenth-century manufacturing techniques? Nollet himself said that repairing instruments should be at the center of a healthy experimental physics, one that demonstrates the alliance of "intellectual sagacity and hands-on skills." Would we have lost much of the 'historical traces,' of the objects' 'historical identity'—authenticity, as defined by Jean-Pierre Cometti, see *infra*—if we had dared reconstruct these reconstructed instruments according to Nollet's own prescriptions in *L'Art des expériences*? Repair (or reconstruction), restoration or conservation, this is the existential question faced by curators and conservators when dealing with a scientific or technological apparatus in a 'state of disrepair.' The following table is a comparison of the CCQ's interventions, or standardized conservation practices, and Nollet's precise construction guidance and recommendations described in *L'Art des expériences*.

43 Replicas and facsimiles have been in the mind of instrument experts for a long time, see Allan Mills, "Facsimiles: Benefits and Worries," *Bulletin of the Scientific Instrument Society* 25 (1990): 1.

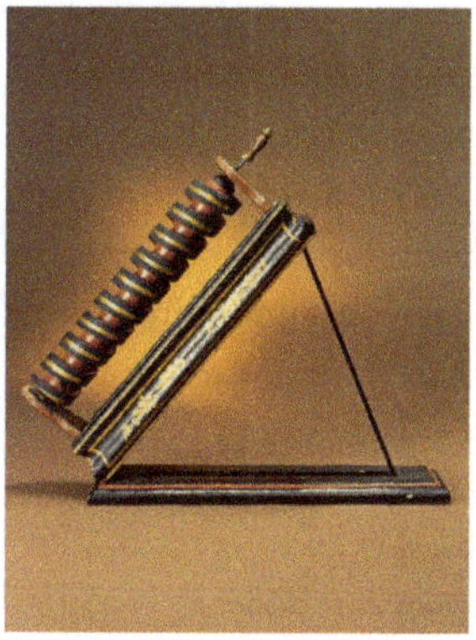

Figure 16
Archimedean screw

NOT from the Collège des Godrans, Dijon, but Nollet-style

Figure 17
Artificial eye

From the Collège des Godrans, Dijon, Nollet-style

Figure 18
Connecting vessels

From the Collège des Godrans, Dijon, but Sigaud de Lafond-style

Under ultraviolet light, the CCQ found on all three instruments a yellow fluorescence, which marks the presence of a varnish containing linseed oil. This is not surprising since one of Nollet's chief formulae for applying color on instruments is to use an "oily varnish [*vernis gras*], suitable for absorbing colors to paint metals." Nollet explains how to make this oily varnish: "In a vessel add two ounces of Venice turpentine, half a pound of karabé or succin [*yellow amber*], & as much of copal gum, both crushed to tiny pieces but not pulverized; *you then add ten ounces of linseed oil* & you stir all the ingredients with an iron spatula after heating on a mediocre fire."[a] We also note that Nollet proposes a varnish based on spirit of wine, rather than linseed oil, to paint on wooden surfaces. If we are dealing here with colors based on an oily varnish, it would have been useful to test other instruments from the Stewart Museum's collection to see how often this occurrence is repeated. The CCQ also established that the varnish on the Archimedean screw is old and dates to before the attack of wood-eating insects. Furthermore, on this same instrument, the screw doesn't show any sign of fluorescence.[b]

a Nollet, *L'Art des expériences*, vol. 1, 451. (emphasis mine) It is interesting to note that as soon as 1720, Filippo Bonanni recommends replacing spirit of wine, turpentine, and fir oil from older and conventional varnish recipes, by linseed oil. Bonanni, *Trattato sopra la vernice detta comunemente cinese* (Rome, 1720). I don't know whether Nollet knew this author and work, though Chinese lacquer was at the center of Nollet's instrument decoration.

b Rapport de restauration du Centre de conservation du Québec (CCQ), number AE-97-03, dated September 2000. Musée McCord Stewart, dossier abbé Nollet. I would like to thank Mrs. Sylvie Dauphin, then collection curator (now retired), for sharing this information with me.

(*cont.*)

Features of the CCQ *conservation:* (1) The general imperfections and tiny holes left by insects were filled with a Lepage sealer [*bouche-pores*] "because it is soluble in water even months after it has hardened." (2) The missing décor was copied from another instrument and added using a stencil. (3) The area was re-tinted with watercolors and the protective varnish was of the Regal Rez 20% type.[c]	*Features of the* CCQ *conservation:* (1) A fissure on the optical tube was filled with a piece of linden cut to size and coated with fish glue. (2) A careful microscopic examination revealed the following sequence of material layers: wood, beige preparation, black lacquer, varnish. Another sequence reads: wood, beige preparation, black lacquer, ocher-red preparation, gold leaf, varnish.[d]	*Features of the* CCQ *conservation:* (1) A stain along one of the glass tubes was mechanically removed. It uncovered some white paint. It was described as an overcoat of varnish. (2) Watercolors were used to re-tint the cleared area. (3) The protective varnish added was of the Regal Rez 40% type, soluble in Stoddard. Several tests were carried out to reproduce the glossy lacquer layer on the back of the instruments, unsuccessfully.[e]
'Reconstruction' suggested according to Nollet: [*numbers in parentheses below are connected to the ones above*] (1) On wood, which requires a coat of varnish, Nollet proposes to plug the 'small cavities' with a "putty made with strong glue & ground chalk." (2) The use of copper foil, as suggested by Nollet for instrument embellishments, would have been preferable. (3) Nollet describes in great details how to prepare and paint the reds and blacks.	*'Reconstruction' suggested according to Nollet:* [*numbers in parentheses below are connected to the ones above*] (1) For their low price and ease of cutting, Nollet recommends the following types of wood: the alder, lime tree, walnut, pear tree, Dutch oak, sorb tree and checker tree. (2) The sequence of material layers discovered under microscopic observation is correct. Nollet favored the choice of reds and blacks for his instruments. Note, however, that the so-called 'beige preparation'	*'Reconstruction' suggested according to Nollet:* [*numbers in parentheses below are connected to the ones above*] (1) Nollet warns of the appropriate way of spreading evenly, without spill, paint and varnish. To maintain the natural whiteness of paper (when used) or white paint, in our case, Nollet suggests using the least colored varnish possible, applying 'the cleanest and whitest' sandarac to small bits of putty. This mix should be

c Ibid.
d Ibid.
e Ibid.

(*cont.*)

For the protective varnish, testing of oily and spirit-of-wine lacquers would have been necessary.[f]	layer is unknown (not described by Nollet). Lastly, instead of gold leaves the abbé recommended copper foil (or *German gold*), bought in booklets from "Quinquaillers & les Marchands de couleurs"; these copper foils were plated with a *mordant* (a preparation absorbed by the wood before it is tainted). The 'ocher' red is based on vermilion.[g]	preferred to that composed of shellac. (2) Why not try using the colors suggested by Nollet? Beyond the reds and blacks, he provides his reader with a series of materials used in producing all sorts of colors. (3) For the protective varnish, testing of oily and spirit-of-wine lacquers would have been necessary.[h]

f Nollet, *L'Art des expériences*, vol. 1, 493, 476–80, 465–6 et 450–56.
g Nollet, *L'Art des expériences*, vol. 1, 3–4 et 465–500.
h Nollet, *L'Art des expériences*, vol. 1, 456, 498–9 et 450–56.

These late-eighteenth-century reconstructions (replicas) of Nollet instruments were regarded by the CCQ as genuine, 'authentic' museum objects, requiring a standard conservation-restoration treatment any other historical object would deserve. Though not a problem per se, was it perhaps a missed opportunity? Instead of commissioning formal conservation care, why not consent to a thorough and well-documented historical reconstruction—one might say a restoration? Nollet's *L'Art des expériences* provides everything we need to contemplate an undertaking often called experimental archaeology. Under this umbrella, the modern reconstruction of Nollet's late Enlightenment reconstructed apparatuses would become a very specific—idiosyncratic—form of conservation-restoration. According to Klaus Staubermann, reconstructing instruments provides, "a new and deeper understanding of the skills, practices and tacit knowledge involved in the making and use of artifacts." The reproduction of original objects becomes "probes in historic skills, performance and culture."[44] Under such considerations, what were the lasting historical and epistemological lessons acquired from conducting the CCQ's 'standard' conservation-restoration on the seven reconstructed Nollet instruments? If nothing of significance came out of it, could the careful reconstruction of reconstructed instruments have been a better solution? How,

44 Staubermann, *Reconstructions*, vi and viii.

moreover, can such a method help us identify fakes and imitations?[45] In other words, could we treat the reconstruction of reconstructed instruments as an adapted—customized—form of conservation-restoration? If one agrees with the ICOM definition of *restoration* spelled out above, the answer would be yes.

It is important to reiterate here that more conversations should be taking place between the field of conservation-restoration, science museums, and the experiential method called 'experimental history of science,' for lack of a better expression. For almost thirty years, numerous scholars have replicated instruments and reworked experiments to great heuristic and intellectual results. These reproductions and reworkings, "which aimed at understanding more deeply the historical actors, their experiments and ideas and their origins," have laid the foundation to a transformative encounter with past practices and scientific knowledge. According to proponents of this way of investigating history of science, experimental reproductions "can help bridge the unavoidable gap that exists between the actions and ideas of historical actors and the textual descriptions or artifactual residues of those actions and ideas that have come down to us."[46] In recent years, for instance, Pamela Smith's major Making and Knowing Project at Columbia University, Sven Dupré's interdisciplinary ARTECHNE: Technique in the Arts 1500–1950 at Utrecht University and University of Amsterdam, William Newman's decades-long reworking of Newton alchemical experiments, Peter Heering and Cibelle Celestino Silva's reevaluation of eighteenth-century Leiden jar operational accounts, and Tom Everett's reconstruction and reworking of the ear phonautograph, originally built in 1874 by Alexander Graham Bell and Clarence J. Blake, have opened

45 Boris Jardine, Joshua Nall, and James Hyslop, "More than Mensing? Revisiting the Question of Fake Scientific Instruments," *Bulletin of the Scientific Instrument Society* 132 (2017): 22–9. Boris Jardine, "Like a Bos: The Discovery of Fake Antique Scientific Instruments at the Whipple Museum," in *The Whipple Museum of the History of Science. Objects and Investigations, to Celebrate the 75th anniversary of R. S. Whipple's Gift to the University of Cambridge*, eds. Joshua Nall, Liba Taub, and Frances Willmoth (Cambridge: Cambridge University Press, 2019), 201–22. Jardine explains Derek Price's study of antique fakes as a move from visibility to legibility, from the collection to international databases. See also Peter de Clercq, ed. *Scientific Instruments: Original and Imitations. The Mensing Collection. Proceedings of a Symposium, Held at the Museum Boerhaave, Leiden, 15–16 October 1999* (Leiden: Museum Boerhaave, 2000).

46 Hjalmar Fors, Lawrence M. Principe, and Otto Sibum, "From the Library to the Laboratory and Back Again: Experiment as a Tool for Historians of Science," *Ambix* 63 (2016): 85–97, quotes on pp. 88–9. This article is a very good summary of this approach, providing a nice bibliography of the most relevant projects up to 2015. For an early discussion of this method, Dominique Pestre, "La pratique de reconstitution des expériences historiques, une toute première réflexion," in *Restating Coulomb. Usages, controverses et réplications autour de la balance de torsion*, eds. Christine Blondel and Matthias Dörries (Florence: Leo S. Olschki, 1994), 17–30.

up perspectives for a fruitful exchange between textual knowledge, hands-on practices, and museum artifacts.[47] Mirroring these examples, reconstructing abbé Nollet's instruments according to *L'Art des expériences* and, wielding the latter, reworking experiments epitomized in his *Leçons de physique expérimentale* would bring new questions and levels of understanding to the devices studied at the CCQ under the restrictive modus operandi of conventional conservation-restoration. By focusing our scholarly attention toward the materiality and experiential behavior of these demonstration machines, we would likely recover what Simon Schaffer so accurately rationalized based on original textual analysis, i.e., the "apparent authority of these devices relies on the gestures performed by the demonstrator and the cultural resources available within the setting of the demonstration. These gestures and resources draw attention away from the workings of the machine and toward the philosophy which the machine is designed to demonstrate."[48] I strongly believe that the philosophy of conservation-restoration need not "draw attention away from the workings of the machine." To flesh out this statement, I follow the intriguing and persuasive teachings of the late philosopher of art Jean-Pierre Cometti.

3 Identity, Integrity, Authenticity: between the Unit of the Total and the Unit of the Whole

In concluding his historical and conservation analysis of the old Cockcroft-Walton particle generator in Lyon, Rémy Geindreau argues that a conservator, holding on too closely to the Brandian tradition, "will be marginalized and

47 Pamela H. Smith and The Making and Knowing Project, "Historians in the Laboratory: Reconstruction of Renaissance Art and Technology in the Making and Knowing Project," *Art History* 39 (2016): 210–33. The online project is found here: https://www.makingandknowing.org/ (accessed 28 May 2020). Smith, *From Lived Experience to the Written Word. Reconstructing Practical Knowledge in the Early Modern World* (Chicago: The University of Chicago Press, 2022). On the pedagogical role of this initiative, Tianna Helena Uchacz, "Reconstructing Early Modern Artisanal Epistemologies and an 'Undisciplined' Mode of Inquiry," *Isis* 111 (2020): 606–13. Sven Dupré's online project: https://artechne.wp.hum.uu.nl/ (accessed 28 May 2020). William Newman, *Newton the Alchemist: Science, Enigma, and the Quest for Nature's 'Secret Fire'* (Princeton: Princeton University Press, 2019). Peter Heering and Cibelle Celestino Silva, "Re-Examining the Early History of the Leiden Jar: Stabilization and Variation in Transforming a Phenomenon Into a Fact," *History of Science* 56 (2018): 314–42. Tom Everett, "Writing Sound with a Human Ear: Reconstructing Bell and Blake's 1874 Ear Phonautograph," *Science Museum Group Journal* 12 (2019), online: http://journal.sciencemuseum.ac.uk/browse/issue-12/writing-sound-with-a-human-ear/ (accessed 28 May 2020).

48 Simon Schaffer, "Machine Philosophy: Demonstration Devices in Georgian Mechanics," *Osiris*, 2nd Series 9 (1994): 157–82, quote on p. 162.

misunderstood" if s/he does not try "to endorse the questions raised by the new conservation-restoration theories" and practices upheld by Paolo Brenni, for instance, regarding scientific instruments.[49] Cesare Brandi is often seen as the founder of the modern conservation-restoration theory. By establishing the Istituto centrale per il restauro (ICR) in Rome in 1939, Brandi laid the groundwork for the field's current professionalization. In the first page of his influential *Teoria del restauro* (1963) Brandi categorically disconnects 'repair' (*risarcimento*) from 'conservation' (*restauro*). Whether, in the former case, the goal is to bring the industrial object (built by artisans or mass-produced) back to its original working conditions (*restituzione in pristino*), in the latter the recovery of the artwork's original function "even if it is part of the restoration, represents ultimately but a secondary or concomitant component, never the primary and fundamental one that considers the work of art as a work of art."[50] From this conclusion naturally derives an ontology divided into two classes: the organic or industrial object (*organico-funzionale*) as a 'unit of the total' (*unità del totale*) and the artwork as a 'unit of the whole' (*unità dell'intero*). For the latter, Brandi argues it is critical to recognize the work of art as such (*"riconoscimento dell'opera d'arte come tale"*) insofar as the restoration aims to recover the potential unity of the work of art (*"ristabilimento dell'unità potenziale dell'opera d'arte"*). The artwork is not composed of distinct parts—the sum of its parts forming a total; its form is conceived rather as indivisible, making up a unique figurative and constitutive unit of a whole: the image *per se*. The impact on the field of conservation-restoration is twofold following this consideration: 1) an artwork physically damaged continues "to exist *potentially* as a whole in each of its fragments" (*dovrà continuare a sussistere potenzialmente come un tutto in ciascuno dei suoi frammenti*); 2) if the same artwork appears to have been materially divided, "we have to search for the original potential unity (*potenziale unità*) contained in each of the fragments, proportionally to the initial form that survives in them."[51] From the concept of 'unit of the whole,' we deduce the three chief conservation-restoration's deontological principles, i.e., *stability* in the choice of material employed on the artwork

49 Geindreau, "Étude préalable à la restauration d'un objet technico-scientifique contemporain," 18. An excellent example of this 'forced' hybridization for science and technology objects is given by Agnès Mirambet-Paris and François Mirambet, "La conservation-restauration du patrimoine technique et industriel dans le cadre de la loi sur les Musées de France, une mission impossible?," *La Lettre de l'OCIM*, 135 (2011): 27–35.

50 Cesare Brandi, *Teoria del restauro* (Turin: Piccola Biblioteca Einaudi, 1977), 4.

51 Ibid., 16–7. See also Cesare Brandi, *Il restoro. Teoria e pratica*, ed. Michele Cordaro (Rome: Editori Riuniti, 1994).

being restored, *reversibility* of actions taken with the latter, and *legibility* of an artwork from a historical point of view.[52]

Putting scientific and technological devices back in working order does not mean ignoring all ethical principles of conservation. Elizabeth Pye, Emerita Professor of Archaeological and Museum Conservation at the University College London's Institute of Archaeology, readily recognizes that the conservation of museum objects requires a rigorous ethic, but it should not be excessively rigid—or inflexible. The operation of complex scientific apparatuses and machines benefit both museum visitors and researchers, who can learn how they work and discover the artisanal or industrial know-how that produced them. If representation is the quintessence of a painting, according to Pye, the quintessence of a mechanical device is its function. The museum must offer visitors the experience of a moving mechanism, considering it implies special conservation-restoration methods and the risk of permanent damage.[53] Though it is not complete 'heresy' to encourage repair or restoration of mechanical devices as a goal itself, it has been, nevertheless, a contentious issue for decades. The safe solution is to call for an in-depth study of the object's status and its general state of conservation. It is better if the object is part of a circumscribed and well-defined scientific and cultural project. Furthermore, it has to benefit from the expertise of an interdisciplinary team made up of historians, restorers, technicians, collection managers, designers and engineers, who will ensure detailed specifications (list of interventions and maintenance measures) and the object's full documentation.[54] But why is it so important to reconstruct the functionality of objects?

---

52 Note, however, that a restoration's reversibility is often questioned. W. A. Oddy, *Reversibility—Does It Exist?* (London: The British Museum Press, 1999), 1–5. For a historical and philosophical context of conservation-restoration, Nicholas Stanley Price, M. Kirby Talley Jr., and Alessandra Melucco Vaccaro, eds., *Historical and Philosophical Issues in the Conservation of Cultural Heritage* (Los Angeles: The Getty Conservation Institute, 1996). For a critical analysis of Brandi's work, Salvador Muñoz Viñas, "'Who is Afraid of Cesare Brandi?' Personal Reflections on the *Teoria del restauro*," *CeROArt* [Online] (June 2015), https://journals.openedition.org/ceroart/4653 (accessed 11 June 2020).

53 Elizabeth Pye, "Challenges of Conservation: Working Objects," *Science Museum Group Journal* [online journal] 6 (2006), http://journal.sciencemuseum.ac.uk/browse/issue-06/challenges-of-conservation/ (accessed 6 March 2020). A rare guide for the conservation-restoration of scientific instruments supports bringing them back in working order. The author argues that "the purpose of a collection is to show current and future generations what the objects on display were, both in their presentation and in their technical functioning." Guy Biraud, *La Restauration et la conservation des appareils scientifiques de collection* (La Rochelle: Édition Biraud, 1987), 15.

54 Bénédicte Rolland-Villemot, "Le traitement des collections industrielles et techniques, de la connaissance à la diffusion," *La Lettre de l'OCIM* 73 (2001): 13–18. George Monger,

Jean Baudrillard suggests in *Le système des objets* that when you take away the elemental attributes and basic use of an object its core identity and distinctiveness are robbed from it. The performing object then becomes a 'pure object,' an object pushed onto an aesthetic plane that no longer belongs to the practical and tangible space of functionality. Baudrillard insists that an "object no longer specified by its function is defined by the subject, but in the passionate abstractness of possession all objects are equivalent."[55] In *Malaise dans les musées*, first and foremost an *attaque en règle* against the controversial Louvre Abu Dhabi project, Jean Clair criticizes this 'passionate abstractness,' which he perceives as the over-embellishment of objects in the exhibitions of the Musée du quai Branly-Jacques Chirac (founded in 2006). He calls the display of objects qua decorative ornaments a '*reductio ad aestheticam*.' The renowned French *conservateur général du patrimoine* condemns the anthropology museum's penchant for dismissing the object's function to emphasize its form. This endeavor, "specific to all art museums," hides a fundamental reality: "only yesterday, [these objects] were primarily objects of worship."[56] The art historian Emerita Svetlana Alpers calls this very act the 'museum effect.' She argues that the museum effect is a way of seeing, or "the tendency to isolate something from its world, to offer it up for attentive looking and thus to transform it into art like our own," i.e., Western art. She finds her conclusion troubling because it means "[i]t is to ourselves, then, that we are representing things in museums." A good exhibition does not simply let visitors marvel at the beauty of objects, tuned in to their own idiosyncratic knowledge; it should provide instead the "contexts and resources that enable audiences to choose to reorganize their knowledge."[57]

"Conservation or Restoration?," *The International Journal of Museum Management and Curatorship* 7 (1988): 375–80. Philippe Tomsin, "Les patrimoines mobiliers scientifique et technique: spécificités de leur restauration, de leur conservation et de leur valorisation," *CeROArt* [online], 1 (2007), http://journals.openedition.org/ceroart/370 (accessed 4 March 2019).

55 Jean Baudrillard, *The System of Objects* (London: Verso, 2005), 92.

56 Jean Clair, *Malaise dans les musées* (Paris: Flammarion, 2007), 110. The philosopher Bernard Deloche draws a similar conclusion in Deloche, *Mythologie du musée. De l'uchronie à l'utopie* (Paris: Le Cavalier Bleu, 2010). Françoise Choay dismisses the quai Branly as a 'Disneyland of exoticism.' Choay, "Un nouveau Luna Park était-il nécessaire?," *Le Débat* 147 (2007): 57–64. On the origins of the quai Branly museum, Sally Price, *Paris Primitive. Jacques Chirac's Museum on the Quai Branly* (Chicago: The University of Chicago Press, 2007). On the emphatic role of art in our society, Lipovetsky and Serroy, *L'Esthétisation du monde*.

57 Svetlana Alpers, "The Museum as a Way of Seeing," in *Exhibiting Cultures: The Poetics and Politics of Museum Displays*, eds. Ivan Karp and Steven D. Lavine (Washington, D.C.: Smithsonian Press, 1991), 25–32, quotes on pp. 27 and 32. Ivan Karp, "Culture and

Objects have always been powerful symbols, requiring prudent dominion by ordinary folks and formal authorities. Caroline Walker Bynum, for instance, reminds us in a remarkable book that "Christian materiality was a problem, as well as an opportunity, for late medieval Christians."[58] Objects of worship were about the senses, the visceral need to be in close proximity to the sacred, in order to gain an immediate capacity of experiencing their powers:

> It was especially wall paintings and three-dimensional sculptures, in all their overt tactility, that in the fourteenth and fifteenth centuries moved, wept, sweated blood—and were later hacked to pieces by those who hated them. The enthusiasm and anxiety such objects produced had everything to do with their materiality.[59]

These images were feared and craved simultaneously, because they "were in some sense living matter." As Bynum explains, "[p]eople behaved as if images were what they represented. To materialize was to animate."[60] Conversely, smashing them meant killing their power once and for all. Such violent and destructive acts of iconoclasm, common in the early modern era, galvanized, especially during the Enlightenment, a sense of commitment toward the preservation of the past. According to Dominique Poulot, the *tabula rasa* insisted upon and decreed by French revolutionaries against the Ancien Régime compelled many officials to rethink how one defined 'art.' The complete break with tradition required by the French Revolution required that history be expunged, wiped clean from the 'art.' According to Poulot, "[t]he protection of the œuvre against vandalism awakened a patriotic cult of talent." Guarding objects crafted under past monarchic regimes demanded to postulate "a permanence of the true and the beautiful, hitherto disdained or misunderstood—and now revealed. It depended on the fact that works could be released from their history."[61] James Simpson, who delved into iconoclasm in the early modern

Representation," in *ibid.*, 11–24, quote on pp. 22–3. A recent example I consider particularly successful is the exhibit on Jean-Paul Riopelle at the Musée des beaux-arts de Montréal, Andréanne Roy, Jacques Des Roches, and Yseult Riopelle, *Riopelle: à la rencontre des territoires nordiques et des cultures autochtones* (Montréal: Musée des beaux-arts de Montréal, 2020).

58 Caroline Walker Bynum, *Christian Materiality. An Essay on Religion in Late Medieval Europe* (New York: Zone Books, 2011), 121.

59 Ibid., 125.

60 Ibid., 122 and 125.

61 Dominique Poulot, "Le musée et le patrimoine: une histoire de contextes et d'origines," in *La Muséologie, champ de théories et de pratiques*, eds. Anik Meunier and Jason Luckerhoff (Montréal: Presses de l'Université du Québec, 2012), 19–39, quotes on pp. 26–7 and 29.

Anglo-American context, likewise argues that the Enlightenment era did not fight against the iconoclastic tradition, inasmuch as adapted to it "by a massive work of image neutralization." Echoing Alpers's 'way of seeing,' Simpson proposes that the role the Enlightenment played in the end was "to drain the object of power, to delimit the 'work of art,' in this sense: to delimit the way art works on us." By getting rid of an image's history and function (regal, ceremonial, devotional) it can be transformed into an inoffensive, neutral object, ready for all to admire without worry. Sans function and purpose come freedom and art: "Focus on the form, ignore the content. Experience freedom, even as you look at religious art. The category of the aesthetic is itself, in sum, a historical product of iconoclasm."[62]

Knowing how we arrived at an almost universal *reductio ab aestheticam* does not mean we have to keep cultivating it. Why does it matter? It matters because material culture studies have repeatedly demonstrated the importance of the shifting and fluid, even unstable use of objects in all socio-cultural, historical, and scientific fields. Hide this lesson behind a veil of aestheticism, shroud it in a learned narrative about emergence and transience and the object—particularly the museum artifact—loses its multiple meanings. *Its function is destroyed.* Jean-Pierre Cometti, who taught philosophy and aesthetic, reminds us that "it is not the material and physical features that define an object—and consequently its identity—but the uses for which it lends itself."[63] In an illuminating book, Cometti proposes a fundamental ontological shift regarding the field of conservation-restoration. No longer a question of exclusively examining the material properties of objects (a strict ontology of the object), one needs to focus his or her "attention to operation and operating conditions." This shift, however, does not alter the Brandian 'unit of the whole.' The object must continue to be considered in its extensiveness. The interest shown in the object's separate components must be "subordinated to

---

See also Poulot, *Surveiller et s'instruire: La Révolution française et l'intelligence de l'héritage historique* (Oxford: Voltaire Foundation [SVEC 344], 1996).

62 James Simpson, *Under the Hammer: Iconoclasm in the Anglo-American Tradition* (Oxford: Oxford University Press, 2011), quotes on pp. 120, 130, and 133 respectively. See also Olivier Christin, "Du culte chrétien au culte de l'art. La transformation du statut de l'image (XVe–XVIIIe siècles)," *Revue d'histoire moderne et contemporaine* 49 (2002): 176–94. This image neutralization and abstraction is not novel to the Enlightenment: it was employed as well in Antiquity and during the Middle Ages. On this, see Jeffrey Abt, "The Origins of the Public Museum," in *A Companion to Museum Studies*, ed. Sharon Macdonald (Oxford: Blackwell Publishing, 2006), 115–34 and Michael Camille, *The Gothic Idol: Ideology and Image-Making in Medieval Art* (Cambridge: Cambridge University Press, 1989), esp. 78–81.

63 Jean-Pierre Cometti, *Conserver/restaurer. L'œuvre d'art à l'époque de sa préservation technique* (Paris: Gallimard, 2015), 71. All translations from this book are mine.

the interest shown in the whole as a significant totality." Here technology "has no priority," because if it were the case, that is to say if the intervention on a specific part "is conceived independently [within the framework of the *unità del totale*], where what matters is the production of an effect by all appropriate means," we would fall back—relapse—straight into the notion of repair.[64]

Starting from the precept of the unit of the whole, Cometti examines the connections federating the idea of identity, integrity, and authenticity tied to an object. He suggests that the notion of identity, which "better supports an unvarnished treatment [*traitement désauratisé*]," is preferable to that of authenticity for attesting to the practice of conservation-restoration. Citing the famous thought experiment involving the ship of Theseus, Cometti recalls that its identity never ceases to be that of the *ship of Theseus*, even when most of the wooden planks and beams composing it were removed and replaced over time. What allows us to affirm it is still and always the same ship? The "guarantee of identity or sameness, if you prefer, lies in a greater or lesser *continuity*, which is that of uses and functioning [*des usages et d'un fonctionnement*]." To avoid 'restoring by imagining,' according to the Brandian formula, it is necessary to locate in the present moment the identity of an object—the ship of Theseus at a precise time t_1 after its inaugural journey. From this derives the notions of integrity (the ontologically defined identity) and authenticity (the historically situated identity), both of which specified by the very object—its identity at an instant t_n. As a result, the "identity of a thing is presupposed by the very idea of a thing, its integrity guarantees its genuineness [*en apporte la garantie*], and the authenticity indicates the conformity of its state at a given time, for example, to a previous state assumed to be original and constitutive of its identity."[65]

Brian Castriota, also seeking to reframe conservation practices, argues that "the presumption that every work has a singular, abiding identity or core … that can be defined at the point of acquisition and maintained through time with conservation oversight is very compelling from the perspective of a

64 Ibid., 31–9.

65 Ibid., 56–74. The concept of time is crucial for Brandi : "Non ci sarà allora bisogno di maggiore insistenza per affermare che l'unico momento legittimo che si offre per l'azione di restauro è quello del presente stesso della coscienza riguardante, in cui l'opera d'arte è nell'attimo ed è presente storico, ma è anche passato e, a costo, altrimenti, di non appartenere alla coscienza umana, è nella storia." Brandi, *Teoria del restauro*, 26–7. On the metaphysics of identity and the ship of Theseus, Stéphane Ferret, *Le bateau de Thésée: Le problème de l'identité à travers le temps* (Paris: Les Éditions de Minuit, 1996).

collecting institution."[66] It is, however, frequently challenged. Like Cometti, he is critical of how authenticity has been framed over time. Castriota maintains an object's future should remain a possibility, i.e., safeguarding it should mean that "its identity may evolve, diverge, or multiply." Objects of cultural heritage are in fact 'processes,' according to Castriota, which "cannot be easily transfigured into contained and eternalised entities." It is necessary to refocus conservation-restoration activities toward a 'discursive and reflective process' that brings out an "object's multiple constructed grounds and centres, and the various means by which tokening links between a work and its instance(s) may be established and sustained."[67]

By emphasizing a philosophy of use, Cometti wants to shift the discourse of conservation-restoration toward a practice which, insofar as it respects the principle of the unit of the whole, seems to favor putting in working condition science and technology devices and machines (to name but these, at the center of our study). The author readily recognizes that the museum environment "tends to ontologize what is conceived in contexts of action modalities." Whether they are mechanical devices or 'recalcitrant' ethnological objects, "the point of view of operation [*le point de vue du fonctionnement*] leads precisely to consider things differently and, as a result, to dismiss the aesthetic point of view."[68] The door is now open for the display of objects in a state of disrepair, where function takes precedence over aesthetic beauty.

4 Reuse and Recycle: Exhibiting the DIY of Scientists and Craftsmen

The 1.5 kW Marconi transmitter (call sign 2LO, London) used by the BBC in 1922 to broadcast its first radio program is nowadays made up, according to Alison Hess, of an amalgam of electromechanical parts that were replaced, transformed, substituted, and repaired over decades of use. Hess asks the question: is it authentic, i.e., is it still the same 2LO transmitter from 1922? Exhibited since 2014 at the Science Museum in London this object is not exactly—ontologically speaking—identical to the original transmitter. Yet, if we consider the metaphor of the ship of Theseus mentioned above, its *identity* remains the same. The transmitter's identity, recalls Hess, "is therefore situated in a process of continuous change. Crafted from materials such as wood, glass, copper and

66 Brian Castriota, "Authenticity, Identity, and Essentialism: Reframing Conservation Practice," in *What is the Essence of Conservation?*, 39–48, quote on p. 41.

67 Ibid., all quotes on p. 45.

68 Cometti, *Conserver/restaurer*, 118–9.

steel, it stands at one point within the shift and flow of a world constructed of and dominated by materials."[69] The 2LO transmitter today is *authentic* in the sense that Cometti understands it, i.e., its identity is historically situated and the material integrity of the whole results from an ontologically defined identity. It is to the general principle of operation (*fonctionnement*), therefore, that one must cling to in order to determine the authenticity of an object, even more than its materiality, which is changing and malleable over time.

A specialist in heritage studies, Siân Jones agrees and proposes that object authenticity—its specific aura—comes from connections between people, places, and things more than materiality of the objects themselves. Authenticity should foremost be acknowledged "as a product of diverse, culturally specific regimes of meaning and value." Together with Thomas Yarrow, her ethnographic study of the experts in charge of conserving the Glasgow Cathedral demonstrated that curators from the Historic Scotland's Cultural Resource Team were specially attuned to preserving the 'Cathedral's fabric,' whereas masons located authenticity in their traditional stone cutting expertise, favoring more interventionist actions—reconstructions. Architects, lastly, saw the building as a conceptual whole, focusing on the overall functionality of the space rather than on individual stones (masons) or its historical fabric (curators). Authenticity, Jones and Yarrow conclude, should be understood as a 'distributed practice.' Authenticity is not inherent, but "'crafted' through different forms of expert practice."[70] Anthropologist Tim Ingold rebels, one might say, against such a far-ranging point of view, alleging that material

69 Alison Hess, "Authenticity, Alterations and Museum Objects: A Close Encounter with 2LO the BBC's First Radio Transmitter," *Journal of Material Culture* 22 (2017): 281–98, quote on pp. 282–3.

70 Siân Jones, "Negotiating Authentic Objects and Authentic Selves: Beyond the Deconstruction of Authenticity," *Journal of Material Culture* 15 (2010): 181–203. She concludes: "Authenticity is not simply a facet of the internal essence of discrete isolated entities as modernist discourses would have us believe, but rather *a product of the relationships between people and things*." (pp. 199–200, author's emphasis) The execution of this concept in the context of built heritage conservation is described in Siân Jones and Thomas Yarrow, "Crafting Authenticity: An Ethnography of Conservation Practice," *Journal of Material Culture* 18 (2013): 3–26, quote on p. 22. Authenticity has also received its share of attention in tourism research, Yvette Reisinger and Carol J. Steiner, "Reconceptualizing Object Authenticity," *Annals of Tourism Research* 33 (2006): 65–86. Boris Jardine comes to a similar conclusion in looking at forgeries and the concept of authenticity, explaining that the latter is not a "condition of the object," but a process "involving networks of objects, scholars, publics, spaces, and techniques, and as such they are subject to the forces of historical change." Jardine, "Like a Bos," 222. Gilles Lipovetsky recently revisited the notion of authenticity in an important book, *Le sacre de l'authenticité* (Paris: Gallimard, 2021).

culture studies—and not only the rather vague concept of materiality—have discarded most references to the very material, *matter*, composing objects. He asks a relevant question: "What academic perversion leads us to speak not of *materials and their properties* but of *the materiality of objects*?"[71] Elsewhere, he speaks of the "duplicity in the comprehension of the material world," i.e., how materiality is understood as both the "raw physicality of the world's 'material character'" and the "socially and historically situated agency of human beings" that gives an object its proper meaning. Though Ingold does not want to ban the use of 'materiality,' he fears the meaning given by humans to artifacts has rendered quasi-invisible the *becoming* of materials, what he describes as "their generative or regenerative potential, indeed their very life."[72]

The field of conservation-restoration, by focusing on the very matter of objects has taken a stance contrary to Ingold's. For Cometti, the materiality of an object is defined by the unit of the whole, which includes but is not limited to the matter composing the object:

> Restoration, considering what we represent as the identity and integrity of an object, is always connected to a materiality, but this materiality is only valid in accordance with attributions of meaning that have obviously nothing specially material, except that these attributions translate into forms of reification that give to materiality the properties to which we demonstrate awareness.[73]

Conservation-restoration, agrees art historian Hanna Hölling, should be seen "as an engagement with materiality, rather than material." Artworks, she continues, "just as their world, are permanently impermanent; they are characterised by relative duration, rather than by snapshots of a reality punctuated by the illusion of material authenticity."[74]

---

71 Tim Ingold, "Materials Against Materiality," *Archaeological Dialogues* 14 (2007): 1–16, quote on p. 3 (author's emphasis). The history of science has recently seen an increased interest in the study of all sorts of matter. See, for instance, Ursula Klein and Emma C. Spary, eds., *Materials and Expertise in Early Modern Europe* (Chicago: The University of Chicago Press, 2010).

72 Tim Ingold, *Making. Anthropology, Archaeology, Art and Architecture* (New York: Routledge, 2013), 27–8.

73 Cometti, *Conserver/restaurer*, 210–11. A good example in the field of science and technology is found in Ben Regel, *et al.*, "Doping at the Science Museum: The Conservation Challenge of Doped Fabric Aircraft in the *Flight* gallery," *Science Museum Group Journal* [En ligne], 6 (2006), http://journal.sciencemuseum.ac.uk/browse/issue-06/doping-at-the-science-museum/ (accessed 13 May 2020).

74 Hanna Hölling, "The *Technique* of Conservation: On Realms of Theory and Cultures of Practice," *Journal of the Institute of Conservation* 40 (2017): 87–96, quotes on p. 87 and 93.

Exhibiting broken objects, or in a *state of disrepair*, makes it possible to accentuate this dichotomy between matter and materiality. In a recent book, Simon Werrett explains how early modern savants used everyday objects to make tools, repair instruments, and carry out experiments—mainly from their homes. Such *thrifty science* highlights the immense adaptability of scientists when it comes to exploring natural phenomena. This type of do-it-yourself (DIY) connects with the notion Claude Lévi-Strauss famously described as *bricolage*, expounding on the 'closed instrumental universe' craftsmen (*bricoleurs*) inhabit. Only one rule exists in such a universe: make do with what lies around. The scientist-*bricoleur* owns "a set of tools and materials which is always finite and is also heterogeneous because what it contains bears no relation to the current project, or indeed to any particular project, but is the contingent result of all the occasions there have been to renew or enrich the stock or to maintain it with the remains of previous constructions or destructions."[75] It is what curator and museologist Glenn Adamson calls 'material intelligence,' i.e., a natural and deep intelligibility of the material world surrounding us, and a disposition—or know-how—that allows us to shape it and give it new forms.[76] Richard Sennett, similarly, coined the term 'material consciousness' to explain how people like to "invest thought in things they can change."[77] In this context, as already mentioned in chapter 1, Werrett suggests that the concept of an 'instrument' might be likened to an evolutionary process rather than a static and fixed definition: "'Instrument' should be thought of as a practice, rather than a thing." Early modern scholars 'instrumented' objects they touched, thus transforming and recycling simple household objects into tools of multiple experimental uses.[78]

Presented simultaneously, Harvard's eighteenth-century pair of globes mentioned in the introduction would allow us to tackle head-on this ordinary and trivial material journey adopted by past savants and today's scientists. By putting aside the supposedly ragged, some might say profaned, globe we weaken

75 Claude Lévi-Strauss, *The Savage Mind* (London: Weidenfeld and Nicolson, 1966), 17. For an analysis of the current age of making, Michel Lallement, *L'Âge du faire. Hacking, travail, anarchie* (Paris: Seuil, 2015).

76 Glenn Adamson, *Fewer, Better Things* (New York: Bloomsbury Publishing, 2018).

77 Richard Sennett, *The Craftsman* (New Haven: Yale University Press, 2008), 119–46, quote on p. 120.

78 Simon Werrett, *Thrifty Science. Making the Most of Materials in the History of Experiment* (Chicago: The University of Chicago Press, 2019), 82. See also Françoise Waquet, *L'Ordre matériel du savoir. Comment les savants travaillent, XVIe–XXIe siècles* (Paris: CNRS Éditions, 2015). Nowadays, beyond science, the role of repair and preventive maintenance is essential, as much for the maintenance of our economy as of our society. Stephen Graham and Nigel Thrift, "Out of Order: Understanding Repair and Maintenance," *Theory, Culture & Society* 24 (2007): 1–25.

the visitor's experience with an unfinished 'biography' of the object. Separated from its unpleasant companion, the celestial globe on exhibit, in accordance with the aesthetic expectations of museums (and, by extension, their visitors), is unable to communicate the full historical and educational trajectory of these instruments dating to the 1760s.[79] Although conservation-restoration has a crucial role to play in preserving the identity of collectibles, it should not become an obstacle or a hindrance toward the historical, educational, and epistemological contributions these objects can provide. Objects in a state of disrepair must accompany, when possible, the display of aesthetically pleasing objects for the sake of a more complete scientific discourse. By focusing the attention of museums' curators and conservators on the concept of operation (*fonctionnement*), as defined by Cometti, we eliminate the deceptively invariable and stationary nature of the object, "for a dynamic theory that integrates time, movement and therefore changes."[80] The broken and dislocated object, hidden in museums' storage rooms, seems to be a privileged vehicle to relay this message.

---

79 On the trajectory of museum objects, Samuel J. M. M. Alberti, "Objects and the Museum," *Isis* 96 (2005): 559–71.

80 Cometti, *Conserver/restaurer*, 70.

CHAPTER 6

Instrument Trajectories: Ways of Knowing the World

Decades ago, although scientific instruments had been introduced into the description and conceptualization of experiment, they were not necessarily granted a life of their own—as theory (always) and experiment (later) were.[1] Antiquarians, collectors, and instrument experts, however, recognized the crucial role played by instruments throughout the evolution—and revolution(s)—of science.[2] Historians, philosophers, and sociologists of science were slower in acknowledging the epistemic nature of instruments. In the late 1980s, Peter Galison described what was—and perhaps still is—one of the most interesting and ambitious programs of instrument research when he wrote:

> We need a history of the material culture of science, but one that is not the dead collection of discarded instruments. In its place we need a history of the way that scientists deploy objects to meet experimental goals whether or not these were set by high theory; a history of instrument-construction linked to the history of technology; a history that encompasses the relation of instruments to forms of demonstration; a history of the laboratory that tracks the development of the organization of scientific work; and a history of the embodiment of theory in hardware.[3]

Going over Galison's argument, the editors of an *Osiris* issue on scientific instruments acknowledged a few years later that "[t]he philosophical debate over whether theory drives experiment or experiment drives theory has tended to obscure the independent role of instruments in science. Instruments come and go, but not necessarily in phase with the vicissitudes of experiment

1 Ian Hacking, *Representing and Intervening: Introductory Topics in the Philosophy of Natural Science* (Cambridge: Cambridge University Press, 1983). Hans Radder, ed., *The Philosophy of Scientific Experimentation* (Pittsburgh: University of Pittsburgh Press, 2003).

2 For instance, the Scientific Instrument Commission recently celebrated its 40th anniversary. A special plenary session entitled "SIC40: Reflecting on Four Decades of Symposia and the Development of Instrument Studies" was organized during the annual meeting held in Athens in September 2022.

3 Peter Galison, "History, Philosophy, and the Central Metaphor," *Science in Context* 2 (1988): 197–212, quote on p. 211.

 | DOI:10.1163/9789004504615_011

and theory. The traditional mix of experiment and theory needs a new ingredient—instruments. It is not just a matter of getting the quantities right; we need an entirely new recipe."[4] Since then, academic chefs have prepared a variety of new scholarly dishes, giving to this *nouvelle cuisine* an instrument flavor we have slowly acquired a taste for.

Museums were concomitantly criticized for not embracing such a forward-thinking account of the material culture of science. Looking at 'scientific toys' over the *longue durée* Gerard L'E. Turner, one of the founders of the Scientific Instrument Commission in the early 1980s, questioned in his 1987 Presidential Address to the *British Society for the History of Science* whether the model of science center and its appeal to *homo ludens* was entirely satisfactory. He argued that in both the old-style science museums and the more recent science centers the "historical element is excluded," therefore concealing inter alia the crucial transmission of craft skills and the economic and social factors on which the making of scientific instruments depends.[5] Though the concept of *ludus* is useful and important, as demonstrated shortly, it can lead museums astray if kept unchecked. For sociologist of science Steven Shapin, the issue was not solely a case of historical amnesia: it was above all an inadequacy in untangling the processes by which science is made. Scientists need not be viewed as magicians, brandishing a magic wand called 'the scientific method'—which provided the solution to every scientific problem. To counteract this erroneous, and potentially dangerous, image of science Shapin believed we needed to show and explain such things as:

> the collective basis of science, which implies that no single scientist knows all of the knowledge that belong to his or her field; the ineradicable role of trust in scientific work, and the consequent vulnerability of good science to bad practices; the contingency and revisability of scientific judgment, and thus the likelihood that what is pronounced true

4 Albert van Helden and Thomas L. Hankins, eds., "Introduction," in *Instruments*, Osiris, 2nd series, 9 (1994): 1–6, quote on p. 6.

5 Gerard L'E. Turner, "Presidential Address: Scientific Toys," *The British Journal for the History of Science* 20 (1987): 377–98. A recent exhibition at the Science Museum in London demonstrates how far museums have come over the past 30 years: *Science City 1550–1800: The Linbury Gallery*. On the same topic, tackled in a more formal way, Jim Bennett and Rebekah Higgitt, "London 1600–1800: Communities of Natural Knowledge and Artificial Practices," *The British Journal for the History of Science* 52 (2019): 183–96. A collection of papers in the *Science Museum Group Journal* 15 (2021) (http://journal.sciencemuseum.ac.uk/issues/spring-2021/, accessed on 5 August 2021) is devoted to this exhibit. On science museums and education, Peter Heering, "Science Museums and Science Education," *Isis* 108 (2017): 399–406. For a general discussion on museums and history of science, see Jim Bennett, "Museums and the History of Science: Practitioner's Postscript," *Isis* 96 (2005): 602–8.

> today may, without culpability, be judged wrong tomorrow; the interpretative flexibility of scientific evidence, and the normalcy of situations in which different good-faith and competent practitioners may come to different assessments of the same evidence.[6]

In the past, science museums have struggled to make exhibitions championing these sweeping precepts. The brave curators (and administrators) who exerted themselves at the task (on rare occasions succeeding) have been caught too often in the 'politics of display,' where knowledge and power—usually understood in a Foucauldian perspective—became powerful twin forces swaying the content and meaning of exhibitions.[7]

Though it is perhaps less obvious today, except for larger institutions such as the Smithsonian Institutions in Washington, D.C., museums historically have been intricately linked to the development of science. In Germany, for instance, M. Norton Wise explains in a recent and thought-provoking book how in the first half of the nineteenth century "the pursuit of experimental science emerged from within the museum." Closely correlated to the neohumanistic ideals of Karl Friedrich Schinkel, the renowned neoclassical architect who construed that "works of art collected in museums" (he designed inter alia the Altes Museum, opened 1830) were analogous "to experiments in the making of history," the collections assembled around the founding of the University of Berlin in 1810 were soon regarded as the core elements of a developing experimental ethos useful to fields like agriculture, industry, and medicine. Embracing a neoclassical aesthetic moved by the polymorphous virtues of the muses, Schinkel's aspiration of a universal *Bildung* was incorporated to all aspects of critical thinking and knowledge. Whether for art, science or industry, museums in this context became, "so to speak, the culture medium for experimentation."[8] In England, 'museums and lecture theaters' formed 'organized ensembles.' Together, they fashioned one's scientific training. But

6 Steven Shapin, "Why the Public Ought to Understand Science-in-the-Making," *Public Understanding of Science* 1 (1992): 27–30, quote on p. 28.

7 Sharon Macdonald, ed., *The Politics of Display: Museums, Science, Culture* (London: Routledge, 1998). Museum exhibition controversies abound, including some early and very visible examples as *Birth and Breeding: The Politics of Reproduction in Modern Britain* (1994) at the Wellcome Institute for the History of Medicine in London, *The Crossroads: The End of World War II, the Atomic Bomb and the Origins of the Cold War* (never built, known as the *Enola Gay* exhibit), and *Science in American Life* (1994) at the Smithsonian Institution in Washington, D.C., respectively discussed in this volume by Ken Arnold and Thomas Gieryn. For a recent assessment, Erminia Pedretti and Ana Maria Navas Iannini, *Controversy in Science Museums. Re-Imagining Exhibition Spaces and Practice* (London: Routledge, 2020).

8 M. Norton Wise, *Aesthetics, Industry & Science. Hermann von Helmholtz and the Berlin Physical Society* (Chicago: The University of Chicago Press, 2018), quotes on pp. xv, 28, 35

as scientific research turned more toward laboratory work in the second half of the nineteenth century, lecture demonstrations were replaced by hands-on rigor at the bench—done every so often side by side with old apparatus hoarded, collected, and displayed in the same physical space. Although collections and museums were being challenged by the 1870s "as crucial site of scientific inquiry," Sophie Forgan reminds us that "[l]aboratories grew up as it were in the shadow of the museum." (Oliver Lodge placed the museum at the center of his ideal physical laboratory in 1890.)[9]

The emergence of a modern scientific practice is closely associated to museums (even more so perhaps for natural history[10]). Today, objects and instruments—*organa*—neatly stored and exhibited in a museum convey a strong and eloquent narrative compelling us to ask questions differently—to review our ways of knowing and seeing the world. The late Princeton historian of science Michael Mahoney, in developing a college course in the 1990s on the history of technology since the Renaissance, struggled to retrieve conventional primary sources until it dawned on him that he was looking in the wrong place:

---

respectively. The author makes a related argument regarding the collections stored in the *Gewerbehaus*, the house of industry, writing that "they played a role similar to what Schinkel proposed for the presentation of works of art at the Altes Museum, as analogues of scientific experiments that looked toward new creations." (p. 75) There is thus a direct connection in circa 1810–1850 between art (aesthetics), science, and industry, all of which being represented in collections/museums from major Berlin institutions, whose end purpose was the growth of knowledge through the performances of experimentation. For an interesting point of view regarding Norton Wise's original argument, Ursula Klein, "Science, Industry, and the German *Bildungsbürgertum*," *Annals of Science* 77 (2020): 366–76.

9 Sophie Forgan, "The Architecture of Display: Museums, Universities and Objects in Nineteenth-Century Britain," *History of Science* 32 (1994): 139–62, quote on p. 155. For a complementary view, Forgan, "Building the Museum: Knowledge, Conflict, and the Power of Place," *Isis* 96 (2005): 572–85. On the role of history and memory in the lab through collections of instruments, Boris Jardine, "The Museum in the Lab: Historical Practice in the Experimental Sciences at Cambridge, 1874–1936," *BJHS: Themes* 4 (2019): 245–71, where he writes that these collections bind the laboratory to society in 'integral networks,' "networks of affect, commerce and authority: the museum in the lab is the space where a form of sociable materiality can coexist with, augment and contrast with a form of (supposedly) asocial materiality." (p. 271) This is exactly what happened with the collection of instruments at Harvard, Jean-François Gauvin, "The Next Level of Play: How Cartesian Devils, Volcanoes, and Quantum Toys Taught General Science at Harvard, 1730–1970," *The Journal of the History of Collections* 31 (2019): 531–46, esp. pp. 535–40.

10 Natural history museums became the hallmark of cutting-edge science in the Western world and in the colonies. On this topic, Susan Sheets-Pyenson, *Cathedrals of Science: The Development of Colonial Natural History Museums During the Late Nineteenth Century* (Montréal: McGill-Queens University Press, 1988).

> What I needed for my students was not a library, but a museum. They shouldn't be reading great books, but examining great things. Or, to put that last point another way, the great ideas we were seeking did not lie in books. They lay in objects. Understanding those ideas meant learning to 'read' in a new way.[11]

This scholarly ambition resonates well with the Harvard University *Tangible Things* initiative, spearheaded by Laurel Ulrich in 2011. The aim was quite simple: "In contrast to more traditional works, our topic is not the evolution of particular objects or forms over time, but rather a method of investigation that begins with a specific artwork, artifact, or specimen and then moves outward in an ever-widening circle." This initiative, which began as a large general education course and an exhibition at the Collection of Historical Scientific Instruments (CHSI) was turned into a published volume a few years later. The chief object-lesson the authors wanted us to remember was that "just about any tangible thing can be pressed into service as primary historical evidence."[12] In this final chapter of the book my goal is to look at three specific types of activities—collecting, playing, digitizing—and focus on the various instrument trajectories in order to demonstrate how effective they are as 'primary historical evidence.' Not meant to be comprehensive, this idiosyncratic account suggests and spurs interest into looking at museum collections from a variety of angles—to probe any object, *organum*, to reveal new ways of knowing the world.

1 Collecting Instruments

The opening of the University of Cambridge's Cavendish Laboratory of experimental physics in 1874 offered a grand vision of the field's future. A closer look inside, however, revealed a somewhat contrasted story. Its first director, James Clerk Maxwell, the grand unifier of electricity and magnetism, who so keenly sought answers to key questions in theoretical physics and metrology, collected—some would say hoarded—old scientific instruments in this state-of-the-art lab. Such an "idiosyncratic approach to the past" was not uncommon in late Victorian Cambridge. It was closely tied "to the construction

11 Michael Mahoney, "Reading a Machine," (1999). https://www.princeton.edu/~hos/h398/readmach/modeltfr.html (accessed on 13 August 2020).

12 Laurel Ulrich, Ivan Gaskell, Sara J. Schechner, Sarah Anne Carter, and Samantha van Gerbig, *Tangible Things. Making History Through Objects* (Oxford: Oxford University Press, 2015), 2. Samuel J. M. M. Alberti, "Objects and the Museum," *Isis* 96 (2005): 559–71.

of institutional identity." As Boris Jardine notes, the physics laboratory was built in the city's center, in a location dedicated to the production of knowledge, where scientific lectures, laboratories, and collections were concentrated: the 'New Museums Site.' Maxwell's 'museum in the lab' fit right in, and unsurprisingly focused on old instruments displaying the history of scientific practice, especially standards of measurement. Despite their initial added cultural value, these historical instruments didn't last long—as was the case elsewhere, typically. The Cavendish collection was eventually dispersed. Yet, even though they were 'transient,' i.e., moving around, the new replacing the old, meaning was closely tied to their emplacement: they "need to be understood in the specific institutional context in which they were amassed."[13] Situatedness, as we will see throughout this chapter, circumscribes the sociohistorical and cultural context of instruments and collections. What remains today of the Cavendish laboratory historical collection is recovered from individual instruments' convoluted trajectories, mapped onto a field of lost and found records, human neglect, remembrance and forgetfulness, and heroic salvaging, reclaiming, and enshrinement—at the Whipple Museum of the History of Science, for instance.

Instrument collections took off in the eighteenth century. Though special instrument cabinets were assembled, curated, and exhibited in princely Renaissance *Wunderkammern*, befittingly regarded as 'sites of knowledge' (such as the court collections in Kassel, Dresden, and Florence), it is during the Enlightenment that the clustering of instruments in one specific place began to have a measurable effect on teaching and experimenting with nature. An aristocrat and polymath like Voltaire, comfortably cloistered in the château de Cirey with the marquise Émilie du Châtelet, spent a handsome amount of money in 1738 to acquire a complete *cabinet de physique* assembled for him by the young abbé Jean-Antoine Nollet. One of Voltaire's inspirations to create his cabinet likely came from a visit in 1735 to the Court of Lorraine, in Lunéville, where he discovered '*un établissement admirable pour les sciences*' filled with machines of all sorts to perform the newest physics experiments. Its caretaker, a simple locksmith turned into a *philosophe*, Philippe Vayringe, was trained in England and, upon his return to France in 1731, was named *professeur de*

13 Jardine, "The Museum in the Lab," 249. On Maxwell, Simon Schaffer, "The Laird of Physics," *Nature* 471 (March 2011): 289–91 and idem, "Late Victorian Metrology and Its Instrumentation: A Manufactory of Ohms," in *Invisible Connections. Instruments, Institutions, and Science*, eds. Robert Bud and Susan E. Cozzens (Bellingham, WA: SPIE Optical Engineering Press, 1992), 23–55.

philosophie expérimentale at the Lunéville Academy. Other renowned cabinets, owned by the Duc de Chaulnes and his brother-in-law, Bonnier de la Mosson, demonstrate in France how didactic enjoyment and serious natural philosophical work could go hand in hand. What were at first described as extravagant spaces and idle curiosities became over the century important loci of scientific training and entertainment (spectacles of nature).[14]

From the 1740s onwards, European academies, colleges and universities began to assemble systematically important instrument cabinets. In England and the Netherlands, makers such as John Desaguliers, George Adams Sr. and his two sons, Willem Jacob 's Gravesande, and the van Musschenbroek brothers, Peter and Jan, equipped schools and royal cabinets with fine instruments—in their respective countries, and beyond. In France, Nollet was for a time the country's most respected instrument maker, providing complete cabinets to the Académie royale de Bordeaux (1741), the Collège de Berne (1750), the Collège de Navarre (1753), the Artillery school of La Fère (1757), the Engineering school of Mézières (1761) and the Artillery school of Bapaume (1765), and teaching experimental physics in some of these same locations. After the Jesuits were expelled from France in 1759, new Catholic colleges asked Nollet for help regarding teaching experimental physics; overwhelmed with similar requests, as he told one correspondent in 1768, Nollet believed only the publication of *L'Art des expériences* (which finally came out in 1770), explaining in detail how to build the essential instruments of a physics cabinet, would relieve him of this drudgery.[15] This is the scenario that eventually played out at

14 Jean-François Gauvin, "Le cabinet de physique du château de Cirey et la philosophie naturelle de Mme du Châtelet et de Voltaire," in *Emilie Du Châtelet: Rewriting Enlightenment Philosophy and Science*, eds. Judith P. Zinsser and Julie Candler Hayes (Oxford: Voltaire Foundation [SVEC 2006: 01], 2006), 165–202. On Vayringe and his attempt to build a calculating machine, Matthew L. Jones, *Reckoning with Matter: Calculating Machines, Innovation, and Thinking about Thinking from Pascal to Babbage* (Chicago: The University of Chicago Press, 2016), 135–9. On the role of spectacle referring to science, Simon Schaffer, "Natural Philosophy and Public Spectacle in the Eighteenth Century," *History of Science* 21 (1983): 1–43. More recently, Bernadette Bensaude-Vincent and Christine Blondel, eds., *Science and Spectacle in the European Enlightenment* (New York: Routledge, 2008). On private cabinets, C. R. Hill, "The Cabinet of Bonnier de la Mosson (1702–1744)," *Annals of Science* 43 (1986): 147–74. On how collecting instruments changed from the Renaissance to the twentieth century, Anthony Turner, "From Mathematical Practice to the History of Science," *Journal of the History of Collections* 7 (1995): 135–50.

15 Jean-François Gauvin, "Eighteenth-Century Entrepreneur: Excerpts from the Correspondence between Jean Antoine Nollet, Etienne-François Dutour, and Jean Jallabert, 1739–1768," in *The Art of Teaching Physics: The Eighteenth-Century Demonstration Apparatus of Jean Antoine Nollet*, eds. Lewis Pyenson and Jean-François Gauvin (Sillery, Qc:

the Collège des Godrans in Dijon toward the end of the century—as discussed in chapter 5.

By the 1740s, experimental instruction became an essential characteristic of natural philosophical training on the other side of the Atlantic, at Harvard College. Taking over for Isaac Greenwood in 1738, John Winthrop's long and successful career as an astronomer and natural philosopher was partly owed to a well-furnished cabinet of instruments, undeniably the best in the American colonies. After the entire collections of instruments and library were lost to fire in 1764, the destruction was met with a wave of donations (books and instruments) from patrons on both side of the Atlantic. Fortunately, Harvard could also count on America's foremost scientist, Benjamin Franklin, who visited London's best instrument maker shops and, over the course of several years, bought state-of-the-art instruments to refurbish Winthrop's cabinet on behalf of the century-old college.[16] Benjamin Martin, one of the most celebrated instrument makers and lecturers of the 1760s, sold to Harvard numerous instruments that are still on display in the permanent gallery of the CHSI.[17] The instruments recreating this *physique amusante* were part of Winthrop's regular—and heretofore expected—instrumentarium in teaching experimental philosophy to undergraduate students.

The number of teaching cabinets and instrument makers exploded in the nineteenth century, on both sides of the Atlantic. As Paolo Brenni explains,

---

Septentrion, 2002), 47–69. On the role of scientific training at the engineering school of Mézières, Sébastien Pautet, "Produire une élite savante et technicienne à l'École du génie de Mézières: dispositions techniques et scientifiques des élèves ingénieurs," *Artefact. Techniques, histoire et sciences humaines* 4 (2016): 119–33. On the evolution of instrument cabinets, see the classic Maurice Daumas, *Les instruments scientifiques aux XVIIe et XVIIIe siècle* (Paris: Presses Universitaires de France, 1953).

16 In a meeting conducted on October 22, 1770, the Harvard Corporation asked Winthrop to officially thank Franklin "for his repeated good services for this College, and particularly in his case of a valuable achromatic telescope lately received thr'o his hands." Harvard University. Corporation. College Books, 1636–1827. College Book 7, 1750–1778. UAI 5.5 Box 5, Harvard University Archives, seq. 272 (p. 212). *Colonial North American Project at Harvard,* http://nrs.harvard.edu/urn-3:HUL.ARCH:10880062 (accessed on 7 November 2022). The telescope's inventory number in the Harvard Collection of Historical Scientific Instruments (CHSI) is 0055a.

17 John R. Millburn, *Benjamin Martin: Author, Instrument-Maker, and 'Country Showman'* (Leiden: Noordhoff International Publishing, 1976). David Wheatland, *The Apparatus of Science at Harvard, 1765–1800* (Cambridge, MA: Harvard University Press, 1968). Martin was also thanked by the Harvard Corporation for 'his repeated kindness' in gifting the college library with several of his works. Harvard University. College Book 7, 1750–1778, seq. 250 (p. 191). Ethan Lasser, ed., *The Philosophy Chamber: Art and Science in Harvard's Teaching Cabinet, 1766–1820* (Cambridge, MA: Harvard Art Museums, 2017).

didactic instruments produced for and bought by European and American colleges were meant for describing and replicating, in a simplified form, historical and fundamental experiments. To do so, cabinets required an abundant list of instruments—easily numbering in the hundreds. According to Brenni, however, they were seriously underused. Not only did they necessitate meticulous manipulations (a good demonstration "is like the trick of a magician"), some experiments (especially related to calorimetry) would take too long during a normal physics lesson. These large cabinets, in short, gathered dust more than they enlightened students over the course of the century. By the 1920s and 1930s, they became irrelevant. Their ultimate fate depended on how much space was required to build new laboratories, their historical (and monetary) value, and more importantly who was still willing to care for them—something Brenni knows first-hand (see chapter 5).[18] While history, philosophy, and sociology of science are experiencing what some scholars dubbed a 'material turn,' in response to the 'linguistic turn' in the social sciences derived in part from Claude Lévi-Strauss's structural anthropology and Jacques Derrida's literary theory, several international organizations have taken a renewed interest in scientific collections—especially in universities. The challenges are plentiful and immense, but as Marta Lourenço and Lydia Wilson demonstrate, "scientific heritage and its associated documentation are an integral part of the scientific infrastructure and therefore relevant sources for research, teaching and outreach." Looking at the trajectory of scientific collections provides an interesting time-dependent cross-section of how the scientific enterprise evolved in a specific context—what Lourenço calls 'time capsules' in the case of significant collections in Portugal, some of which laid dormant for the better part of two hundred years or more.[19] Let us return for a moment to the Harvard collection.

---

18 Paolo Brenni, "The Evolution of Teaching Instruments and Their Use between 1800 and 1930," *Science & Education* 21 (2012): 191–226. Brenni, *Il gabinetto di fisica dell'Istituto tecnico toscano. Guida alla visita* (Florence: Edizioni Polistampa, 2009). He also created a YouTube channel dedicated to replicating demonstrations using original apparatuses—showcasing his skills as a 'magician.' (see chapter 5) On teaching collections and pedagogy in general, Elizabeth Cavicchi and Peter Heering, eds., *Historical Scientific Instruments in Contemporary Education* (Leiden: Brill, 2022).

19 Marta C. Lourenço and Lydia Wilson, "Scientific Heritage: Reflections on Its Nature and New Approaches to Preservation, Study and Access," *Studies in History and Philosophy of Science* 44 (2013): 744–53, quote on p. 752. As the Chair of the International Council of Museums Committee for University Museums and Collections (UMAC), Lourenço is keenly aware of the importance and work necessary to increase awareness on this issue. See also Dominique Ferriot and Lourenço, "De l'utilité des musées et collections des universités," *Lettre de l'OCIM* 93 (2004): 4–16. Lourenço and José Pedro Sousa Dias, "'Time

In the early history of the Cavendish laboratory, the past was built alongside the new; at Harvard, the past was kept at a distance, pushed aside to let the modern and leading-edge take over. The Harvard Physical Cabinets, partly going back to the eighteenth-century philosophy chamber, went into a profound transformation in the later 1800s. After the apparatuses were "safely moved from Harvard Hall, and systematically arranged in appropriate rooms" in the newly built (1884) Jefferson Physical Laboratory, they continued for a short time to be employed mainly as didactic instruments, demonstrating the basic laws of physics and nature.[20] In 1890, a complete survey was made, where "the cases were numbered, and each piece of apparatus was stamped with a number and the number of the case" and then entered into a record book. (Fig. 19L–R) A card-based loan system was put in place and, as John Trowbridge (hired in 1870) remarked, "no abuses have resulted from this policy."[21] More importantly, the policy of the department "changed radically in regard to the expenditures for apparatus." Whereas in the past, Harvard, and most other American colleges, spent large amount of money on apparatus made in England, France, and Germany, new instruments were now being developed for teaching experimental physics in the Jefferson Laboratory. The designs were then sent to Boston companies for production (and for profit to them)—for the benefit of Harvard and other institutions of learning. If, between 1870 and 1880, Harvard ordered an average of $1,500 per year of foreign apparatuses, that amount was reduced to less than $100 between 1887 and 1890. According to Trowbridge, this was "due to the increase in the laboratory method of instruction in physics and the practical abandonment of the lecture method."[22] Progress was significant, and as reported in 1892, "students

Capsules' of Science: Museums, Collections, and Scientific Heritage in Portugal," *Isis* 108 (2017): 390–98.

20 Harvard University, *Annual Reports of the President and Treasurer of Harvard College, 1884–85* (Cambridge, MA: Harvard University, 1886), 155. The collection of instruments was moved out of Holden Chapel and into the University Hall before 1850, when the Chapel was yet again redesigned to house an anatomical museum and theater. By 1870, the collection of apparatus outgrew the rooms given to it. The Corporation decided to move its caretaker, Joseph Lovering, to the first floor of Harvard Hall to provide him with a bigger lecture room, recitation room, and an "apparatus-room large enough to accommodate the collections of the department of Physics for some years to come." Harvard University, *Annual Reports of the President and Treasurer of Harvard College, 1869–70* (Cambridge, MA: Harvard University, 1871), 24–5.

21 Harvard University, *Annual Reports of the President and Treasurer of Harvard College, 1889–90* (Cambridge: Harvard University, 1891), 167.

22 Harvard University, *Annual Reports of the President and Treasurer of Harvard College, 1890–91* (Cambridge, MA: Harvard University, 1892), 169–70. Wallace Sabine, for instance, designed a new spectroscope for elementary courses on light, "by means of

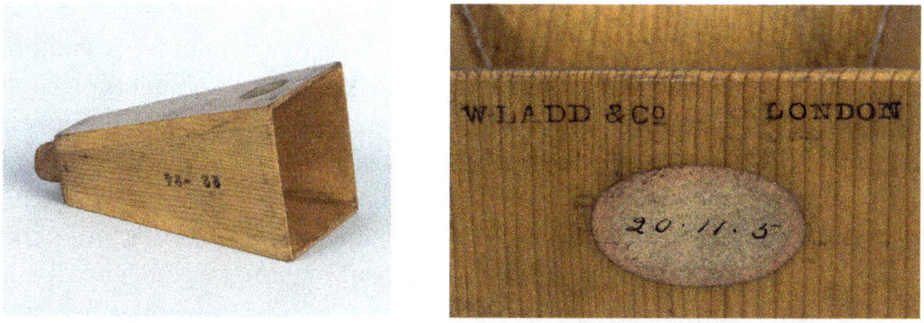

FIGURE 19.L–R A wooden prismatic resonator from ca. 1876. Several objects in the CHSI collection bear two sets of identification numbers: stamped (22–24)—seen left; a pasted oval piece of paper (20.11.5)—seen right. They appear to be the result of two separate inventories conducted by the physics department. © COLLECTION OF HISTORICAL SCIENTIFIC INSTRUMENTS, HARVARD UNIVERSITY, 1998-1-0218B

who have lately returned from Germany inform me [Trowbridge] that there is no laboratory in Europe which is now superior to the Jefferson Physical Laboratory in regard to material equipment."[23] The establishment of an instrument workshop within Jefferson Laboratory, which made possible extended research by professors and original work by graduate students, also placed the Harvard lab "in the first rank of American laboratories."[24]

New instruments entering the department of physics were added to the Laboratory, no longer to the physics cabinet. Trowbridge believed that "physical cabinets mark an era of the past; and the laboratory of the future must have, so to speak, live capital in the shape of working apparatus rather than inactive capital in pieces of apparatus which have merely an historical interest."[25] In another report, almost a decade later, he commented similarly: "The growth, therefore, of the physical cabinets in the future promises to be

which measurements are possible which formerly could be made only with very expensive instruments." Harvard University, *Annual Reports of the President and Treasurer of Harvard College*, 1891–92 (Cambridge, MA: Harvard University, 1893), 175.

23 Harvard University, *Annual Reports of the President and Treasurer of Harvard College*, 1891–92, 175. For the development of physics teaching in Germany, Katherine Olesko, *Physics as a Calling: Discipline and Practice in the Königsberg Seminar for Physics* (Cornell: Cornell University Press, 1991).

24 Harvard University, *Annual Reports of the President and Treasurer of Harvard College*, 1897–98 (Cambridge, MA: Harvard University, 1899), 243.

25 Harvard University, *Annual Reports of the President and Treasurer of Harvard College*, 1892–93 (Cambridge, MA: Harvard University, 1894), 193. Harvard's story is similar to many other European universities, as discussed above.

a very small one; for the new apparatus enters instantly into everyday use in the special work rooms of the Laboratory."[26] As more room for laboratory work was required owing to the ever-increasing number of students electing courses in physics, Trowbridge warned the Corporation that "with the growth of the Physical Department, more of the space now devoted to historical and antiquated apparatus will be utilized for laboratory purposes." Fortunately, the more interesting instruments, such as the "Benjamin Franklin machines, the Orreries, [and] the first types of dynamos, have been placed together in a room open to visitors on the first floor of the building."[27] Still, Trowbridge believed this 'inactive capital' could prove to be useful to students: "for the unused and historical pieces of apparatus are before the student while he is engaged in testing the physical laws which they illustrate, an old apparatus, which is apt to be lifeless in its influence to all except an antiquarian, becomes alive and speaking to the working student."[28] While students in 1885 could still find entertainment with experimental lectures conducted by and for one another, the same apparatuses in 1902 had become evocative objects—not unlike the collection of 'relics' preserved at the Royal Society of London, discussed recently by Rebekah Higgitt.[29] Trowbridge was proud to write to Harvard President Charles Eliot in 1902 that the "laboratory has now the best equipment of any Physical Laboratory either in this country or abroad; and its scientific work last year certainly was more important than that of any American laboratory. Our mental equipment should keep pace with our mechanical equipment; and therefore I advocate bringing more to the front young instructors."[30] The work in the laboratory, at the bench, took precedence over lecture demonstrations and old iconic instruments.

During the first few decades of the twentieth century, the Harvard historical instruments (and those that would become historical) moved around, were hidden in closets, cannibalized, stolen (or borrowed …), sent to the Smithsonian and, more often than not, neglected and forgotten. They were

26 Harvard University, *Annual Reports of the President and Treasurer of Harvard College, 1901–02* (Cambridge, MA: Harvard University, 1903), 255.

27 Ibid.

28 Harvard University, *Annual Reports of the President and Treasurer of Harvard College, 1902–03* (Cambridge, MA: Harvard University, 1904), 239.

29 Harvard University, *Annual Reports of the President and Treasurer of Harvard College, 1885–86* (Cambridge, MA: Harvard University, 1887), 133. Sherry Turkle, *Evocative Objects: Things We Think With* (Cambridge, MA: MIT Press, 2011). Rebekah Higgitt, "Instruments and Relics: The History and Use of the Royal Society's Object Collections c.1850–1950," *Journal of the History of Collections* 31 (2019): 469–85.

30 Harvard University Archives, UAI 5.150, box 68, *Records of the President of Harvard University Charles William Eliot. Trowbridge, John, 1893–1903*, Trowbridge to Eliot, 3 March 1902.

eventually 'rediscovered' and saved from oblivion by David P. Wheatland (1898–1993), who founded the Collection of Historical Scientific Instruments (CHSI) in the late 1940s. Henceforth, Wheatland rummaged through laboratories, corridors, and attics in Harvard science buildings in search of historical and important instruments, saving thousands of them from destruction due to the standard practices of researchers, driven by their perennial thirst for spare parts. (Fig. 20) Filling holes in the core collection, Wheatland also bought hundreds of instruments from antiquarians and auction houses.[31] The CHSI's collection continues to grow today with the collaboration of numerous physics professors, from Gerald Holton (who gave some of his personal instruments related to the Harvard Project Physics he famously co-created in the 1960s) to Eric Heller, who stumbled in 2014 on a box of 'quantum toys' that became, for a while, a personal obsession (see section *infra*).

The history of collections is one of material ebb and flow. Objects go in and out of collections, they appear and disappear over time and space. Collections not only begin and mature, they end as well—and are sometimes even reborn. It is through the lens of ending that we see best the refracted light of collection practices, a light that "makes visible the perpetual care, assessment, de-accessioning and editing needed for collections to persist." For the editors of a special *BJHS Themes* number, collections are not kept in conditions of stasis and permanence, but rather are continuously rearranged by "the twin forces of circulation and maintenance." The end of a collection as one knows it does not imply the end of its sociohistorical and scientific utility. Looking at various kinds of possible endings (desecration, recycling, marginalization, mercantile) "can tell us specifically about scientific collections, their meanings and their power."[32] With any collecting, there is the expectation of ending—and between the incessant motion, dislocation, relocation, and disappearance of

31 Sara Schechner, "David P. Wheatland (1898–1993)," online https://chsi.harvard.edu/chsi-history (accessed 27 August 2020). On Wheatland in general, William J. H. Andrewes, "The Life and Work of David Pingree Wheatland (1898–1993)," *Journal of the History of Collections* 7 (1995): 261–8. On Wheatland's purchases, Andrewes, "The Legacy of David Wheatland," *Nuncius* 16 (2001): 687–701.

32 Boris Jardine, Emma Kowal, and Jenny Bangham, "How Collections End: Objects, Meaning and Loss in Laboratories and Museums," *BJHS Themes* 4 (2019): 1–27, quotes on pp. 2, 4–5. On the concept of active ending and its heuristic role, Ann Laura Stoler, ed., *Imperial Debris. On Ruins and Ruination* (Durham: Duke University Press, 2013), where we learn that the word 'ruin' is both a noun and a verb, "the state of a thing and a process affecting it." Thomas Laqueur, *The Work of the Dead: A Cultural History of Mortal Remains* (Princeton: Princeton University Press, 2015), on the universal significance of the dead body and how we dispose of it. Gregg Mitman, Marco Armiero, and Robert S. Emmet, *Future Remains. A Cabinet of Curiosities for the Anthropocene* (Chicago: The University of Chicago Press, 2018), on what to collect on the eve of our foreseeable extinction.

FIGURE 20 Cloud chamber used by J. C. Street and E. C. Stephenson at Harvard for their discovery of the muon. Believed lost for a long time, it was later found in Edward Purcell's office in Lyman Laboratory, as identified by Robert Pound in 1998.

the objects themselves. Objects also have a life of their own, a 'biography' that often plays a determining role in whether a collection is kept intact or taken apart and obliterated. Studying how they were made, and what they were made for, provides another epistemic layer to why they were collected first—or why they were forgotten or simply tossed. The whole process of rediscovery can open up novel ways of thinking about instruments—and it can be fun!

2 Knowing through Playing

A few years ago, when I was still working at Harvard, I found myself playing a lot during office hours. The reason was simple: I had stumbled on a set of unassuming old-school metallic cubes nicknamed 'quantum toys' in the early 1960s by their inventor, Costas (Cos) Papaliolios, a Harvard physicist known for his novel contributions to optical ground-based astronomy. (Fig. 21) These devices consisting of open-ended boxes, large and small, into which are inserted light

FIGURE 21 Quantum toys
© COLLECTION OF HISTORICAL SCIENTIFIC INSTRUMENTS, HARVARD UNIVERSITY, 2014-1-0231

polarizers, were given to the Collection of Historical Scientific Instruments (CHSI) late in the autumn of 2014. Marked with three variants of Dirac's quantum mechanics bra-ket notation system, their function initially remained curious. Why would one use such juvenile contraptions to survey the intricacies of microphysics? What pedagogical lesson was embedded in them? Were they designed to calculate anything or simply to help students acquire a basic interpretation of quantum mechanics? How does one 'play' with them? Delivered to the CHSI in a package simply identified as 'Julian Schwinger, Phys 251a,' the cubes' connection to a famous Harvard physicist and Nobel Prize laureate caught my attention.[33] Papaliolios's quantum toys are a form of haptic representation of quantum mechanical operators, the reification of Schwinger's basic measurement symbols. Papaliolios developed the idea circa 1962, merely two years after completing Physics 251a with Schwinger. They were designed,

33 A full article came out of this research. See Jean-François Gauvin, "Playing with Quantum Toys: Julian Schwinger's Measurement Algebra and the Material Culture of Quantum Mechanics Pedagogy at Harvard in the 1960s," *Physics in Perspective* 20 (2018): 8–42. The reader should refer to this article to understand how these 'Pauli cubes,' as they were also known, were used and crafted.

we learn, as a teaching aid for quantum mechanics in a flagship General Education course at Harvard titled Natural Sciences 2. The intent behind the 'toys,' according to Papaliolios, was "to teach some of the basic aspects of Quantum Mechanics to non-physics students."[34] The main heuristic feature of the novel pedagogical tools comes from the fact that Papaliolios replaced Schwinger's paradigmatic Stern-Gerlach simple two-state system with another one established on the property of light called helicity—or polarization. That shift made it easier to manufacture a tangible and hands-on device from which the basis of a quantum mechanics formalism *à la* Schwinger could naturally arise. The quantum toys' creation illustrates Papaliolios's ability to convert an abstract idea into a leisurely demonstration apparatus easy to manipulate and appreciate. Historically, they are no different than, say, the early modern orreries that displayed to a wider audience the intricacies of the celestial sphere. They bring down (or up) to a performative level that which is inscrutable to most. It is the operation of the device (the *habitus*) that makes the abstract notion real and fathomable. It became Papaliolios's tactic to reform basic teaching of quantum mechanics.

In the published account of these quantum toys, I did not attempt to provide a 'thick description,' nor do I believe they would qualify as a 'thing knowledge.'[35] The hours I spent playing with these toys, tangible things in their own right, opened up an intriguing path into the history of quantum mechanics instruction at Harvard in the 1960s. It also led me to ask seriously the question: What if playing, as an epistemic category, was as powerful an instrument of science and knowledge production as reasoning and making? What if *homo ludens* as a defining concept was as meaningful and elemental as *homo sapiens* and *homo faber*? Johan Huizinga made just such an assertion in a book first published in German in 1944. Huizinga argued that 'play' was not necessarily a matter of psychology or biology, but "a function of culture proper." It is found everywhere "as a well-defined quality of action which is different from 'ordinary' life." Play is a free action—it is in fact freedom. It is a way to step outside reality and to dive into a pretend one. It can easily become a serious affair, unexpectedly ensnaring someone when her devotion to play passes

34 Harvard University Archives, Costas Papaliolios Papers (unprocessed accessions 14811), Box 21, Folder 'Quantum Toys,' Papaliolios to Dr. John Fowler, 5 January 1968.

35 Ken Alder, ed., "Focus: Thick Things," *Isis* 98 (2007): 80–142. Alder first introduced this notion in his *Engineering the Revolution: Arms & Enlightenment in France*, 1763–1815 (Chicago: The University of Chicago Press, 1997). Davis Baird, *Thing Knowledge: A Philosophy of Scientific Instruments* (Berkeley: University of California Press, 2004). Liba Taub, ed., "Focus: The History of Scientific Instruments," *Isis* 102 (2011): 689–729. Another useful concept to think with is Bill Brown, "Thing Theory," *Critical Inquiry*, 28 (2001): 1–22.

into rapture. Because it is done under one's own volition, play is beautiful; it is enchanting and captivating, casting a spell on us: "Really to play, a man must play like a child."[36] Huizinga's cultural approach to play resonates with current studies on the production of scientific knowledge.[37]

Sherry Turkle, director of the MIT Initiative on Technology and Self, takes a somewhat similar approach by investigating the intellectual and emotional functions of childhood objects in the emergence of students' and scientists' love for science. She argues that "giving science its best chance means guiding children to objects they love." This line of reasoning is influenced by the work of Donald W. Winnicott, a psychoanalyst who developed the notion of *transitional objects*, i.e., objects that help us make the transition between childhood and adulthood, and between our inner self and the outside world. Growing up, we are continuously in search of objects we can love, objects that will define us in opposition to the realm of reality. According to Winnicott, however, it is the intermediate zone brought by the object, the transitional space, that is creative and expressive. It is where science happens, according to Turkle.[38] And where craft originates, following Richard Sennett: "Craftsmanship draws on what children learn in play's dialogue with physical materials, the discipline of following rules, the advance of complexity in making rules." The craft of play, work, and innovation are intimately interconnected, perhaps going back to prehistoric societies, since "the rhythm of routine in craftsmanship draws on childhood

36 Johan Huizinga, *Homo ludens: A Study of the Play-Element in Culture* (London: Routledge & Kegan Paul, 1949), 4–9.

37 Turner, "Presidential Address: Scientific Toys." Peter Heering and Roland Wittje, eds., *Learning by Doing. Experiments and Instruments in the History of Teaching* (Stuttgart: Franz Steiner Verlag, 2011). Melanie Keene, *Science in Wonderland: The Scientific Fairy Tales of Victorian Britain* (Oxford: Oxford University Press, 2015). Sofie Lachapelle, *Conjuring Science. A History of Scientific Entertainment and Stage Magic in Modern France* (New York: Palgrave Macmillan, 2015). A fascinating case study is Anke Te Heesen, *The World in a Box: The Story of an Eighteenth-Century Picture Encyclopedia* (Chicago: The University of Chicago Press, 2002).

38 Sherry Turkle, ed., *Falling for Science: Objects in Mind* (Cambridge, MA: MIT Press, 2008), 5, 37, and 279. See also her other books on the topic of affect and material culture: Turkle, *Evocative Objects*, and Turkle, ed., *The Inner History of Devices* (Cambridge, MA: MIT Press, 2008). On how regular people deal with stuff, from an anthropological point of view, Daniel Miller, *The Comfort of Things* (Cambridge: Polity Press, 2008). On a theoretical approach to 'stuff' see Miller, *Stuff* (Cambridge: Polity Press, 2010). For more on transitional and cognitive objects, linking childhood play to the transition of practical to theoretical knowledge, see Robin A. Hodgkin, "Cognitive Objects," *Oxford Review of Education* 14 (1988): 353–62.

experience of play."[39] Making things and producing science have deeper (puerile) roots than what is commonly believed. Since our long-forgotten childhood, the objects we love are habitually the things we play with.

Historical examples abound: Isaac Newton's prisms in the 1660s were foremost entertainment and decorative objects before they became in Robert Boyle's words, as already quoted in chapter 1, "the usefullest Instrument Men have yet imploy'd about the Contemplation of Colours." During the same period, glass-drops (also known as *larmes de verre* or Prince Rupert's drops) were fashionable and equally fascinated the French peripatetic Balthasar de Monconys and natural philosophers such as Robert Hooke and Jacques Rohault, who after careful investigations considered these trinkets 'new scientific objects.' In London, exploring the transient quality and features of common soap bubbles, Victorian physicists such as Michael Faraday and William Thomson combined "childlike enthusiasm and sober microphysics" in searching for clues of nature's underlying structure. And mechanical molecular models in the middle of the twentieth century, made of simple balls and sticks, were more than a visual and didactic support displayed and constructed in a classroom. (Fig. 22) They were legitimate research tools used to "augment and extend the modeler's cognitive capacities and performances." It was the manipulations of—the playing with—various components of mechanical models that enhanced cognitive performances.[40]

Since the late 1960s, science museums have harnessed the power of game and play in their display. The Exploratorium in San Francisco, founded by Frank Oppenheimer, was—and still is—a pioneer and a leader in the field. Oppenheimer saw no problem with, and actually encouraged, children running and freely roaming the whole museum on their path to discovery. Exploration, he believed, should not be encumbered by an inflexible set of rules; to do so was to disrespect greatly the visitor. It is often said that Oppenheimer got angry

---

39 Richard Sennett, *The Craftsman* (New Haven: Yale University Press, 2008), 268–74, quotes on pp. 273 and 268 respectively. Felix Riede, *et al.*, "The Role of Play Objects and Object Play in Human Cognitive Evolution and Innovation," *Evolutionary Anthropology* 27 (2018): 46–59.

40 Simon Schaffer, "Glass Works: Newton's Prisms and the Uses of Experiment," in *The Uses of Experiment: Studies in the Natural Sciences*, eds. David Gooding, Trevor Pinch, and Simon Schaffer (Cambridge: Cambridge University Press 1989), 67–104, quote on p. 76. Mihnea Dobre, "On Glass-Drops: A Case Study of the Interplay Between Experimentation and Explanation in Seventeenth-Century Natural Philosophy," *Journal of Early Modern* Studies 2 (2013): 105–24. Simon Schaffer, "A Science Whose Business is Bursting: Soap Bubbles as Commodities in Classical Physics," in *Things that Talk: Object Lessons from Art and Science*, ed. Lorraine Daston (New York: Zone Books, 2004), 147–92. Mathieu Charbonneau, "The Cognitive Life of Mechanical Molecular Models," *Studies in the History and Philosophy of Biological and Biomedical Sciences* 44 (2013): 585–94.

FIGURE 22 Mechanical molecular model made in May 1965 by a Harvard student, Philip L. Berneburg. It represents the mineral montmorillonite, a member of the smectite group.

whenever an exhibit couldn't be made to 'misbehave.' Experimental failures and exhibit breakdowns were all part of having fun with science.[41] Though adults may sometimes feel unwelcome in science museums,[42] research has shown how significant the pedagogy of play is toward learning science. From John Locke's *Some Thoughts Concerning Education* (1693) to Jean-Jacques Rousseau's *Émile* (1762) and, more recently, to Montessori schools educators have long-since discerned a core concept, i.e., "the more natural and unstructured the environment is, the richer is children's play as an educational activity."[43] And this is true whether one, such as a museum, sets up an engineering environment or a virtual one.[44] This participatory standpoint is in constant evolution and has found a novel outlet in what is commonly known as 'citizen science.' It is no longer enough to learn how scientists produce knowledge—even by inviting them and their laboratory into a museum.[45] People want to feel they are part of the process: it is, most argue, a way of democratizing science, to tear down the elitist barrier between scientists and the public. The notion is popular and not without its problems. To whom is it addressed, for instance, and who can enjoy it despite the grandstanding call of democratization? Leaving these social and cultural difficult questions aside, citizen science is popular because projects are designed so the public find enjoyment in looking at data. The public is invited to 'play' with large amount of data, whether it deals with astronomy, synthetic biology or the environment.[46]

---

41 K. C. Cole, *Something Incredibly Wonderful Happens: Frank Oppenheimer and His Astonishing Exploratorium* (Chicago: The University of Chicago Press, 2009). One among many recent case studies from the National Science Museum in Thailand, Peeranut Kanhadilok and Mike Watts, "Youth at Play: Some Observations from a Science Museum," *International Journal of Adolescence and Youth* 22 (2017): 179–94.

42 Winifred Kehl, "I Have a Confession: I'm an Adult. How Can Science Museums Welcome Grown-Ups?" *Museum* (July–August 2013), 46–51. Steven Conn, *Do Museums Still Need Objects?* (Philadelphia: University of Pennsylvania Press, 2010), 138–71. The chapter is entitled "Where Have All the Grown-Ups Gone?"

43 Roberto Farné, "Pedagogy of Play," *Topoi* 24 (2005): 169–81, quote on p. 173. On the role of games to apprehend and make sense of culture, social connection, politics, and economic change, see Silvia Lindtner and Paul Dourish, "The Promise of Play: A New Approach to Productive Play," *Games and Culture* 6 (2011): 453–78.

44 Pirko Tõugu, *et al.*, "Connecting Play Experiences and Engineering Learning in a Children's Museum," *Journal of Applied Developmental Psychology* 53 (2017): 10–19. Jonathan P. Rowe, *et al.*, "Play in the Museum: Design and Development of a Game-Based Learning Exhibit for Informal Science Education," *International Journal of Gaming and Computer-Mediated Simulations* 9 (2017): 96–113.

45 Morgan Meyer, "Researchers on Display: Moving the Laboratory into the Museum," *Museum Management and Curatorship* 26 (2011): 261–72.

46 For a general introduction to citizen science, Susanne Hecker, *et al.*, *Citizen Science. Innovation in Open Science, Society and Policy* (London: UCL Press, 2018). To unpack the

Participatory science thus goes beyond the simple forms of communication of knowledge we commonly encounter in museums of science and technology. We need only actively capitalize on a range of epistemological games such as the classic Fold It and Eyewire, and we could open up a unique dynamic window on the world of scientific research. The function of such games is not so much to contribute to the acquisition of theoretical knowledge (*homo sapiens*) as to increase practical and epistemological skills (*homo faber*). To adopt the terminological distinctions made by Michel de Certeau, epistemological games enable players to develop tactics that are not limited to particular environments or skills; they are the very antithesis of strategies, which are predicated on an assessment of balances of power for a subject with a separate and identifiable capacity to choose and to act. More to the point, tactics are not tied to any particular environment (often a predefined one). As a result, players can seize opportunities, and opt for the unexpected, for tricky moves and crafty schemes which can almost always be learned only by carrying out actions and discovering them by chance or of necessity. In other words, the art of adopting successful tactics means being able to act without a defined and dominant structure (power). It should not be surprising then to learn that designers of games draw primarily on the human brain's capacity for pattern recognition and problem-solving. Pride of place is given to players' intellectual savoir-faire and epistemological resourcefulness—and that is what creates the basis for scientific discovery.[47]

In many respects, a similar pedagogical philosophy reigned at Harvard from the 1730s onwards. Based on archival materials and on the Collection of Historical Scientific Instruments (CHSI) holdings, I argue elsewhere that science teaching at Harvard was time and again a matter of learned play and recreation.[48] From lecture demonstrations delivered in the Philosophy Chamber in the 1760s to experimental work performed in the Jefferson Physical Laboratory in the 1880s, students were trained—and entertained—through their hands-on interaction with scientific instruments. (As Liba Taub reminded us recently, the concept of 'scientific instrument' is somewhat fluid and embraces

concept, especially as regards STS, Bruno Strasser, *et al.*, "'Citizen Science'? Rethinking Science and Public Participation," *Science and Technology Studies* 32 (2018): 52–76.

47 Michel de Certeau, *L'Invention du quotidien. 1- L'art de faire* (Paris: Gallimard, 1990). Jean-François Gauvin, "When Museums of Science Create Citizen Expertise: Participatory Science and the *Gamification of Techne*," *Thema: La revue des Musées de la civilisation* 2 (2015): 28–38. https://thema.mcq.org/index.php/Thema/issue/view/5 (accessed 27 August 2020).

48 Gauvin, "The Next Level of Play."

a broad variety of definitions.[49] See also chapter 1.) It is by means of the 'magical' effects of a Cartesian devil or a solar microscope, the dangerous and deafening explosion of an artificial volcano, the building (and breaking) of a measuring apparatus, and the playing with mathematical and quantum toys that vast areas of scientific knowledge were diligently imparted. Not unlike the original physics collection built in the late 1890s by and for the Science Museum in London,[50] the collection of instruments preserved at the CHSI is a testament to pedagogical advancements in teaching science. Over time, questions regarding what to collect in the rapidly evolving field of 'modern' physics, largely targeting research prowess and industrial triumphs over teaching methods, proved to be a recurring challenge for curators at the Science Museum—and to a lesser extent at the CHSI when it was founded in the late 1940s. Instruments come in all shapes and sizes. Whether they are considered low-cost productions or classified as high-end artworks, they should not be offhandedly dismissed, but rather carefully studied in order to understand their specific cultural and pedagogical values. Not only do they demonstrate effectively that playfulness is integral to science teaching, they propose an interesting angle of inquiry into scientific museum collections. At a time when science museums and science centers are driven toward ever more fun and interactive displays, it is necessary to understand that this core notion of play is embedded within the history and philosophy of scientific instrument design and construction.

3 Digitizing Collections

David Pantalony, a curator at the Canada Science and Technology Museum, makes the case for what he calls "the value of an honest approach to material culture." Scholars, he argues, should study objects (scientific or not) for what they are, not for what we want them to be. Not only are collections of objects "a valuable window into humankind's relationship with the material world"—they provide social, scientific, and historical information impossible to corral in more conventional paper-based sources—they offer as well

49 Liba Taub, "What is a Scientific Instrument, Now?," *Journal of the History of Collections* 31 (2019): 453–67.

50 Alison Boyle, "Modern Physics in the Museum: Shaping a UK National Collection in the Twentieth Century," *Journal of the History of Collections* 31 (2019): 487–502.

a "unique source of diversity in education, culture, research, and knowledge economy."[51] Instruments are data in need of analysis.

Pantalony is not the only one believing that: Many have been mobilized to the cause over the last few decades.[52] If the scholarly value of individual objects is not in question here—was it ever?—what can one say about their collective relevance, i.e., 'collections as data,' when thousands of objects in aggregated collections are no longer seen as distinct epistemic entry points for the production of knowledge but rather assimilated and processed as an integrated (and sizable) amount of raw data? With the rapid growth of Internet use and digital information and communication technologies (ICTs), museums have attempted to reinvent themselves, on an organizational and structural level as well as on the presentation and display of their collections (the COVID-19 pandemic has painfully drawn attention to that fact).[53] Jeffrey Schnapp, known for his work at Harvard in the domains of media, knowledge design, digital arts, and humanities, contends, though, that museums in general failed in adapting to the digital age. (He focuses on art museums, but the argument is valid everywhere.) For instance, he notes that predigital inventory catalogs designed with museum staff in mind did not mutate into inventive and inspired online catalogs adapted to a public view: their style and strategy remain "top down, tethered to standardized schemes, [and] based upon atomized approaches to objects." Either made with tongue in cheek or intended as provocative, Schnapp

---

51 David Pantalony, "What Remains: The Enduring Value of Museum Collections in the Digital Age," *Journal of the History of Science and Technology* 14 (2020): 160–82, quotes on p. 181. The five case studies included here are fascinating, ranging from tuning forks to a cyclotron. On the concept of 'object energy,' Nina Robbins, "Why We Still Need Collections—Museums in the Business of Originality," *ICOFOM Study Series* 48–1 (2020): 183–92.

52 For instance, van Helden and Hankins, "Instruments." Liba Taub, ed., "On Scientific Instruments," *Studies in History and Philosophy of Science Part A* 40 (2009): 337–438. Taub, "Focus: The History of Scientific Instruments." Samuel J. M. M. Alberti, ed., "Shaping Scientific Instrument Collections," *Journal of the History of Collections* 31 (2019): 445–562. Monographies have as well been influential in shaping the field. One that rocked my world was Peter Galison, *Image & Logic. A Material Culture of Microphysics* (Chicago: The University of Chicago Press, 1997).

53 The call for change is not new, Darren Peacock, "Making Ways for Change: Museums, Disruptive Technologies and Organisational Change," *Museum Management and Curatorship* 12 (2008): 333–51. For a recent analysis of digital culture and museum, see Florence Andreacola, ed., "Musées et mondes numériques," *Culture & Musées* 35 (2020): 11–209. For collections as data, S. L. Ziegler, "Open Data in Cultural Heritage Institutions: Can We Be Better Than Data Brokers," *DHQ: Digital Humanities Quarterly* 14.2 (2020), online: http://digitalhumanities.org:8081/dhq/vol/14/2/000462/000462.html (accessed 2 August 2021).

in a telling example compares the digital identities of a Bialetti Moka Express (coffee maker) collected 1) at the New York Museum of Modern Art (MoMA) and sold 2) on Amazon.com. His analysis and conclusion are fascinating: "The Amazon description is richer, more accurate, and opens up a space for engagement and collaborative description. ... [It] treats the Moka as much a *network* as a *product.*" How, he inquires, can we thus describe, "the things and, in turn, the aggregates of things known as collections ... in ways that make them not just accessible to computational analysis but also engaging, if not sensorially, then in other ways?"[54] Thanks to the digital age, collections of objects are ripe for constructing and sharing alternative stories if one knows how to probe and deconstruct them.[55]

Mitchell Whitelaw, associate professor in the School of Art and Design at the Australian National University, agrees. Online GLAM catalogs (Galleries, Libraries, Archives, and Museums), based on simple keyword searches *à la* Google, with fields often partially documented, are what he dubbed 'ungenerous interfaces.' Though the amount of data found on the web for cultural collections is indeed very generous, counting in millions of entries, their presentation in general is not. This situation is particularly important "because the interface plays an inescapable role in mediating digital heritage." According to Whitelaw,

> As an interface, search fails to match the ample abundance of our digital collections and the generous ethos of the institutions that hold them. A more generous interface would do more to represent the scale and richness of its collection. It would open the doors, tear down the drab lobby; instead of demanding a query it would offer multiple ways in, and support exploration as well as the focused inquiry where search excels. In revealing the complexity of digital collections, a generous interface would also enrich interpretation by revealing relationships and structures within a collection.[56]

---

54 Jeffrey Schnapp, "The Intimate Lives of Cultural Objects," in *The Routledge Companion to Media Studies and Digital Humanities*, ed. Jenteray Sayers (New York: Routledge, 2018), 423–32, quotes on pp. 426–8 (emphasis original).

55 See, for example, Bernadette Biedermann, "'Virtual Museums' as Digital Collection Complexes. A Museological Perspective Using the Example of Hans-Gross-Kriminalmuseum," *Museum Management and Curatorship* 32 (2017): 281–97. For a more general and encompassing analysis, Tula Giannini and Jonathan P. Bowen, eds., *Museums and Digital Culture. New Perspectives and Research* (Cham: Springer, 2019).

56 Mitchell Whitelaw, "Generous Interfaces for Digital Cultural Collections," *DHQ: Digital Humanities Quarterly* 9.1 (2015), online: http://www.digitalhumanities.org/dhq/vol/9/1/000205/000205.html (accessed 14 August 2020). For more examples, consult his website: http://mtchl.net/tag/generousinterfaces/ (accessed 14 August 2020).

Schnapp speaks of the "expressive potential of the metadata and media associated with objects and collections … that make up the *sensorium* of information."[57] To him, simply showing collections on the web is insufficient: "it is by making them usable with analytical and interpretive tools, which render them viewable and navigable as both aggregates and individual items, that we can begin to unlock their potential as engines of cultural argument on the micro, meso, and macro scale."[58]

With colleagues from the Université de Montréal, I decided a few years ago to analyze the data provided online by the Waywiser database, containing the description of more than 11,000 scientific instruments from the Collection of Historical Scientific Instruments (CHSI) at Harvard.[59] The search engine is an 'ungenerous interface,' per Whitelaw's description. What then, I asked, can we learn from the overall CHSI collection beyond the individually cataloged objects? Unsurprisingly, as depicted in figure 23, we see that the vast majority of early modern instruments comes from Europe. From around 1800, Europe is losing ground rapidly to the United States, up until 1870 when the latter becomes the dominant producer of instruments preserved in the CHSI. It is not a trend that should surprise anyone. However, it is worth noting the 1870 inflection point corresponds perfectly, as described above, to Harvard hiring John Trowbridge, who orchestrated the shift toward laboratory work and local purchases of instruments. Regarding the 1960 discrepancy in the graph, when a sharp peak of European instruments appears, it results from a single instrument, an M20 KGS microscope produced by the Swiss company Wild Heerbrugg Ltd. A further investigation reveals that the instrument and its comprehensive set of accessories (more than 160 of them), bought on 2 September 1960 by David Wheatland for the handsome amount of $6,000, were all given separate accession numbers, hence artificially increasing the number of 'instruments'

57 Schnapp, "The Intimate Lives of Cultural Objects," 426 (emphasis original).

58 Ibid, 428.

59 Maxime Sainte-Marie, Jean-François Gauvin, and Vincent Larivière, "Les collections muséales au service de la science. Une analyse de la Collection of Historical Scientific Instruments de l'université Harvard," *Document numérique* 2–3 (2017): 137–58. The CHSI online database is found here: http://waywiser.fas.harvard.edu/. For a similar analysis on a much larger collection, Isobel MacDonald, "Counting When, Who and How: Visualizing the British Museum's History of Acquisition Through Collection Data, 1753–2019," *Journal of the History of Collections*, 2022, fhac034, https://doi.org/10.1093/jhc/fhac034. For a similar analysis on another instrument collection, Tacye Phillipson, "Collections Development in Hindsight: A Numerical Analysis of the Science and Technology Collections of National Museums Scotland since 1855," *Science Museum Group Journal* 12 (2019), online: http://dx.doi.org/10.15180/191205/001 (accessed 4 November 2022).

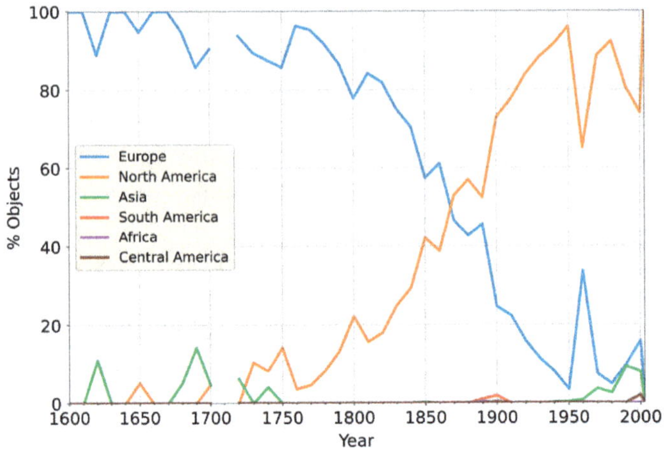

FIGURE 23 Instrument acquisitions over time at the Collection of Historical Scientific Instruments (CHSI), Harvard University. The online Waywiser catalog, with over 11,000 entries, analyzed here from the perspective of the country of origins and the year of manufacture. The gap for 1710 simply means there is no object from between 1701 and 1710. Notice the turning point around 1870.
GRAPH DRAWN BY MAXIME SAINTE-MARIE

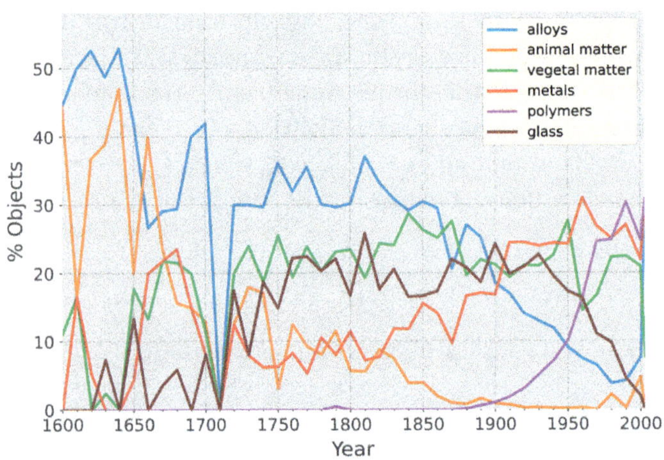

FIGURE 24 Instrument materials over time at the Collection of Historical Scientific Instruments (CHSI), Harvard University. The online Waywiser catalog, with over 11,000 entries, analyzed here from the perspective of the materials and the year of manufacture.
GRAPH DRAWN BY MAXIME SAINTE-MARIE

from Europe in the database on that date.[60] The impact of such curatorial and registrar decisions, very much local, is important to notice and acknowledge, as we will see shortly.

What the Waywiser data reveals regarding the materials employed in making the CHSI instruments is worth a brief mention as well. Figure 24 shows, on the one hand, that the use of organic-based matter (mostly wood) was, on average, rather stable over the centuries (unlike animal-based matter [ivory, leather], which continually decreased over the same period). On the other hand, we see a sharp dip in the utilization of alloys (brass, bronze, etc.) and glass, offset by a similar rise in the mobilization of polymers (Bakelite, hard rubber, and plastic) beginning in the 1900s. The correlation coefficients between glass and polymers actually are very high: indeed, close to 90% of the variation noticed for glass can be explained by the variation seen for polymers, and vice versa.[61] This correlation, however, would not exist if every instrument preserved in the CHSI collection was included in Waywiser: most importantly the 5,000 vacuum tubes made of a glass receptacle inside which is kept the electronics. This added number of objects made with glass material would completely modify the data visually displayed on the above graph. 'All data are local,' and they require analyses beyond the numbers.

More recently, I studied the Louis-Marie Herbarium at Université Laval—where I currently work. Considering the discussion in chapter 1 regarding what an *organum* is, these botanical specimens can no doubt be considered 'instruments.' One of the most important herbaria in Canada, the Herbarium's collection of botanical specimens, going back to the foundation of the university in 1852, has a long and rich history in research and pedagogy. It prioritizes Arctic-alpine, subarctic, and boreal species from Canada and the northern hemisphere.[62] The data of over 145,000 specimens (from circa 800,000), covering mainly the province of Québec, was examined in order to find novel ways of visualizing it. The data analysis provides inter alia a better sense of

60 Wheatland said of this purchase: "It is intended to use all these [accessories] for comparison, demonstration, and some for investigation. Eventually perhaps other basic instruments may be ordered and the accessories distributed among them. We are anxious therefore to have the list as complete as possible to make all the tests and optical experiments." Quote taken from Waywiser, in the historical attributes of the microscope's entry (inv. number, 1997-1-0722).

61 Saint-Hilaire, Gauvin, and Larivière, "Les collections muséales au service de la science," 155.

62 Robert Gauthier, "L'Herbier Louis-Marie," *Cap-aux-diamants* 72 (2003): 54–6. For a general sense of this collection, see Kim Damboise, Herbier Louis-Marie (QFA)—Collection de plantes vasculaires. Version 10.6. Université Laval. Occurrence dataset https://doi.org/10.5886/3p8ltbg7 (accessed via GBIF.org, 20 August 2020). I would like to thank Kim Damboise for sharing the data.

the collection to a general audience, which cannot expect much visual help from the current online and 'ungenerous interface' called Hercul vasculaires.[63] Figure 25, for instance, displays where our selected set of specimens were collected geographically. The collecting density (represented in the darker red color) around the St. Lawrence River is unsurprising, considering how diverse the vegetation is along the river and, of course, how easy it is to reach. A quick look also reveals the many specimens collected in the northern part of Québec, following the main hydrographic basins, rivers, and shores of major water bodies, especially James Bay, Hudson Bay, and Ungava Bay. This flora is presently being studied for an important publishing project.[64]

Beyond the geography of specimen collecting, one might be interested in the sorts of plants collected and when during the year. Figure 26 provides an answer to this question. It will not surprise anyone to learn that most specimens are collected during the summer, in the months of July and August. Though June and September see some activity, the rest of the year remains marginal (winter is coming ...). Such visual information enables anyone to see and interpret the data differently, and to ask questions that would perhaps be overlooked in a regular keyword database *à la* Google. Looking at the number of specimens per family collected per annum (Fig. 27) provides another set of questions worth thinking about. For instance, we see a burst of collecting activity between 1855 and 1864, so how do we explain this? (It is the work of a famous Québec naturalist, abbé Léon Provancher. Though mostly known as an entomologist, his study of native plants led to an important publication, the two-volume *Flore canadienne* (1862), which remained unsurpassed until the *Flore laurentienne* (1935) of brother Marie-Victorin). Additionally, we can distinctly observe which families of plants are sought after and when they were mainly collected. Why is that the case? Is there a historical and epistemic explanation? We could ask as well why some years have seen more collecting activities than others, and who are behind these? Data-driven visualizations cannot answer such questions; actually, it is not their role. The function of data-driven visualizations is to help discern patterns simply impossible to ascertain in the context of ubiquitous keyword-searchable interfaces. The CHSI and Louis-Marie Herbarium databases illustrate the key argument behind the relevancy of investigating data provided by historical and cultural collections. It

63 See it online: http://hercul.herbier.ulaval.ca/ (accessed 11 November 2022). The database now contains close to 200,000 entries.

64 Serge Payette, ed., *Flore nordique du Québec et du Labrador*, 3 vols. (Québec: Presses de l'Université Laval, 2013–2018).

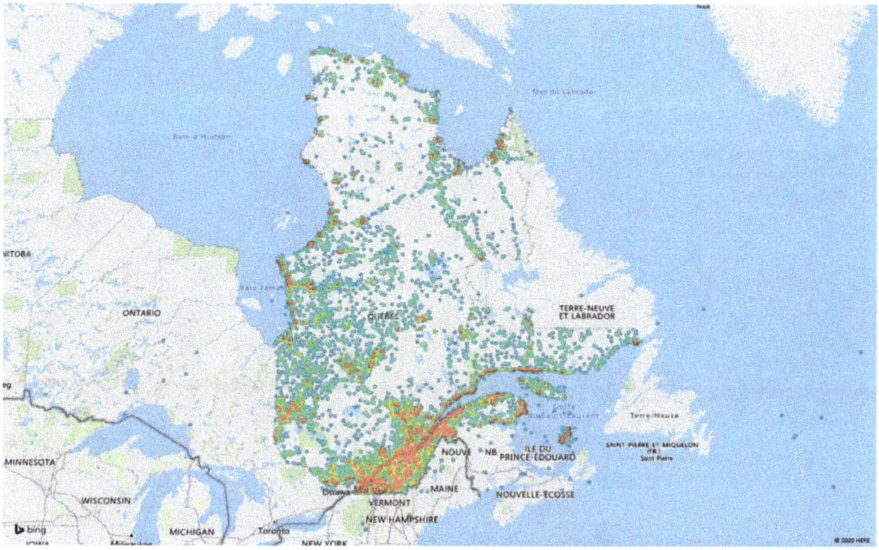

FIGURE 25 A map of the province of Québec showing where botanical specimens were collected. The gradient represents the least (green) to most (red) collected specimens.
GRAPH GENERATED BY MY RESEARCH ASSISTANT, ZHENGYANG ZHANG

provides new tools to probe the past, present, and future of collecting. The goal of data-driven analyses for cultural collections is not to replace object-driven research, or 'object lessons,' but to further it and make it more comprehensive. As Bruno Strasser brilliantly demonstrates, there is a profound connection between the use and in-depth examination of scientific data collections, or databases, and museum collections. Computers, databases, and the high flow of raw data have had a deep, transformative impact on the production of knowledge in the biological sciences over the past half-century. The use and management of modern databases—a new kind of 'instrument'—are "best understood as part of a much longer tradition of collecting, comparing, and classifying objects in nature." Strasser stresses that "today's databases are to the contemporary experimental life sciences what museums have been to natural history: repositories of things and knowledge, as well as key tools for their further production."[65] Yet, one might ask, what is the nature of these data? Is the context of their creation irrelevant to the new knowledge they produce?

---

65 Bruno Strasser, *Collecting Experiments. Making Big Data Biology* (Chicago: The University of Chicago Press, 2019), 8–9.

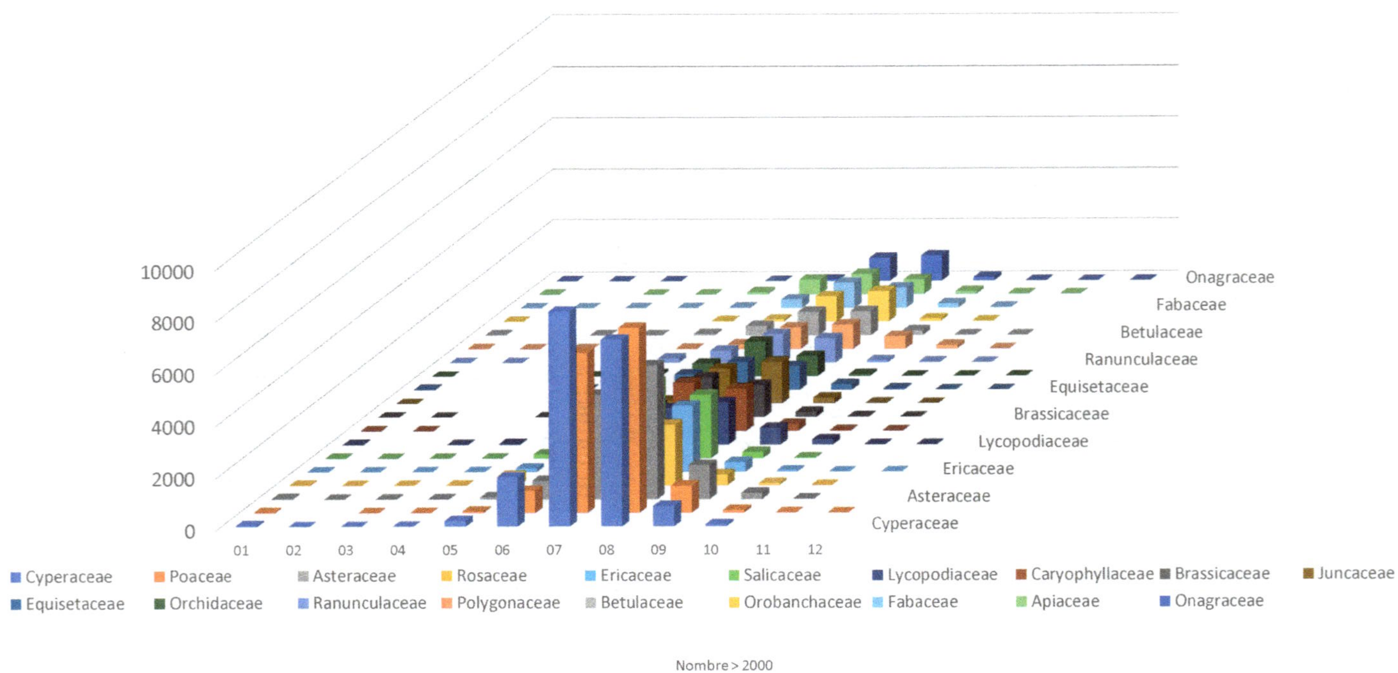

FIGURE 26 A graph showing when botanical specimens were collected. A 3D illustration displaying the most recurring families collected, the number of specimens for each and the month they were collected.
GRAPH GENERATED BY MY RESEARCH ASSISTANT, ZHENGYANG ZHANG

For Yanni Loukissas, a digital media professor at Georgia Institute of Technology, 'all data are local.' Inspired by Clifford Geertz's concept of 'local knowledge,' Loukissas contends that even 'big data' are "grounded in and inseparable from their social, material, and spatial conditions." Databases ought to be understood "as situated and partial because of their attachments to distributed places." Taking as example the Harvard University Arnold Arboretum and its close to 72,000 accessions, Loukissas critically analyzes and uses original methods of information visualization to explore this historical collection dating to 1872. Based on digital and theoretical thinking, his visually compelling images illustrate well "the manifold relationships between data and place." The goal is to display how significantly database practices rely on local knowledge and, especially, to communicate that "data as cultural forms" are "situated socially and technologically, but also in place." It would be a mistake to believe that data are 'out there' and completely disconnected from their locus of creation. On the contrary, Loukissas maintains that "When taken out of place, data can come to be seen as either the view from nowhere or nothing more than data dirt." This is the reason why one must learn "to look *at data*, to investigate how they are made and embedded in the world, before you can look *through data*."[66]

Data visualization of cultural heritage collections is still in its infancy—albeit growing quickly.[67] Its potential is tremendous. On the one hand, it offers opportunities to redefine completely how we look at object collections by proposing novel research questions essentially driven by the visual reconfiguration of data. On the other hand, it favors a conversation with the broader public, who

---

66 Yanni Alexander Loukissas, "A Place for Big Data: Close and Distant Readings of Accessions Data from the Arnold Arboretum," *Big Data & Society* 3 (July-December 2016): 1–20, quotes on pp. 3–4. He developed the concept and gave more examples in Loukissas, *All Data are Local: Thinking Critically in a Data-Driven Society* (Cambridge, MA: MIT Press, 2019), 53, and xv–xvi (emphasis original). The Arnold Arboretum is discussed in chapter 2. The visualization is accessible on the web: http://www.lifeanddeathofdata.org/ (accessed 25 August 2020). Schnapp contends the same thing in his "The Intimate Lives of Cultural Objects," 424–5. Another good case for contextualizing and situating data and datasets is found in Matthew Lavin, "Why Digital Humanists Should Emphasize Situated Data Over Capta," *DHQ: Digital Humanities Quarterly* 15.2 (2021), online http://www.digitalhumanities.org/dhq/vol/15/2/000556/000556.html (accessed 2 August 2021).

67 An interesting analysis of how the field stands today is found in Florian Windhager, *et al.*, "Visualization of Cultural Heritage Collection Data: State of the Art and Future Challenges," *IEEE Transactions on Visualization and Computer Graphics* 25 (2019): 2311–30. One of the chief players is certainly Marian Dörk, who has developed a creative research lab on the topic. See, for instance, Mark-Jan Bludau, Marian Dörk, and Frank Heidmann, "Relational Perspectives as Situated Visualizations of Art Collections," *Digital Scholarship in the Humanities* 36, issue Supplement_2 (2021): ii17–ii29.

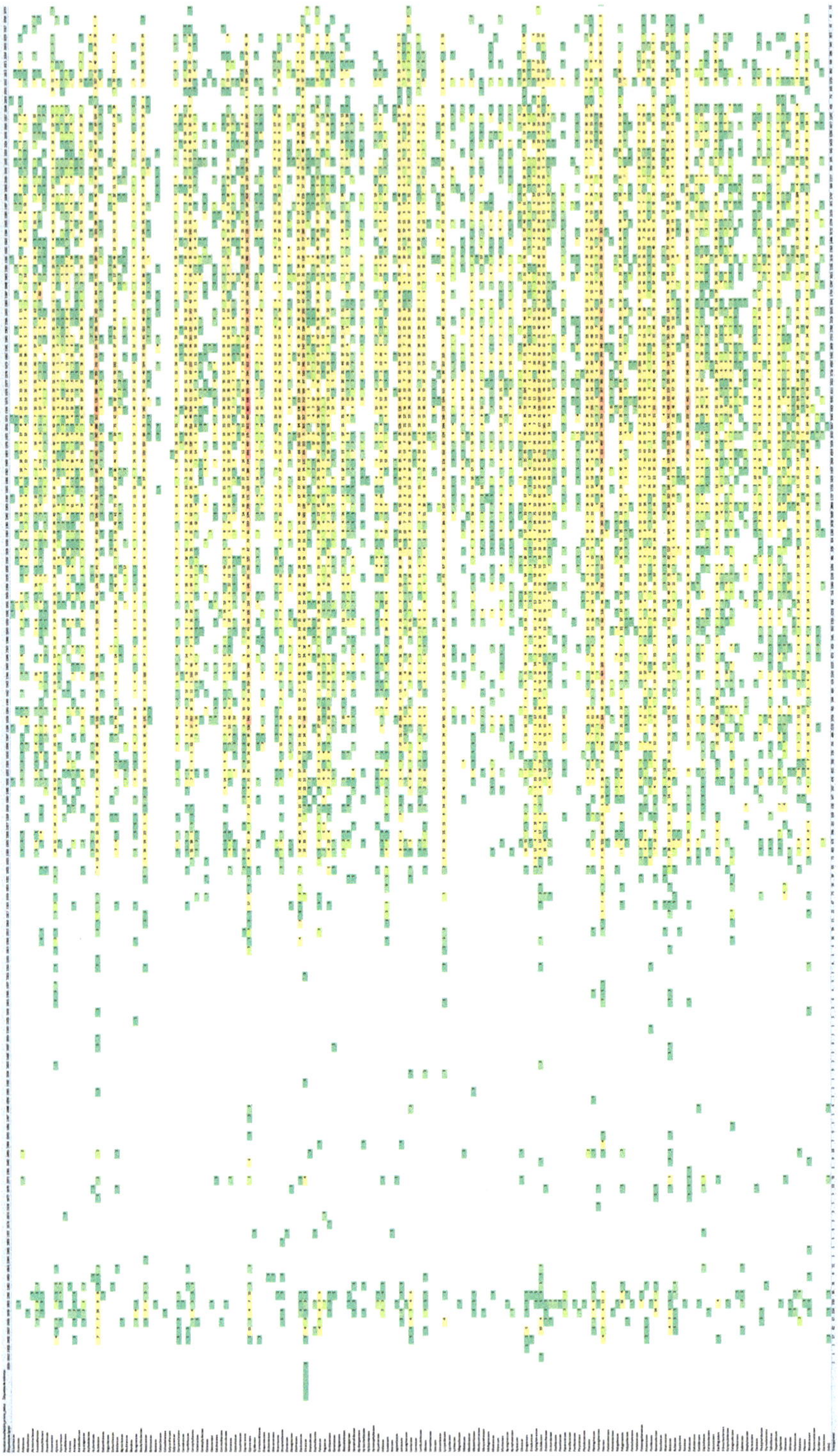

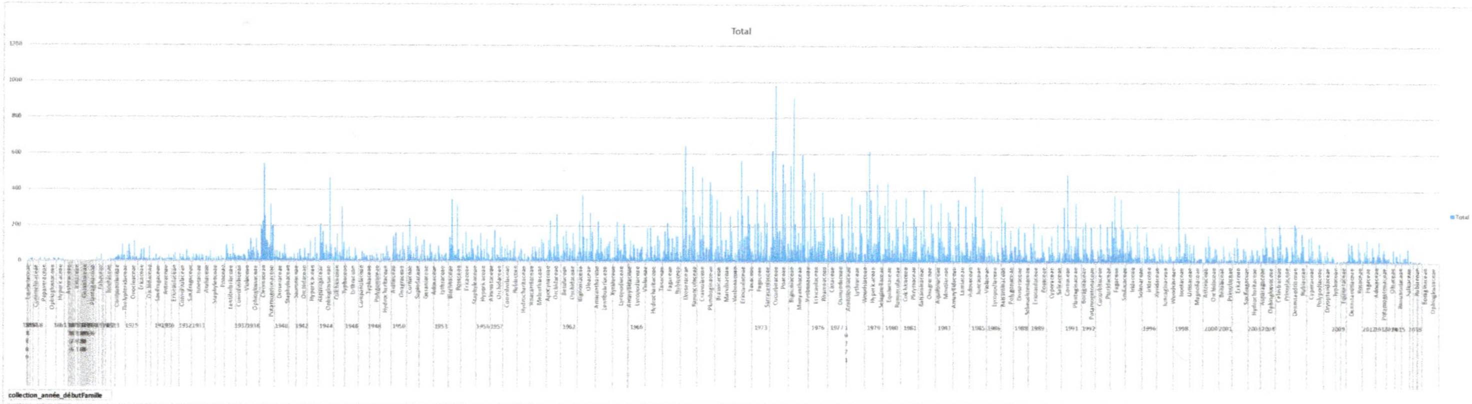

FIGURE 27.A–B Two graphs showing different representations of the number of collected specimens per annum. Above, an illustration depicting on the y-axis the names of the family of plants collected and, on the x-axis, the year they were collected, from 1820 to 2019. The number of collected specimens for each family, per annum, is depicted by the colored squares. Again, the gradient represents the least (green) to the most (red) collected specimens. Below, the same numbers represented in a more conventional graph.

GRAPHS GENERATED BY MY RESEARCH ASSISTANT, ZHENGYANG ZHANG

is not necessarily interested in specific objects or information but would rather like to get the 'big picture' of collection content and significance—and thus grasp why such cultural heritage collections need preserving. For scholars as well as cultural enthusiasts, the digital mastery and display of large amount of data usher in an exciting new era of knowledge-production possibilities available to all for interpretation and pure enjoyment.

Epilogue

In a recent historiographical essay, historian of science Jérôme Lamy observes (as others do periodically) the breadth, significance, and growth of instrument studies since the early 2000s.[1] Looking in turn at instrument practices, at the market economy of instruments, at their politics, their visual and theoretical representations, and the multiple challenges that poses the preservation of this scientific heritage, Lamy provides an interesting framework suitable to address a field no longer at the far periphery of the history of science. This book is another such contribution that underscores the value of examining what an 'instrument' means to our understanding of knowledge production. Set around the concepts of early modern *organum* and *habitus*, it investigates how natural philosophers perceived and recognized the tools and devices of their scientific trade and practices, a practice they also analyzed and deconstructed to determine the strength and failure of their ways of knowing and doing. I argue that the early modern terms and definitions corresponding to 'instrument' and 'practices,' or *organum* and *habitus*, are likely the more accurate and fitting to explain how natural philosophers conceived and self-reflected upon their scientific activities. Together, *organum* and *habitus* form a dyad that represents the importance and central position taken by instrumental practices in the early modern period.

As it might have become obvious while reading the book (I did not try very hard to hide it), my professional career considerably influenced the academic one—and vice versa. The third part of the book, *museum*, is an effort to bring this concept into a dialogue with the first two parts, to generate a triptych composed of *organum, habitus*, and *museum*. As a structured, well-studied, and sometimes controversial institution, the museum has been for centuries a place for preserving, documenting, and exhibiting (among other things) objects recovered at a certain moment of their active or concluded life. Once safely stored in a museum, how can we reclaim the core meaning of an 'instrument' as an *organum*, the multiple uses and dispositions of its *habitus*? By shifting the point of view of a conventional conservation-restoration from an ontology solely focused on the materiality of an object to an ontology emphasizing an object's operation and its operating conditions. Part 3 explores this shift from a theoretical perspective, adding numerous examples taken for the most part from a personal museum experience.

1 Jérôme Lamy, "Le renouveau de l'histoire des instruments scientifiques: Essai bibliographique," *Artefact. Techniques, histoire et sciences humaines* 17 (2022): 9–38. En ligne: https://journals.openedition.org/artefact/13018.

 | DOI:10.1163/9789004504615_012

I want incidentally to end the book with a reflection based on one such personal museum experience. I believe it encompasses the overall essence of my argument, although the historical context is contemporary. A few months after being hired as Director of Administration / Lecturer for the Collection of Historical Scientific Instruments (CHSI) at Harvard, Peter Galison and I met with Thomas Rockwell, Senior Creative Director of Global Collaborations for the San Francisco Exploratorium (October 18, 2011). Over the following 2 ½ years, we discussed the feasibility of recreating at the Exploratorium the CHSI's successful *X-rays of the Soul: Rorschach & the Projective Test* exhibit (2011–2012), besides conducting a weeklong workshop involving students from the Department of the history of science at Harvard and professional staff from the CHSI. Though the exhibit was not duplicated owing to various logistical and budgetary complications, it appeared clear the workshop would be an easy and rewarding opportunity to cross-fertilize our museum philosophies and intellectual practices.

As Frank Oppenheimer, founder of the Exploratorium, used to say: "No one ever flunked a museum." Though the CHSI's and Exploratorium's undertakings and practices appeared at separate ends of a methodological and philosophical spectrum (something Michael John Gorman wants to fundamentally change with his *idea collider* concept and development process[2]), our respective visitors enjoyed what they saw in each location. There was nothing to fail, only gains to make. The workshop asked one essential question: Was a mutual methodological formula spanning both philosophies possible? Convinced it was possible, the stimulating discussions and hands-on activities in San Francisco in June 2014 revealed the many rewards—and obstacles—of building such a bridge. For the dozen of CHSI participants, who felt we could contribute to the structural foundation of that bridge, one important thing we learned from the workshop was the tremendous value of spending concentrated time hashing out important questions and issues. We all felt we accomplished more during the weeklong workshop than if we'd met a couple of hours per week over a longer period. The rapid prototyping concept we explored at the Exploratorium was novel and illuminating to us. We discovered that when a group openly discusses an idea, from all possible sides and angles, it provides greater enthusiasm and multiple fresh contributions. We learned to be bolder, not to be afraid of failure. To reach the full potential of the CHSI, the *laboratory of laboratories*, as Galison likes to call it, we recognized we ought to contemplate and experiment with the Exploratorium's innovative practices.

2 Michael John Gorman, *Idea Colliders: The Future of Science Museums* (Cambridge, MA: MIT Press, 2020).

It was in this spirit and environment that I developed a thought experiment during the workshop. Named *Bursting Waves: Feedback Loops and the American Technoscience Complex*, the exhibit concept covered the period between 1940 and 1970, the rise (and full-swing expansion) of electronics in the Cold War industrial-military complex. The goal was to see whether I could embrace the notion of a multimedia exhibit and, at the same time, find if and how I could juxtapose—intricately mesh—in one space the visitor's experience of an art gallery, a science center, and a history of science museum. I needed something strong, something evocative at the core of the project. I found it in the design artwork by Robert Gerard Pietrusko, Associate Professor of landscape architecture at the Weitzman School of Design, University of Pennsylvania. Bobby, who was at Harvard at the time and made crucial contributions to the *X-rays of the Soul* exhibit, was developing an interactive installation called *Six Microphones*, described as "a site-determined composition of audio feedback which explores the mutually constitutive relationships among sound, space, and audience."[3] (Fig. 28) The artwork itself was not generating any sound; ambient sound was amplified and reorganized by the visitor's own movement within the gallery. The artist characterized it as 'an act of organized listening:' the artwork "transforms the gallery from a space in which sound simply propagates, to the medium required to bring sound into being. While not site-specific, *Six Microphones* is fully site-determined: no sound is generated, it is only amplified." Around this installation, which used microphones pointed at loudspeakers to produce the feedback, and where bodies crashed against sound waves, historical objects from the CHSI collections would be displayed in support of the overarching narrative focused on the production, understanding, and deployment of electronic-generated waves. Computers, synthesizers, radios, and wave generators—black boxes—from the Cold War era would coalesce in one storyline about technology, war, art, and physics. Helped by movie clips, hands-on exercises, and foremost by the interactive artwork, visitors would have discovered an invisible world. It would have been a feast for the ears—and for the mind.

At the core of the exhibit, Bobby Pietrusko's sound installation would have been set up at the entrance of the space. It was the art experience (the unsettling sensation) visitors would have to negotiate before being introduced to the science and sociocultural ideas in the second part of the room. Feel first, explain and understand later: Linking art to science. Dark gray melamine foam acoustic tiles would have covered the walls in the front half of the exhibit

3 Quotes and information are found on Pietrusko's website: http://www.warning-office.org/wo-six-microphones (accessed on 3 September 2021).

FIGURE 28 Six Microphones installation
CREDIT: ROBERT GERARD PIETRUSKO, 'SIX MICROPHONES.'
© JEFF ELSTONE, 2015

space to somewhat soundproof the acoustic area, where the sound installation would be experienced by visitors. Moreover, these tiles would help introduce—and explain—the work of Leo Beranek (mentioned in the permanent exhibition of the Putnam Gallery, one floor below), who designed the first anechoic chamber in the early 1940s. The second part of the exhibit was divided by a central wall, on which graphics and projections would bring out the Cold War technoscience complex. That section (Fig. 29) was composed of several instruments from the CHSI collection—including a vast array of iconic vacuum tubes and old radios—in addition to banner-like projection screens, LCD's display screens arranged in tiles, and a hands-on station. A reverse view of the exhibit displays on that side of the wall a white board, where visitors would draw electronics sketches, equations, and mention what they like about the exhibit—and whatever else they wished to express. A creative white board that went naturally with the close-by hands-on station. The latter, designed by the Exploratorium professionals, would become a place where people could recreate feedback loops, play with and learn from basic electronic components and radios, which were so important to so many people from the 1930s onward, including several future famous scientists. In just one week, I managed to generate a complete vision for the project, which thrilled me by its originality and

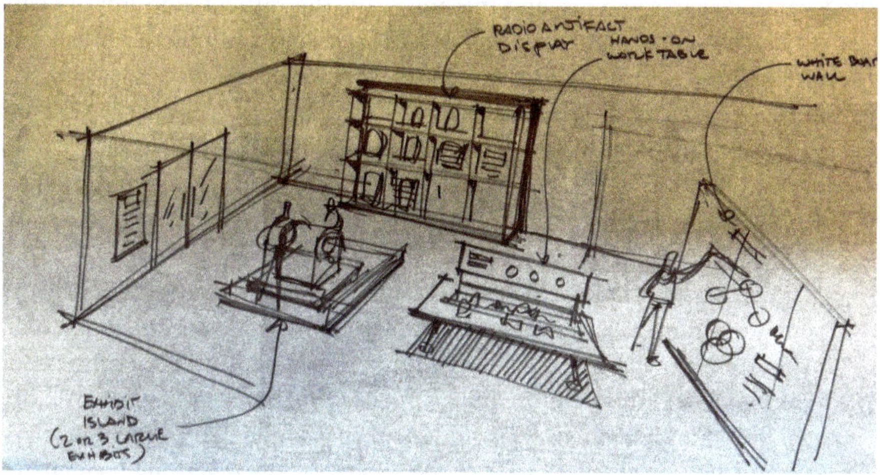

FIGURE 29 A reverse-view of the exhibit project's second part using rapid prototyping technique.
SKETCH DRAWN BY ADAM PROST, EXPLORATORIUM

its contrast with everything I had ever done at the CHSI. Though the exhibit was never created, it opened my eyes to the endless possibilities a creative team brings to a project.

This work-based anecdote is important as regards the content and development of this book. For the past twenty years, I have kept one foot in academia and the other in the museum world. Crossing regularly back and forth between these two universes, I discovered how little attention each pays to the other—most of the time. Though remarkable scholarly and professional initiatives and productions have seen the light of day over the past thirty years, I still wonder why a rift exists between the fields of the history of science and scientific heritage. Teaching with and learning through objects, for one thing, is more difficult than with books and archival materials. Objects are clumsy, sometimes large and heavy, and are difficult to access (and handle) owing to extensive (I would dare say overwhelmingly restrictive) conservation guidelines and protocols. Objects, to borrow Chris Caple's apt description, are in many ways 'reluctant witnesses to the past.'[4] Although they can be digitized, as are paper archives, doing so still prevents scholars from performing deep and careful analyses. One cannot 'read' an object as she would a (reproduced or not) piece of paper; an object offers a myriad of subtle hints once it lands in our hands: the weight, the craftsmanship of materials and its chemical analysis,

4 Chris Caple, *Objects. Reluctant Witnesses to the Past* (London: Routledge, 2006).

the wear and tear, scratches and missing parts, the intricate motion of gears and levers, the strategy and tactics of use, the previous add-ons and calculated reconstruction performed to fix it, etc. (Historians of the book would argue it is likewise for print materials ...) I'm always reminded of Michael Wright's description of the Antikythera mechanism, of which he did such a remarkable reconstruction many years ago. Discussing the virtue of a hands-on reproduction, he wrote:

> Computer animation has advantages in communication with the public. ... However, the virtual model has no friction and suffers no wear. There are no questions of mass, inertia, weight, strength, stiffness or working loads. It is no more than a moving diagram. Without the validity conferred on it by the working of the physical model, it would be of very little value, and it lacks much of the appeal of "the real thing:" which, we find, the physical model does most definitely possess.[5]

The appeal of the 'real thing' inspires us to look beyond the semiotics of objects and instruments. The 'real thing' entices us to investigate its provenance, its use and ruin (disrepair), and its treasured rebirth in cultural institutions dedicated to preservation, research and education. Without the 'real thing' entire scholarly fields (archaeology and paleontology easily come to mind) would not exist. Our history as a species would be filled with gaping holes and unanswerable questions. It is, I believe, the same for the history of science.

Instruments of Knowledge is thus a modest attempt at emphasizing the role of the object, the 'real thing,' toward the production of knowledge and *scientia*. To achieve that goal, *organum*, *habitus*, and *museum*, the three key concepts structuring this book, were chosen to ground my study in relevant theoretical and practical issues based on ongoing academic and museum experience. What is an instrument, how it is wielded and used, and why it is safeguarded *ad aeternam* when the function it accomplished is heretofore considered obsolete, are basic questions that have motivated my work over the last two decades. *Organum*, *habitus*, and *museum* are intellectual hubs on my personal road map, an itinerary that has helped me think about the role and meaning of instruments in the history of science and in my professional career as a museum curator and administrator. The short Exploratorium workshop mentioned above, including the presentation of the exhibit thought experiment I

5 Michael Wright, "The Antikythera Mechanism: Reconstruction as a Medium for Research and Publication," in *Reconstructions. Recreating Science and Technology of the Past*, ed. Klaus Staubermann (Edinburgh: National Museums Scotland, 2011), 1–20, quote on p. 18.

conducted, is a good example of my creative approach regarding 'the dead collection of discarded instruments' (Galison, see chapter 6), preserved in scores of museums, private collections, colleges, and universities across the world. The field is growing and the attention toward scientific heritage is reaching an unprecedented apogee. This book, attempting to link scholarship about early modern instruments to an original museum practice is my small contribution to the building of a bridge embracing both sides of the (hopefully narrowing) divide.

Bibliography

Abt, Jeffrey. "The Origins of the Public Museum." In *A Companion to Museum Studies*, edited by Sharon Macdonald, 115–34. Oxford: Blackwell Publishing, 2006.

Ackermann, Robert J. *Data, Instruments, and Theory: A Dialectical Approach to Understanding*. Princeton: Princeton University Press, 1985.

Adamson, Glenn. *Fewer, Better Things*. New York: Bloomsbury Publishing, 2018.

Agamben, Giorgio. *What is an Apparatus? And Other Essays*, translated by David Kishik and Stefan Pedatella. Stanford: Stanford University Press, 2009.

Aguila, Emmanuel. *Conservation-restauration du "Tellurium ou loxocosme" (n⁰ 07255-0000) du Conservatoire national des arts et métiers. Étude de l'usure des pivots dans un mécanisme d'horlogerie classique*. Paris: Institut national du patrimoine, mémoire de fin d'études, 2015.

Aguila, Emmanuel. "La mécanique céleste. La restauration du tellurium du CNAM." *Patrimoines. Revue de l'Institut national du patrimoine* 12 (2016): 104–9.

Alberti, Samuel J. M. M. "Objects and the Museum." *Isis* 96 (2005): 559–71.

Alberti, Samuel J. M. M., ed. "Shaping Scientific Instrument Collections." *Journal of the History of Collections* 31 (2019): 445–562.

Alder, Ken. *Engineering the Revolution: Arms & Enlightenment in France, 1763–1815*. Chicago: The University of Chicago Press, 1997.

Alder, Ken. "Making Things the Same: Representation, Tolerance, and the End of the Old Regime in France." *Social Studies of Science* 28 (1998): 499–545.

Alder, Ken, ed. "Focus: Thick Things." *Isis* 98 (2007): 80–142.

Alpers, Svetlana. *The Art of Describing: Dutch Art in the Seventeenth Century*. Chicago: The University of Chicago Press, 1983.

Alpers, Svetlana. "The Museum as a Way of Seeing." In *Exhibiting Cultures: The Poetics and Politics of Museum Displays*, editedby Ivan Karp and Steven D. Lavine, 23–32. Washington, D.C.: Smithsonian Press, 1991.

André, Louis. "Scientific Instrument Restoration: Problems and Practice at the Musée des arts et métiers and in France." In *Proceedings of the Workshop Held in Florence, December 14–15, 1998*, edited by Laura Manetti, 23–30. Florence: Le Lettere, 2000.

Andreacola, Florence, ed. "Musées et mondes numériques." *Culture & Musées* 35 (2020): 11–209.

Andrewes, William J. H. "The Life and Work of David Pingree Wheatland (1898–1993)." *Journal of the History of Collections* 7 (1995): 261–8.

Andrewes, William J. H. "The Legacy of David Wheatland." *Nuncius* 16 (2001): 687–701.

Antognazza, Maria Rosa. *Leibniz. An Intellectual Biography*. Cambridge: Cambridge University Press, 2009.

Appadurai, Arjun, ed. *The Social Life of Things. Commodities in Cultural Perspective.* Cambridge: Cambridge University Press, 1986.

Aristotle. *The Complete Works of Aristotle,* edited by Jonathan Barnes, 2 vols., revised edition. Princeton: Princeton University Press, 1984.

Arnou, René, S.J. *L'Homme a-t-il le pouvoir de connaître la vérité? Réponse de Saint Thomas: La connaissance par habitus.* Rome: Presses de l'Université Grégorienne, 1970.

Ash, Eric H., ed. *Expertise. Practical Knowledge and the Early Modern State.* Chicago: The University of Chicago Press [*Osiris*, vol. 25], 2010.

Aufrère, Sydney. *La Momie et la tempête: Nicolas-Claude Fabri de Peiresc et la curiosité égyptienne en Provence au début du XVIIe siècle.* Avignon: Editions A. Barthélemy, c1990.

Augarde, Jean-Dominique. "La fabrication des instruments scientifiques du XVIIIe siècle et la corporation des fondeurs." In *Studies in the History of Scientific Instruments,* edited by Christine Blondel, *et al.*, 53–72. London: Rogers Turner Books Ltd, 1989.

Austern, Linda Phyllis. "'Tis Nature's Voice': Music, Natural Philosophy and the Hidden World in Seventeenth-Century England." In *Music Theory and Natural Order from the Renaissance to the Early Twentieth Century,* edited by S. Clark and A. Rehding, 30–67. Cambridge: Cambridge University Press, 2001.

Bacon, Francis. *The New Organon,* edited by Lisa Jardine and Michael Silverthorne. Cambridge: Cambridge University Press, 2000 [1620].

Bailhache, Patrice. "L'Harmonie universelle: La musique entre les mathématiques, la physique, la métaphysique et la religion." *Les Études philosophiques* 1/2 (1994): 13–24.

Baillet, Adrien. *La Vie de Monsieur Des-Cartes,* 2 vols. Paris, 1691.

Baird, Davis. *Thing Knowledge: A Philosophy of Scientific Instruments.* Berkeley: University of California Press, 2004.

Bakhtin, Mikhail. *Rabelais and his World,* translated by Helene Iswolsky. Bloomington: Indiana University Press, 1984.

Baltes, John. *The Empire of Habit. John Locke, Discipline, and the Origins of Liberalism.* Rochester: University of Rochester Press, 2016.

Barbieri, Patrizio. "Musiche e strumenti 'alla greca' nel seicento Italiano." In *Francesco Buti tra Roma e Parigi: Diplomazia, poesia, teatro,* edited by Francesco Luisi, 349–64. Roma: Torre d'Orfeo Editrice, 2009.

Barbieri, Patrizio. *Physics of Wind Instruments and Organ Pipes 1100–2010.* Rome: Il Levante Libreria Editrice, 2013.

Barbour, J. Murray. *Tuning and Temperament: A Historical Survey.* New York: Da Capo Press, 1972.

Barclay, Robert. "Instruments, Conservation, Restoration, Copying of." Grove Music Online. https://www-oxfordmusiconline-com.ezp-prod1.hul.harvard.edu/grovemusic/view/10.1093/gmo/9781561592630.001.0001/omo-9781561592630-e-0000046526.

Baron, Roger. *Science et sagesse chez Hugues de Saint-Victor*. Paris: Lethielleux, 1957.

Bates, Don. "*Machina ex Deo*. William Harvey and the Meaning of Instrument." *Journal of the History of Ideas* 61 (2000): 577–93.

Baudrillard, Jean. *The System of Objects*. London: Verso, 2005.

Beeckman, Isaac. *Journal tenu par Isaac Beeckman de 1604 à 1634*, edited by Cornélis de Waard, 4 vols. The Hague: M. Nijhoff, 1939–53.

Beguin, René. *Restauration des horloges, montres et pendules*. Fribourg: Office du livre S. A., 1979.

Belgioioso, Giulia. "Descartes e gli artigiani." In *La Biografia intellettuale di René Descartes attraverso la* Correspondance, edited by Jean-Robert Armogathe, Giulia Belgioioso, and Carlo Vinti, 113–65. Naples: Vivarium, 1999.

Bennett, Jim. "The Mechanic's Philosophy and the Mechanical Philosophy." *History of Science* 24 (1986): 1–28.

Bennett, Jim. "Geometry in Context in the Sixteenth Century: The View from the Museum." *Early Science and Medicine* 7 (2002): 214–30.

Bennett, Jim. "Shopping for Instruments in Paris and London." In *Merchants & Marvels: Commerce, Science and Art in Early Modern Europe*, edited by Pamela H. Smith and Paula Findlen, 370–95. New York: Routledge, 2002.

Bennett, Jim. "Presidential Address: Knowing and Doing in the Sixteenth Century: What Were Instruments For?" *The British Journal for the History of Science* 36 (2003): 129–50.

Bennett, Jim. "Museums and the History of Science: Practitioner's Postscript." *Isis* 96 (2005): 602–8.

Bennett, Jim. "The Mechanical Arts." In *The Cambridge History of Science. Vol. 3 Early Modern Science*, edited by Katharine Park and Lorraine Daston, 673–95. Cambridge: Cambridge University Press, 2008.

Bennett, Jim, and Sofia Talas, eds. *Cabinet of Experimental Philosophy in Eighteenth-Century Europe*. Leiden: Brill, 2013.

Bennett, Jim, and Rebekah Higgitt. "London 1600–1800: Communities of Natural Knowledge and Artificial Practices." *The British Journal for the History of Science* 52 (2019): 183–96.

Bennett, Tony. "The Exhibitionary Complex." *New Formations* 4 (1988): 73–102.

Bensaude-Vincent, Bernadette, and Christine Blondel, eds. *Science and Spectacle in the European Enlightenment*. New York: Routledge, 2008.

Berg, Maxine. *Luxury & Pleasure in Eighteenth-Century Britain*. Oxford: Oxford University Press, 2010.

van Berkel, Klaas. "Descartes' Debt to Beeckman: Inspiration, Cooperation, Conflict." In *Descartes' Natural Philosophy*, edited by Stephen Gaukroger, John Schuster, and John Sutton, 46–59. London: Routledge, 2000.

Bertoloni Meli, Domenico. "The Role of Numerical Tables in Galileo and Mersenne." *Perspectives on Science* 12 (2004): 164–90.

Bertoloni Meli, Domenico. *Thinking with Objects: The Transformation of Mechanics in the Seventeenth Century*. Baltimore: The Johns Hopkins University Press, 2006.

Bertoloni Meli, Domenico. *Mechanism, Experiment, Disease: Marcello Malpighi and Seventeenth-Century Anatomy*. Baltimore: The Johns Hopkins University Press, 2011.

Bertoloni Meli, Domenico. *Mechanism: A Visual, Lexical, and Conceptual History*. Pittsburg: University of Pittsburg Press, 2019.

Bertucci, Paola. *Artisanal Enlightenment. Science and the Mechanical Arts in Old Regime France*. New Haven: Yale University Press, 2017.

Bewer, Francesca G. *A Laboratory for the Art. Harvard's Fogg Museum and the Emergence of Conservation in America, 1900–1950*. New Haven: Yale University Press, 2010.

Biagioli, Mario. *Galileo's Instruments of Credit: Telescopes, Images, Secrecy*. Chicago: The University of Chicago Press, 2006.

Biasiol, Pietro. "Christophe de Villiers: Un *savant* di provincia nella corrispondenza di M. Mersenne." *Rivista Critica di Storia della Filosofia* 36 (1981): 379–400.

Biedermann, Bernadette. "'Virtual Museums' as Digital Collection Complexes. A Museological Perspective Using the Example of Hans-Gross-Kriminalmuseum." *Museum Management and Curatorship* 32 (2017): 281–97.

Biraud, Guy. *La Restauration et la conservation des appareils scientifiques de collection*. La Rochelle: Édition Biraud, 1987.

Blair, Ann. *Too Much to Know. Managing Scholarly Information Before the Modern Age*. New Haven: Yale University Press, 2010.

Blake, Ralph M. "The Rôle of Experience in Descartes' Theory of Method (I)." *The Philosophical Review* 38 (1929): 125–43.

Blake, Ralph M. "The Rôle of Experience in Descartes' Theory of Method (II)." *The Philosophical Review* 38 (1929): 201–18.

Blay, Michel. "Mersenne expérimentateur: Les études sur le mouvement des fluides jusqu'en 1644." *Les Études philosophiques* 1/2 (1994): 69–86.

Bludau, Mark-Jan, Marian Dörk, and Frank Heidmann. "Relational Perspectives as Situated Visualizations of Art Collections." *Digital Scholarship in the Humanities* 36, issue Supplement_2 (2021): ii17–ii29.

de Boer, Bas, Hedwig te Molder, and Peter-Paul Verbeek. "Understanding Science-in-the-Making by Letting Scientific Instruments Speak: From Semiotics to Postphenomenology." *Social Studies of Science* 51, no. 3 (2021): 392–413.

Boerhaave, Hermann. *Elements of Chemistry* …, 2 vols. London, 1735.

Boethius, Anicius Manlius Severinus. *Fundamentals of Music*, translated by Calvin M. Bower. New Haven: Yale University Press, 1989.

Bortolotto, Chiara, ed. *Le Patrimoine culturel immatériel: enjeux d'une nouvelle catégorie*. Paris: Maison des Sciences de l'Homme, 2011.

Bourdieu, Pierre. *Le sens pratique*. Paris: Les Éditions de Minuit, 1980.

Bourdieu, Pierre. *Raisons pratiques. Sur la théorie de l'action*. Paris: Seuil, 1994.

Boyle, Alison. "'Not for Their Beauty': Instruments and Narratives at the Science Museum, London." In *Scientific Instruments on Display*, edited by Silke Ackermann, Richard Kremer, and Mara Miniati, 37–60. Leiden: Brill, 2014.

Boyle, Alison. "Modern Physics in the Museum: Shaping a UK National Collection in the Twentieth Century." *Journal of the History of Collections* 31 (2019): 487–502.

Boyle, Robert. *The Christian Virtuoso Shewing that by Being Addicted to Experimental Philosophy, a Man is Rather Assisted than Indisposed to be a Good Christian*. London, 1690.

Brandi, Cesare. *Teoria del restauro*. Turin: Piccola Biblioteca Einaudi, 1977.

Brandi, Cesare. *Il restoro. Teoria e pratica*, edited by Michele Cordaro. Rome: Editori Riuniti, 1994.

Brenni, Paolo. "Strumenti scientifici: il restauro del meteorografo Secchi." *OPD Restauro* 5 (1993): 99–107.

Brenni, Paolo. "Il meteorografo di padre Angelo Secchi." *Nuncius* 8 (1993): 197–247.

Brenni, Paolo. "Better than New? Scientific Instrument Restoration in Italy." In *Proceedings of the Workshop Held in Florence, December 14–15, 1998*, edited by Laura Manetti, 51–6. Florence: Le Lettere, 2000.

Brenni, Paolo. "Quelques considérations sur la restauration des instruments scientifiques." *CoRé* 17 (2006): 8–11.

Brenni, Paolo. *Il gabinetto di fisica dell'Istituto tecnico toscano. Guida alla visita*. Florence: Edizioni Polistampa, 2009.

Brenni, Paolo. "The Fraunhofer's Refractor of Tartu (Dorpstat) and Its Restoration." *Bulletin of the Scientific Instrument Society* 113 (2012): 2–7.

Brenni, Paolo. "The Evolution of Teaching Instruments and Their Use between 1800 and 1930." *Science & Education* 21 (2012): 191–226.

Brenni, Paolo. "Filming Nineteenth-Century Physics Demonstrations with Historical Instruments." In *Historical Scientific Instruments in Contemporary Education*, edited by Elizabeth Cavicchi and Peter Heering, 34–49. Leiden: Brill, 2022.

Brentjes, Sonja. *Peiresc's Interests in the Middle East and Northern Africa in Respect to Geography and Cartography*. Berlin: Max Planck Institute for the History of Science [preprint 269], 2004.

Briggs, Robin. "The Académie des sciences and the Pursuit of Utility." *Past and Present* 131 (1991): 38–88.

Brown, Bill. "Thing Theory." *Critical Inquiry* 28 (2001): 1–22.

Brugmans, Henri-L. *Le séjour de Christian Huygens à Paris et ses relations avec les milieux scientifiques français, suivi de son* Journal de voyage à Paris et à Londres. Paris: Impressions Pierre André, 1935.

Brundtland, Terje. "From Medicine to Natural Philosophy: Francis Hauksbee's Way to the Air-Pump." *The British Journal for the History of Science* 41 (2008): 209–40.

Bryden, D. J. "Evidence from Advertising for Mathematical Instrument Making in London, 1556–1714." *Annals of Science* 49 (1992): 301–36.

Bub, Daniel, and Michael Masson. "Gestural Knowledge Evoked by Objects as Part of Conceptual Representations." *Aphasiology* 20, no. 9 (2006): 1112–24.

Bucciantini, Massimo, Michele Camerota, and Franco Giudice. *Galileo's Telescope: A European Story,* translated by Catherine Bolton. Cambridge, MA: Harvard University Press, 2015.

Buccolini, Claudio. "Mersenne et la philosophie baconienne en France à l'époque de Descartes." In *Bacon et Descartes. Genèse de la modernité philosophique*, edited by Elodie Cassan. Paris: ENS Editions, 2014.

Buchwald, Jed. "Descartes's Experimental Journey Past the Prism and Through the Invisible World to the Rainbow." *Annals of Science* 65 (2008): 1–46.

Burkitt, Ian. "Technologies of the Self: Habitus and Capacities." *Journal for the Theory of Social Behaviour* 32 (2002): 219–37.

Burnett, Charles. "European Knowledge of Arabic Texts Referring to Music: Some New Material." *Early Music History* 12 (1993): 1–17.

Camic, Charles. "The Matter of Habit." *The American Journal of Sociology* 91 (1986): 1039–87.

Camille, Michael. *The Gothic Idol: Ideology and Image-Making in Medieval Art.* Cambridge: Cambridge University Press, 1989.

Canguilhem, Georges. "Descartes et la technique [1937]." In *Œuvres complètes tome 1: Écrits philosophiques et politiques (1926–1939)*, edited by Jean-François Braunstein and Yves Schwartz, 490–98. Paris: Vrin, 2011.

Caple, Chris. *Objects. Reluctant Witnesses to the Past.* London: Routledge, 2006.

Cardano, Girolamo. *Hieronymus Cardanus (1501–1576): Writings on Music*, translated and edited by Clement A. Miller. [n. p.]: American Institute of Musicology, 1973.

Cardon, Dominique. *La Draperie au Moyen Âge. Essor d'une grande industrie européenne.* Paris: CNRS, 1999.

Carlyle, Margaret. "Collecting the World in her Boudoir: Women and Scientific Amateurism in Eighteenth-Century Paris." *Early Modern Women: An Interdisciplinary Journal* 11 (2016): 149–61.

Carreras, Juan José, and Bernardo García García, eds. *The Royal Chapel in the Time of the Habsburgs: Music and Ceremony in Early Modern European court*, translated by Yolanda Acker with English edition by Tess Knighton. Woodbridge, UK: Boydell Press, 2005.

Carter, Tim, and John Butt, eds. *The Cambridge History of Seventeenth-Century Music.* Cambridge: Cambridge University Press, 2005.

Castriota, Brian. "Authenticity, Identity, and Essentialism: Reframing Conservation Practice." In *What is the Essence of Conservation? Materials for a Discussion*, edited by François Mairesse and Renata F. Peters, 39–48. Paris: ICOFOM, 2019.

Cavaillé, Jean-Pierre. "Descartes stratège de la destination." *XVIIe siècle* 177 (1992): 551–59.

Cavaillé, Jean-Pierre. "'Le plus éloquent philosophe des derniers temps': les stratégies d'auteur de René Descartes." *Annales: histoire, sciences sociales* 49 (1994): 349–67.

Cavicchi, Elizabeth, and Peter Heering, eds. *Historical Scientific Instruments in Contemporary Education*. Leiden: Brill, 2022.

Cazaux, Christelle. *La Musique à la cour de François Ier*. Paris: École Nationale des Chartes, 2002.

Cazeaux, Isabelle. *French Music in the Fifteenth and Sixteenth Centuries*. Oxford: Basil Blackwell, 1975.

de Certeau, Michel. *L'Invention du quotidien. 1—L'art de faire*. Paris: Gallimard, 1990.

Charbonneau, Mathieu. "The Cognitive Life of Mechanical Molecular Models." *Studies in the History and Philosophy of Biological and Biomedical Sciences* 44 (2013): 585–94.

Choay, Françoise. "Un nouveau Luna Park était-il nécessaire?" *Le Débat* 147 (2007): 57–64.

Christin, Olivier. "Du culte chrétien au culte de l'art. La transformation du statut de l'image (XVe–XVIIIe siècles)." *Revue d'histoire moderne et contemporaine* 49 (2002): 176–94.

Clair, Jean. *Malaise dans les musées*. Paris: Flammarion, 2007.

Clark, Elizabeth A. *History, Theory, Text. Historians and the Linguistic Turn*. Cambridge, MA: Harvard University Press, 2004.

Clarke, Desmond M. *Descartes's Philosophy of Science*. Manchester: Manchester University Press, c1982.

de Clercq, Peter, ed. *Scientific Instruments: Original and Imitations. The Mensing Collection. Proceedings of a Symposium, Held at the Museum Boerhaave, Leiden, 15–16 October 1999*. Leiden: Museum Boerhaave, 2000.

Cohen, Sarah R. *Art, Dance, and the Body in French Culture of Ancien Régime*. Cambridge: Cambridge University Press, 2000.

Cole, K. C. *Something Incredibly Wonderful Happens: Frank Oppenheimer and His Astonishing Exploratorium*. Chicago: The University of Chicago Press, 2009.

Collins, Harry. *Tacit & Explicit Knowledge*. Chicago: The University of Chicago Press, 2010.

Collins Judd, Cristle. *Reading Renaissance Music Theory: Hearing with the Eyes*. Cambridge: Cambridge University Press, 2000.

Cometti, Jean-Pierre. *Conserver/restaurer. L'œuvre d'art à l'époque de sa préservation technique*. Paris: Gallimard, 2015.

Conein, Bernard, Nicolas Dodier, and Laurent Thévenot, eds. *Les Objets dans l'action. De la maison au laboratoire*. Paris: Éditions de l'École des Hautes Études en Sciences Sociales, 1993.

Conn, Steven. *Do Museums Still Need Objects?* Philadelphia: University of Pennsylvania Press, 2010.

Cook, Harold C. *Matters of Exchange. Commerce, Medicine, and Science in the Dutch Golden Age*. New Haven: Yale University Press, 2007.

Cook, Harold C. *Young Descartes*. Chicago: The University of Chicago Press, 2018.

Cosgrove, D. E. "Should We Take It All so Seriously? Culture, Conservation, and Meaning in the Contemporary World." In *Durability and Change: The Science, Responsibility, and Cost of Sustaining Cultural Heritage*, edited by W. E. Krumbein, *et al.*, 259–66. London: Wiley, 1994.

Cottingham, John, ed. *Descartes' Conversation with Burman*. Oxford: Clarendon Press, 1976.

Crenn, Gaëlle, and Jean-Christophe Vilatte, eds. "L'émotion dans les expositions." *Culture et Musées* 36 (2020): 1–196.

Cudworth, Ralph. *The True Intellectual System of the Universe, Wherein All the Reason and Philosophy of Atheism is Confuted and its Impossibility Demonstrated*. London, 1678.

Cypess, Rebecca. *Curious & Modern Inventions. Instrumental Music as Discovery in Galileo's Italy*. Chicago: The University of Chicago Press, 2016.

Darge, Rolf. *Habitus per actus cognoscuntur. Die Erkenntnis des Habitus und die Funktion des moralischen Habitus im Aufbau des Handlung nach Thomas von Aquin*. Bonn: Bouvier Verlag, 1996.

Daston, Lorraine, ed. *Biographies of Scientific Objects*. Chicago: The University of Chicago Press, 2000.

Daston, Lorraine, ed. *Things that Talk: Object Lessons from Art and Science*. New York: Zone Books, 2004.

Daston, Lorraine. "On Scientific Observation." *Isis* 99 (2008): 97–110.

Daston, Lorraine, and Katharine Park. *Wonders and the Order of Nature, 1150–1750*. New York: Zone Books, 1998.

Daston, Lorraine, and Peter Galison. *Objectivity*. New York: Zone Books, 2007.

Daston, Lorraine, and Elizabeth Lunbeck, eds. *Histories of Scientific Observation*. Chicago: The University of Chicago Press, 2011.

Daumas, Maurice. *Les instruments scientifiques aux XVIIe et XVIIIe siècle*. Paris: Presses Universitaires de France, 1953.

Dear, Peter. *Mersenne and the Learning of the Schools*. Ithaca: Cornell University Press, 1988.

Dear, Peter. "Mersenne et l'expérience scientifique." *Les Études philosophiques* 1/2 (1994): 53–67.

Dear, Peter. *Discipline and Experience: The Mathematical Way in the Scientific Revolution*. Chicago: The University of Chicago Press, 1995.

Dear, Peter. "A Mechanical Microcosm: Bodily Passions, Good Manners, and Cartesian Mechanism." In *Science Incarnate: Historical Embodiments of Natural Knowledge*, edited by Christopher Lawrence and Steven Shapin, 51–82. Chicago: The University of Chicago Press, 1998.

Dear, Peter. "What is the History of Science the History *of*? Early Modern Roots of the Ideology of Modern Science." *Isis* 96 (2005): 390–406.

Dear, Peter. *The Intelligibility of Nature. How Science Makes Sense of the World*. Chicago: The University of Chicago Press, 2006.

Debord, Guy. *La société du spectacle*. Paris: Gallimard, 1992 [1967].

Deferrari, Roy J., *et al. A Lexicon of St. Thomas Aquinas Based on the* Summa Theological *and Selected Passages of his Other Works*. Washington, D.C.: Catholic University of America Press, ca. 1948.

Dekker, Elly, and Kristen Lippincott. "The Scientific Instruments in Holbein's Ambassadors: A Re-Examination." *Journal of the Warburg and Courtauld Institutes* 62 (1999): 93–125.

Delbourgo, James. *Collecting the World. Hans Sloane and the Origins of the British Museum*. Cambridge, MA: Harvard University Press, 2017.

Deloche, Bernard. *Mythologie du musée. De l'uchronie à l'utopie*. Paris: Le Cavalier Bleu, 2010.

Denis, Jean. *Traité de l'accord de l'espinette*, new introduction by Alan Curtis. New York: Da Capo Press, 1969 [second ed. of 1650].

Dennis, Michael Aaron. "Graphic Understanding: Instruments and Interpretation in Robert Hooke's *Micrographia*." *Science in Context* 3 (1989): 309–64.

Derolez, Séverine. "La patrimonialisation des objets scientifiques contemporains et leurs contextes de valorisation: Cas de l'accélérateur de particules Cockroft-Walton." *Culture & Musées* 29 (2017): 147–50.

Descartes, René. *Règles utiles et claires pour la direction de l'esprit en la recherche de la vérité*, edited and translated by Jean-Luc Marion. The Hague, Nijhoff, 1977.

Descartes, René. *The Philosophical Writings of Descartes*, translated by John Cottingham, Robert Stoothoff, and Dugald Murdoch, 3 vols. Cambridge: Cambridge University Press, 1984–1991.

Descartes, René. *Œuvres de Descartes*, edited by Charles Adam and Paul Tannery, new edition, 11 vols. Paris: Vrin, 1996.

Descartes, René. *Abrégé de musique*, translated and edited by Frédéric de Buzon. Paris: Presses universitaires de France, 1987.

Des Chene, Dennis. *Spirits and Clocks: Machine and Organism in Descartes*. Ithaca: Cornell University Press, 2001.

Descola, Philippe. "The Two Natures of Lévi-Strauss." In *The Cambridge Companion to Lévi-Strauss*, edited by Boris Wiseman, 103–17. Cambridge: Cambridge University Press, 2009.

Detienne, Marcel, and Jean-Pierre Vernant. *Les Ruses de l'intelligence. La mètis des Grecs*. Paris: Flammarion, 1974.

Devie, Dominique. *Le Tempérament musical. Philosophie, histoire, théorie et pratique*. Béziers: Société de musicologie de Languedoc, 1990.

Dew, Nicholas. "'Vers la ligne': Circulating Measurements Around the French Atlantic." In *Science and Empire in the Atlantic World*, edited by James Delbourgo and Nicholas Dew, 53–72. New York: Routledge, 2008.

Dew, Nicholas. *Orientalism in Louis XIV's France*. Oxford: Oxford University Press, 2009.

Deyon, Pierre. "Variations de la production textile aux XVIe et XVIIe siècles." *Annales: ESC* 18 (1963): 921–55.

DiMeo, Michelle. *Lady Ranelagh: The Incomparable Life of Robert Boyle's Sister*. Chicago: The University of Chicago Press, 2021.

Dobre, Mihnea. "On Glass-Drops: A Case Study of the Interplay Between Experimentation and Explanation in Seventeenth-Century Natural Philosophy." *Journal of Early Modern* Studies 2 (2013): 105–24.

Dobre, Mihnea, and Tammy Nyden, eds. *Cartesian Empiricisms*. Dordrecht: Springer, 2013.

Dostrovsky, Sigalia. "Early Vibration Theory: Physics and Music in the Seventeenth Century." *Archive for the History of Exact Sciences* 14 (1975): 169–218.

Dournon, Geneviève. "Instrumentariums et classifications." *Revue de musicologie* 79 (1993): 197–207.

Dubois, Jacques. "An Eighteenth-Century Collection of Scientific Instruments at Dijon." In *The Art of Teaching Physics: The Eighteenth-Century Demonstration Apparatus of Jean-Antoine Nollet,* edited by Lewis Pyenson and Jean-François Gauvin, 99–118. Sillery, QC: Les Éditions du Septentrion, 2002.

Dubourg Glatigny, Pascal, and Hélène Vérin, eds. *Réduire en art: La technologie de la Renaissance aux Lumières*. Paris: Éditions de la Maison des sciences de l'Homme, 2008.

Dudley, Sandra, ed. *Museum Objects: Experiencing the Properties of Things*. New York: Routledge, 2012.

Dufay, Bruno, *et al.* "L'atelier parisien de Bernard Palissy." *Revue de l'Art* 78 (1987): 33–60.

Duncan, David Allen. *The Tyranny of Opinions Undermined: Science, Pseudo-Science and Scepticism in the Musical Thought of Marin Mersenne*. PhD diss., Vanderbilt University, 1981.

Duncan, David Allen. "Persuading the Affections: Rhetorical Theory and Mersenne's Advice to Harmonic Orators." In *French Musical Thought, 1600–1800*, edited by Georgia Cowart, 148–75. Ann Arbor: U.M.I. Research Press, 1989.

Dunn, Richard. "'More Artistic than Scientific': Exhibiting Instruments as Decorative Arts in the Victoria and Albert Museum." In *Scientific Instruments on Display*, edited by Silke Ackermann, Richard Kremer, and Mara Miniati, 61–76. Leiden: Brill, 2014.

Dupré, Sven. "Ausonio's Mirrors and Galileo's Lenses: The Telescope and Sixteenth-Century Practical Optical Knowledge." *Galilaeana: Journal of Galilean Studies* 2 (2005): 145–80.

Dupré, Sven. "Doing it Wrong: The Translation of Artisanal Knowledge and the Codification of Error." In *The Structures of Practical Knowledge*, edited by Matteo Valleriani, 167–88. Cham: Springer International Publishing, 2017.

Dupré, Sven, and Michael Korey. "Inside the *Kunstkammer*: The Circulation of Optical Knowledge and Instruments at the Dresden Court." *Studies in History and Philosophy of Science* 40 (2009): 405–20.

Durant, John. "The Challenge and the Opportunity of Presenting 'Unfinished Science'." In *Creating Connections. Museums and the Public Understanding of Current Research*, edited by David Chittenden, Graham Farmelo, and Bruce Lewenstein, 47–60. Walnut Creek: AltaMira Press, 2004.

Eamon, William. *Science and the Secrets of Nature: Books of Secrets in Medieval and Early Modern Culture*. Princeton: Princeton University Press, 1994.

Eamon, William. "Markets, Piazzas, and Villages." In *The Cambridge History of Science, Volume 3: Early Modern Science*, edited by Katharine Park and Lorraine Daston, 206–23. Cambridge: Cambridge University Press, 2008.

Edwards, David. *Artscience. Creativity in the Post-Google Generation*. Cambridge, MA: Harvard University Press, 2008.

Edwards, David. *The Lab. Creativity and Culture*. Cambridge, MA: Harvard University Press, 2010.

Elliott, W. S., and A. E. Sale. "The Computer Conservation Society—Machines of the 1950's Restored to Life in the Science Museum, London." *Fifth National Conference on Engineering Heritage 1990: Interpreting Engineering Heritage; Preprints of Papers*, 39–44.

Even-Ezra, Ayelet. *Lines of Thought: Branching Diagrams and the Medieval Mind*. Chicago: The University of Chicago Press, 2021.

Everett, Tom. "Writing Sound with a Human Ear: Reconstructing Bell and Blake's 1874 Ear Phonautograph." *Science Museum Group Journal* 12 (2019). http://journal.sciencemuseum.ac.uk/browse/issue-12/writing-sound-with-a-human-ear/.

Fabbri, Natacha. *De l'utilité de l'harmonie: Filosofia, scienza e musica in Mersenne, Descartes e Galileo*. Pisa: Edizioni della Normale, 2008.

Fabre, Daniel, ed. *Émotions patrimoniales*. Paris: Éditions de la Maison des sciences de l'homme, 2013.

Farné, Roberto. "Pedagogy of Play." *Topoi* 24 (2005): 169–81.

Farr, James R. *Hand of Honor: Artisans and Their World in Dijon, 1550–1650*. Ithaca: Cornell University Press, 1988.

Fatet, Jérôme, ed. *A quoi servent les instruments scientifiques? Réflexions et études de cas sur les rôles et les fonctions des instruments dans la pratique scientifique*. Limoges: Presses universitaires de Limoges, 2016.

Fenlon, Iain. "Isabella d'Este e i suoi contemporanei. Musica e mecenatismo presso le corti dell'Italia settentrionale." In *Bernardo Clesio e il suo tempo*, edited by Paolo Prodi, 607–37. Rome: Bulzoni, 1986.

Ferret, Stéphane. *Le bateau de Thésée: Le problème de l'identité à travers le temps*. Paris: Les Éditions de Minuit, 1996.

Ferreyrolles, Gérard. *Les Reines du monde. L'Imagination et la coutume chez Pascal*. Paris: Honoré Champion Editeur, 1995.

Ferriot, Dominique, and Martha C. Lourenço. "De l'utilité des musées et collections des universités." *Lettre de l'OCIM* 93 (2004): 4–16.

Field, Judith V. "What is Scientific About a Scientific Instrument?" *Nuncius* 3 (1988): 3–26.

Findlen, Paula. "The Museum: Its Classical Etymology and Renaissance Genealogy." *Journal of the History of Collections* 1 (1989): 59–78.

Findlen, Paula. *Possessing Nature: Museums, Collecting, and Scientific Culture in Early Modern Italy*. Berkeley: University of California Press, 1994.

Findlen, Paula, ed. *Early Modern Things: Objects and Their Histories, 1500–1800*. New York: Routledge, 2013.

Finocchiaro, Maurice A. "Vincenzo Galilei's Musicology and Galileo's Science: Methodological Comparison and Contrast." *Isis* 111 (2020): 740–58.

Floris Cohen, H. *Quantifying Music: The Science of Music at the First Stage of the Scientific Revolution, 1580–1650*. Dordrecht: D. Reidel Publishing Company, 1984.

Floris Cohen, H. and Jacomien Prins. "Music, Philosophy, and Science." In *The Cambridge History of Sixteenth-Century Music*, edited by Iain Fenlon and Richard Wistreich, 439–71. Cambridge: Cambridge University Press, 2019.

Fontenelle, Bernard Le Bovier de, *Histoire de l'Académie royale des sciences*, 2 vols. Paris, 1733.

Forgan, Sophie. "The Architecture of Display: Museums, Universities and Objects in Nineteenth-Century Britain." *History of Science* 32 (1994): 139–62.

Forgan, Sophie. "Building the Museum: Knowledge, Conflict, and the Power of Place." *Isis* 96 (2005): 572–85.

Fors, Hjalmar, Lawrence M. Principe, and Otto Sibum. "From the Library to the Laboratory and Back Again: Experiment as a Tool for Historians of Science." *Ambix* 63 (2016): 85–97.

Friedman, Michael. "Joachim Jungius and the Transfer of Knowledge from Weaving to Mathematics in the 17th Century." *Cahiers François Viète* III–13 (2022): 43–82.

Furetière, Antoine. *Dictionnaire universel, contenant généralement tous les mots françois tant vieux que modernes, et les termes de toutes les sciences et des arts* ... The Hague, 1690.

Gal, Ofer, and Raz Chen-Morris. "Empiricism Without the Senses: How the Instrument Replaced the Eye." In *The Body as Object and Instrument of Knowledge: Embodied Empiricism in Early Modern Science*, edited by Charles T. Wolfe and Ofer Gal, 121–47. Dordrecht: Springer, 2010.

Galilei, Galileo, *Sidereus nuncius or The Sidereal Messenger*, translated by Albert van Helden. Chicago: The University of Chicago Press, 1989.

Galilei, Galileo, *Sidereus nuncius or The Sidereal Messenger*, translated by Albert van Helden. *Discorsi e dimostrazioni matematiche, intorno à due nuove scienze*. Leiden, 1638.

Galilei, Vincenzo. *Dialogue on Ancient and Modern Music*, translated by Claude V. Palisca. New Haven: Yale University Press, 2003.

Galison, Peter. "History, Philosophy, and the Central Metaphor." *Science in Context* 2 (1988): 197–212.

Galison, Peter. *Image & Logic. A Material Culture of Microphysics*. Chicago: The University of Chicago Press, 1997.

Galluzzi, Paolo. *The Italian Renaissance of Machines*. Cambridge, MA: Harvard University Press, 2020.

Gamba, Enrico, and Roberto Mantovani. "Gli strumenti scientifici di Guidobaldo del Monte." In *Guidobaldo del Monte (1545–1607): Theory and Practice of the Mathematical Disciplines from Urbino to Europe*, edited by Antonio Becchi, Domenico Bertoloni Meli, and Enrico Gamba. Berlin: Max Planck Research Library for the History and Development of Knowledge, Proceedings 4, 2013.

Garber, Daniel. *Descartes's Metaphysical Physics*. Chicago: The University of Chicago Press, 1992.

Garber, Daniel. *Descartes Embodied: Reading Cartesian Philosophy Through Cartesian Science*. Cambridge: Cambridge University Press, 2001.

Garber, Daniel. "On the Frontlines of the Scientific Revolutions: How Mersenne Learned to Love Galileo." *Perspectives on Science* 12 (2004): 135–63.

Garcia, Tristan, and Vincent Normand, eds. *Theater, Garden, Bestiary. A Materialist History of Exhibitions*. Berlin: Sternberg Press, 2019.

Garrioch, David, ed. *The Republic of Skill. Artisan Mobility, Innovation, and the Circulation of Knowledge in Premodern Europe*. Leiden: Brill, 2022.

Gaukroger, Stephen. *Descartes: An Intellectual Biography*. Oxford: Clarendon Press, 1995.

Gaukroger, Stephen. *The Emergence of a Scientific Culture: Science and the Shaping of Modernity, 1210–1685*. Oxford: Oxford University Press, 2006.

Gaukroger, Stephen, and John Schuster. "The Hydrodynamic Paradox and the Origins of Cartesian Dynamics." *Studies in the History and Philosophy of Science* 23 (2002): 535–72.

Gauthier, Robert. "L'Herbier Louis-Marie." *Cap-aux-diamants* 72 (2003): 54–6.

Gauvin, Jean-François. "Eighteenth-Century Entrepreneur: Excerpts from the Correspondence Between Jean-Antoine Nollet, Étienne-François Dutour and Jean-Jallabert, 1739–1768." In *The Art of Teaching Physics: The Eighteenth-Century Demonstration Apparatus of Jean-Antoine Nollet,* edited by Lewis Pyenson and Jean-François Gauvin, 47–69. Sillery, Qc: Les Éditions du Septentrion, 2002.

Gauvin, Jean-François. "Artisans, Machines, and Descartes's *Organon.*" *History of Science* 44 (2006): 187–216.

Gauvin, Jean-François. "Le cabinet de physique du château de Cirey et la philosophie naturelle de Mme du Châtelet et de Voltaire." In *Emilie Du Châtelet: Rewriting Enlightenment Philosophy and Science,* edited by Judith P. Zinsser and Julie Candler Hayes, 165–202. Oxford: Voltaire Foundation [SVEC 2006: 01], 2006.

Gauvin, Jean-François. "Instruments of Knowledge." In *The Oxford Handbook of Philosophy in Early Modern Europe,* edited by Desmond Clarke and Catherine Wilson, 315–37 (Oxford: Oxford University Press, 2011).

Gauvin, Jean-François. "The Instrument That Never Was: Inventing, Manufacturing, and Branding Réaumur's Thermometer During the Enlightenment." *Annals of Science* 69 (2012): 515–49.

Gauvin, Jean-François. "When Museums of Science Create Citizen Expertise: Participatory Science and the *Gamification of Techne.*" *Thema: La revue des Musées de la civilisation* 2 (2015): 28–38. https://thema.mcq.org/index.php/Thema/issue/view/5.

Gauvin, Jean-François. "Functionless: Science Museums and the Display of 'Pure Objects'." *Science Museum Group Journal,* vol. 5 (Spring 2016). http://journal.sciencemuseum.ac.uk/browse/issue-05/functionless/.

Gauvin, Jean-François. "Playing with Quantum Toys: Julian Schwinger's Measurement Algebra and the Material Culture of Quantum Mechanics Pedagogy at Harvard in the 1960s." *Physics in Perspective* 20 (2018): 8–42.

Gauvin, Jean-François. "The Next Level of Play: How Cartesian Devils, Volcanoes, and Quantum Toys Taught General Science at Harvard, 1730–1970." *The Journal of the History of Collection* 31 (2019): 531–46.

Gauvin, Jean-François. "Huygens's *Carriole*: Urban Commodity, Authorship, and the Intelligence of Carriage Design." *Nuncius* 38 (2023): 72–102.

Gauvin, Jean-François. "Réparation, restauration, exposition: valoriser les instruments disloqués des musées de sciences et de technologie." In *Les Réparations dans l'histoire. Cultures techniques et savoir-faire dans la longue durée,* edited by Gianenrico Bernasconi, Guillaume Carnino, Liliane Hilaire-Pérez, and Olivier Raveux, 259–69. Paris: Presses des mines, 2022.

Gauvin, Jean-François, Maxime Sainte-Marie, and Vincent Larivière. "Les collections muséales au service de la science: Une analyse de la Collection of Historical Scientific Instruments de l'université Harvard." *Document numérique* 20 (2–3) (2017): 137–58.

Geffroy, Anne-Marie. *Conservation-restauration d'un instrument scientifique: un théodolite de la fin du XIXe siècle du musée des arts et métiers. Étude comparative du comportement au vieillissement artificiel de trois vernis sur laiton*. Paris: Institut national du patrimoine, mémoire de fin d'études, 2004.

Geindreau, Rémy. "Étude préalable à la restauration d'un objet technico-scientifique contemporain. Le générateur Cockcroft-Walton de l'Institut de Physique Nucléaire de Lyon." *CeROArt* [online], EGG 5 (2016). http://journals.openedition.org/ceroart/4933.

Geoghegan, Hilary, and Alison Hess. "Object-Love at the Science Museum: Cultural Geographies of Museum Storerooms." *Cultural Geographies* 22 (2015): 445–65.

Gessner, Samuel. "The Journals and the Instrument Maker: Visuality of Butterfield's Instruments in the *Journal des Sçavans* and the *Philosophical Transactions* around 1680." *Nuncius* 30, no. 3 (2015): 610–36.

Gewirtz, Alan. "Experience and the Non-Mathematical in the Cartesian Method." *Journal of the History of Ideas* 2 (1941): 183–210.

Giannini, Tula, and Jonathan P. Bowen, eds. *Museums and Digital Culture. New Perspectives and Research*. Cham: Springer, 2019.

Giglioni, Guido. "How Bacon Became Baconian." In *The Mechanization of Natural Philosophy*, edited by Sophie Roux and Daniel Garber, 27–54. Dordrecht: Springer [Boston Studies in the Philosophy of Science], 2013.

Gilson, Etienne. *Index scolastico-cartésien*. PhD diss., Université de Paris, 1912.

Goclenius, Rudolph. *Lexicon philosophicum quo tanquam clave philosophiae fores aperiuntur*. Hildesheim: Georg Olms Verlag, 1964 [Frankfurt, 1613].

Goldstein, Catherine. "Routine Controversies: Mathematical Challenges in Mersenne's Correspondence." *Revue d'histoire des sciences* 66 (2013): 249–73.

Gorman, Michael John. *Idea Colliders: The Future of Science Museums*. Cambridge, MA: MIT Press, 2020.

Gossman, Lionel. *Men and Masks: A Study of Molière*. Baltimore: The Johns Hopkins University Press, 1963.

Goubert, Pierre. *Beauvais et le Beauvaisis de 1600 à 1730: contribution à l'histoire sociale de la France du XVIIe siècle*. Paris: S.E.V.P.E.N., 1960.

Gouk, Penelope. *Music, Science and Natural Magic in Seventeenth-Century England*. New Haven: Yale University Press, 1999.

Graham, Stephen, and Nigel Thrift. "Out of Order: Understanding Repair and Maintenance." *Theory, Culture & Society* 24 (2007): 1–25.

Graham Burnett, D. *Descartes and the Hyperbolic Quest: Lens Making Machines and Their Significance in the Seventeenth Century*. Philadelphia: Transactions of the American Philosophical Society, new series, vol. 95, No. 3, 2005.

Granato, Marcus, and Françoise Le Guet Tully. "Les principes de la restauration d'instruments scientifiques: le cas du cercle méridien Gautier de l'Observatoire de Rio de Janeiro." *In Situ* [online], 10 (2009). http://journals.openedition.org/insitu/3865.

Grant, Florence. "Mechanical Experiments as Moral Exercise in the Education of George III." *The British Journal for the History of Science* 48 (2015): 195–212.

Grant, Ruth W. "John Locke on Custom's Power and Reason's Authority." *The Review of Politics* 74 (2012): 607–29.

Grosslight, Justin. "Small Skills, Big Networks: Marin Mersenne as Mathematical Intelligencer." *History of Science* 51 (2013): 337–74.

Guerrini, Anita. *The Courtiers' Anatomists. Animals and Humans in Louis XIV's Paris*. Chicago: The University of Chicago Press, 2015.

Guillin, Vincent. "Les études cartésiennes de G. Canguilhem." *Études philosophiques* 114 (2008): 65–84.

Gumbrecht, Hans Ulrich. *Production of Presence. What Meaning Cannot Convey*. Stanford: Stanford University Press, 2003.

Hacking, Ian. *Representing and Intervening: Introductory Topics in the Philosophy of Natural Science*. Cambridge: Cambridge University Press, 1983.

Hacking, Ian. "The Disunities of Science." In *The Disunity of Science: Boundaries, Contexts, and Power*, edited by P. Galison and D. J. Stump, 37–74. Stanford: Stanford University Press, 1996.

Hacking, Ian. *Historical Ontology*. Cambridge, MA: Harvard University Press, 2002.

Hamou, Philippe. *La Mutation du visible. Essai sur la portée épistémologique des instruments d'optique au XVIIe siècle*, 2 vols. Villeneuve D'Ascq (Nord): Presses universitaires du septentrion, 1999.

Hahn, Roger. *The Anatomy of a Scientific Institution: The Paris Academy of Sciences, 1666–1803*. Berkeley: University of California Press, 1971.

Hankins, Thomas L., and Robert J. Silverman. *Instruments and the Imagination*. Princeton: Princeton University Press, 1995.

Harkness, Deborah. "Managing an Experimental Household: The Dees of Mortlake and the Practice of Natural Philosophy." *Isis* 88 (1997): 247–62.

Harkness, Deborah. *The Jewel House: Elizabethan London and the Scientific Revolution*. New Haven: Yale University Press, 2007.

Harvard University. Corporation. College Books, 1636–1827. *Colonial North American Project at Harvard*. https://colonialnorthamerica.library.harvard.edu/.

Harvard University. *Annual Reports of the President and Treasurer of Harvard College*. Cambridge, MA: Harvard University, various years.

Hecker, Susanne, *et al. Citizen Science. Innovation in Open Science, Society and Policy.* London: UCL Press, 2018.

Heering, Peter. "Science Museums and Science Education." *Isis* 108 (2017): 399–406.

Heering, Peter, and Roland Wittje, eds. *Learning by Doing. Experiments and Instruments in the History of Teaching.* Stuttgart: Franz Steiner Verlag, 2011.

Heering, Peter, and Cibelle Celestino Silva. "Re-Examining the Early History of the Leiden Jar: Stabilization and Variation in Transforming a Phenomenon Into a Fact." *History of Science* 56 (2018): 314–42.

Heesen, Anke Te. *The World in a Box: The Story of an Eighteenth-Century Picture Encyclopedia.* Chicago: The University of Chicago Press, 2002.

Heilbron, John. *Electricity in the 17th and 18th Centuries: A Study of Early Modern Physics.* Berkeley: University of California Press, 1979.

Heilbron, John. *The Sun in the Church. Cathedrals as Solar Observatories.* Cambridge, MA: Harvard University Press, 1999.

Heinich, Nathalie. *Des valeurs. Une approche sociologique.* Paris: Gallimard, 2017.

van Helden, Albert. "The Birth of the Modern Scientific Instrument, 1550–1700." In *The Uses of Science in the Age of Newton*, edited by J. G. Burke, 49–84. Berkeley: University of California Press, 1983.

van Helden, Albert, and Thomas L. Hankins. "Introduction." In *Instruments*, Osiris, 2nd series, 9 (1994): 1–6.

Heller, Henri. *Labour, Science and Technology in France, 1500–1620.* Cambridge: Cambridge University Press, 1996.

Heller-Roazen, Daniel. *The Fifth Hammer: Pythagoras and the Disharmony of the World.* New York: Zone Books, 2011.

Hennion, Antoine. "Those Things That Hold Us Together: Taste and Sociology." *Cultural Sociology* 1 (2007): 97–114.

Hesketh, Ian. "Technologies of the Scientific Self: John Tyndall and His Journal." *Isis* 110 (2019): 460–82.

Hess, Alison. "Authenticity, Alterations and Museum Objects: A Close Encounter with 2LO the BBC's First Radio Transmitter." *Journal of Material Culture* 22 (2017): 281–98.

Higgitt, Rebekah. "Challenging Tropes: Genius, Heroic Invention, and the Longitude Problem in the Museum." *Isis* 108 (2017): 371–80.

Higgitt, Rebekah. "Instruments and Relics: The History and Use of the Royal Society's Object Collections c.1850–1950." *Journal of the History of Collections* 31 (2019): 469–85.

Hilaire-Pérez, Liliane. *L'Invention technique au siècle des Lumières.* Paris: Albin Michel, 2000.

Hilaire-Pérez, Liliane. "Technology as a Public Culture in the Eighteenth Century: The Artisan's Legacy." *History of Science* 45 (2007): 135–53.

Hilaire-Pérez, Liliane. *La pièce et le geste. Artisans, marchands et savoir technique à Londres au XVIIIe siècle.* Paris: Albin Michel, 2013.

Hilaire-Pérez, Liliane, and Catherine Verna. "Dissemination of Technical Knowledge in the Middle Ages and the Early Modern Era. New Approaches and Methodological Issues." *Technology and Culture* 47 (2006): 536–65.

Hill, C. R. "The Cabinet of Bonnier de la Mosson (1702–1744)." *Annals of Science* 43 (1986): 147–74.

Hill, John Walter. *Baroque Music: Music in Western Europe, 1580–1750*. New York: W. W. Norton & Company, 2005.

Hill, Katherine. "'Juglers or Schollers?': Negotiating the Role of a Mathematical Practitioner." *The British Journal for the History of Science* 31 (1998): 253–74.

Hobbes, Thomas. *The English Works of Thomas Hobbes* [electronic resource], 16 vols. Charlottesville, V.A.: InteLex Corporation [Past Masters Collection], 1995.

Hodgkin, Robin A. "Cognitive Objects." *Oxford Review of Education* 14 (1988): 353–62.

Hölling, Hanna. "The *Technique* of Conservation: On Realms of Theory and Cultures of Practice." *Journal of the Institute of Conservation* 40 (2017): 87–96.

Holmes, Frederic L., and Trevor H. Levere, eds. *Instruments and Experimentation in the History of Chemistry*. Cambridge, MA: MIT Press, 2000.

Hübners, Johann. *Curieuses und Eales Natur- Kunst- Berg- Gewerck- und Handlungs-Lexicon …*, 7th ed. Leipzig, 1736.

Huizinga, Johan. *Homo ludens: A Study of the Play-Element in Culture*. London: Routledge & Kegan Paul, 1949.

Hume, David. *A Treatise of Human Nature: Being an Attempt to Introduce the Experimental Method of Reasoning into Moral Subjects*, 3 vols. London, 1739–1740.

Hume, David. *Philosophical Essays Concerning Human Understanding*. London, 1748.

Ihde, Don. *Instrumental Realism: The Interface Between Philosophy of Science and Philosophy of Technology*. Bloomington: Indiana University Press, 1991.

Ingold, Tim. "Materials Against Materiality." *Archaeological Dialogues* 14 (2007): 1–16.

Ingold, Tim. "Making Culture and Weaving the World." In *Museum Objects. Experiencing the Properties of Things*, edited by Sandra H. Dudley, 368–84. New York: Routledge, 2012.

Ingold, Tim. *Making. Anthropology, Archaeology, Art and Architecture*. New York: Routledge, 2013.

Isacoff, Stuart. *Temperament: The Idea that Solved Music's Greatest Riddle*. New York: Alfred A. Knopf, 2001.

Isherwood, Robert M. *Music in the Service of the King: France in the Seventeenth Century*. Ithaca: Cornell University Press, 1973.

Israel, Giorgio. "Des *Regulæ* à la *Géométrie*." *Revue d'histoire des sciences* 51 (1998): 183–236.

Jackson, Myles W. "Music and Science During the Scientific Revolution." *Perspectives on Science* 9 (2001): 106–15.

Jacob, Christian, ed. *Lieux de savoir, 2: Les mains de l'intellect.* Paris: Albin Michel, 2011.

Jacomy, Bruno. "Du cabinet au conservatoire: Les instruments scientifiques du Conservatoire des arts et métiers à Paris." *Journal of the History of Collections* 7, no. 2 (1995): 227–33.

Jardine, Boris, Joshua Nall, and James Hyslop. "More than Mensing? Revisiting the Question of Fake Scientific Instruments." *Bulletin of the Scientific Instrument Society* 132 (2017): 22–9.

Jardine, Boris. "Like a Bos: The Discovery of Fake Antique Scientific Instruments at the Whipple Museum." In *The Whipple Museum of the History of Science. Objects and Investigations, to Celebrate the 75th anniversary of R. S. Whipple's Gift to the University of Cambridge*, edited by Joshua Nall, Liba Taub, and Fraces Willmoth, 201–22. Cambridge: Cambridge University Press, 2019.

Jardine, Boris. "The Museum in the Lab: Historical Practice in the Experimental Sciences at Cambridge, 1874–1936." *BJHS: Themes* 4 (2019): 245–71.

Jardine, Boris, Emma Kowal, and Jenny Bangham. "How Collections End: Objects, Meaning and Loss in Laboratories and Museums." *BJHS Themes* 4 (2019): 1–27.

Johns, Adrian. "Coffeehouses and Print Houses." In *The Cambridge History of Science, Volume 3: Early Modern Science*, edited by Katharine Park and Lorraine Daston, 320–40. Cambridge: Cambridge University Press, 2008.

Johnston, Stephen. "Mathematical Practitioners and Instruments in Elizabethan England." *Annals of Science* 48 (1991): 319–44.

Jones, Matthew L. "Descartes's Geometry as Spiritual Exercise." *Critical Inquiry* 28 (2001): 40–71.

Jones, Matthew L. *The Good Life in the Scientific Revolution: Descartes, Pascal, Leibniz, and the Cultivation of Virtue.* Chicago: The University of Chicago Press, 2006.

Jones, Matthew L. *Reckoning with Matter. Calculating Machines, Innovation, and Thinking about Thinking from Pascal to Babbage.* Chicago: The University of Chicago Press, 2016.

Jones, Siân. "Negotiating Authentic Objects and Authentic Selves: Beyond the Deconstruction of Authenticity." *Journal of Material Culture* 15 (2010): 181–203.

Jones, Siân, and Thomas Yarrow. "Crafting Authenticity: An Ethnography of Conservation Practice." *Journal of Material Culture* 18 (2013): 3–26.

de Jonge, Henk Jan. "Erasmus's Translation of the New Testament: Aim and Method." *The Bible Translator* 67 (2016): 29–41.

Juler, Edward, and Alistair Robinson, eds. *Post-Specimen Encounters Between Art, Science and Curating. Rethinking Art Practice and Objecthood Through Scientific Collections.* Bristol and Chicago: Intellect, 2020.

Julien, Marie-Pierre, and Jean-Pierre Warnier, eds. *Approches de la culture matérielle. Corps à corps avec l'objet.* Paris: L'Harmattan, 1999.

Jurgens, Madeleine. *Documents du Minutier central concernant l'histoire de la musique (1600–1650)*, 2 vols. Paris: S.E.V.P.E.N., 1967; La Documentation Française, 1974.

Kanhadilok, Peeranut, and Mike Watts. "Youth at Play: Some Observations from a Science Museum." *International Journal of Adolescence and Youth* 22 (2017): 179–94.

Kant, Immanuel. *Lectures on Logic*, translated and edited by J. Michael Young. Cambridge: Cambridge University Press, 1992.

Kant, Immanuel. *Critique of Pure Reason*, translated and edited by Paul Guyer and Allen W. Wood. Cambridge: Cambridge University Press, 1998.

Karp, Ivan. "Culture and Representation." In *Exhibiting Cultures: The Poetics and Politics of Museum Displays*, edited by Ivan Karp and Steven D. Lavine, 11–24. Washington, D.C.: Smithsonian Press, 1991.

Kasser, Michel. "Picard et le renouvellement de l'art du nivellement à la fin du XVIIe siècle." In *Jean Picard et les débuts de l'astronomie de précision au XVIIe siècle*, edited by Guy Picolet, 255–63. Paris: Editions du CNRS, 1987.

Keene, Melanie. *Science in Wonderland: The Scientific Fairy Tales of Victorian Britain*. Oxford: Oxford University Press, 2015.

Kehl, Winifred. "I Have a Confession: I'm an Adult. How Can Science Museums Welcome Grown-Ups?" *Museum* (July–August 2013), 46–51.

Kim, Mi Gyung. "Chemical Analysis and the Domains of Reality: Wilhelm Homberg's *Essays de chimie*, 1702–1709." *Studies in the History and Philosophy of Science* 31 (2000): 37–69.

Klein, Ursula. "The Laboratory Challenge: Some Revisions of the Standard View of Early Modern Experimentation." *Isis* 99 (2008): 769–82.

Klein, Ursula. "Artisanal-Scientific Experts in Eighteenth-Century France and Germany." *Annals of Science* 69 (2012): 303–6.

Klein, Ursula. "Hybrid Experts." In *The Structures of Practical Knowledge*, edited by Matteo Valleriani, 287–306. Cham: Springer International Publishing, 2017.

Klein, Ursula. "Science, Industry, and the German *Bildungsbürgertum*." *Annals of Science* 77 (2020): 366–76.

Klein, Ursula, and Wolfgang Lefèvre. *Materials in Eighteenth-Century Science: A Historical Ontology*. Cambridge, MA: MIT Press, 2007.

Klein, Ursula, and Emma C. Spary, eds. *Materials and Expertise in Early Modern Europe: Between Market and Laboratory*. Chicago: The University of Chicago Press, 2010.

Koganzon, Rita. "'Contesting the Empire of Habit': Habituation and Liberty in Lockean Education." *American Political Science Association* 111 (2016): 547–58.

Konoval, Brandon. "Pythagorean Pipe Dream? Vincenzo Galilei, Marin Mersenne, and the Pneumatic Mysteries of the Pipe Organ." *Perspectives on Science* 26 (2018): 1–51.

Kottick, Edward L. "Building a 15th-Century Lute." *The Galpin Society Journal* 26 (1973): 72–83.

Kühn, Hermann. "The Restoration of Historic Technological Artefacts, Scientific Instruments and Tools." *The International Journal of Museum Management and Curatorship* 8 (1989): 389–405.

Lachapelle, Sofie. *Conjuring Science. A History of Scientific Entertainment and Stage Magic in Modern France*. New York: Palgrave Macmillan, 2015.

de Lafond, Sigaud. *Description et usage d'un cabinet de physique expérimentale*, 2 vols. Paris: P. Fr. Gueffier, 1775.

Lallement, Michel. *L'Âge du faire. Hacking, travail, anarchie*. Paris: Seuil, 2015.

Lamarque, Peter. *Work & Object: Explorations in the Metaphysics of Art*. Oxford: Oxford University Press, 2010.

Lamy, Jérôme. "Le renouveau de l'histoire des instruments scientifiques: Essai bibliographique." *Artefact. Techniques, histoire et sciences humaines* 17 (2022): 9–38.

Laqueur, Thomas. *The Work of the Dead: A Cultural History of Mortal Remains*. Princeton: Princeton University Press, 2015.

Lasser, Ethan, ed. *The Philosophy Chamber: Art and Science in Harvard's Teaching Cabinet, 1766–1820*. New Haven: Yale University Press, 2017.

Latour, Bruno. *Science in Action: How to Follow Scientists and Engineers Through Society*. Cambridge, MA: Harvard University Press, 1987.

Laursen, John Christian. "David Hume on Custom and Habit and Living with Skepticism." *Daimon. Revista internacional de filosofia* 52 (2011): 87–99.

Lavin, Matthew. "Why Digital Humanists Should Emphasize Situated Data Over Capta." *DHQ: Digital Humanities Quarterly* 15.2 (2021). http://www.digitalhumanities.org/dhq/vol/15/2/000556/000556.html.

Lawrence, Christopher, and Steven Shapin, eds. *Science Incarnate. Historical Embodiments of Natural Philosophy*. Chicago: The University of Chicago Press, 1998.

Lazar, Horia. "L'Habitude de penser: de la machine arithmétique à l'"abêtissement'." *Chronique: Institut catholique de Toulouse* (1995): 57–70.

Le Cerf, G., ed. (with the collaboration of E. R. Labande). *Les Traités d'Henri-Arnaut de Zwolle et de divers anonymes (ms B. N. Latin 7295)*. Paris, 1932.

Lefèvre, Wolfgang, ed. *Picturing Machines, 1400–1700*. Cambridge, Mass.: The MIT Press, 2004.

Le Guet Tully, Françoise, Joël Françoise, William Mourey, Jacques Rebière, and Jean Davoigneau. "L'Observatoire de Nice Côte d'Azur, site de Nice. Projets et réalisations." In *Métal. Actes de la conférence internationale sur la conservation des métaux*, edited by William Mourey and Luc Robbiola, 22–5. London: James & James, 1998.

Leibniz, Gottfried Wilhelm. *Die philosophischen Schriften von Gottfried Wilhelm Leibniz*, edited by C. J. Gerhardt, 7 vols. Berlin: Weidmannsche Buchhandlung, 1875.

Lenoble, Robert. *Mersenne, ou la naissance du mécanisme*. Paris: Vrin, 1943.

de Lespinasse, René. *Les Métiers et corporations de la ville de Paris*, 3 vols. Paris: Imprimerie Nationale, 1886–1897.

Lesure, François. "La facture instrumentale à Paris au seizième siècle." *The Galpin Society Journal* 7 (1954): 11–52.

Lesure, François. "Pierre Trichet's *Traité des instruments de musique*: supplément." *The Galpin Society Journal* 15 (1962): 70–81.

Lévi-Strauss, Claude. "Structural Analysis in Linguistics and in Anthropology." In *Structural Anthropology*, 31–54. New York: Basic Books, 1963.

Lévi-Strauss, Claude. *The Savage Mind*. London: Weidenfeld and Nicolson, 1966.

Lévi-Strauss, Claude. "Structuralism and Ecology." *Social Science Information* 12 (1973): 7–23.

Lindley, Mark. "Mersenne on Keyboard Tuning." *Journal of Music Theory* 24 (1980): 166–203.

Lindtner, Silvia, and Paul Dourish. "The Promise of Play: A New Approach to Productive Play." *Games and Culture* 6 (2011): 453–78.

Lipovetsky, Gilles, and Jean Serroy. *L'Esthétisation du monde. Vivre à l'âge du capitalisme artiste*. Paris: Gallimard, 2013.

Lipovetsky, Gilles. *Le sacre de l'authenticité*. Paris: Gallimard, 2021.

Locke, John. *An Essay Concerning Human Understanding*. London, 1690.

Locke, John. "Practice and Habit." *The Chautauquann; A Weekly News Magazine* 3 (1883): 317–8.

Locke, John. *Of the Conduct of the Understanding*, edited by Paul Schuurman. PhD diss., University of Keele, 2000.

Long, Pamela O. "Power, Patronage, and the Authorship of *Ars*: From Mechanical Know-How to Mechanical Knowledge in the Last Scribal Age." *Isis* 87 (1997): 1–41.

Long, Pamela O. *Openness, Secrecy, Authorship: Technical Arts and the Culture of Knowledge from Antiquity to the Renaissance*. Baltimore: The Johns Hopkins University Press, 2001.

Loriferne, Hubert. "L'influence de Picard dans les travaux d'alimentation en eau du château de Versailles sous Louis XIV." In *Jean Picard et les débuts de l'astronomie de précision au XVII*[e] *siècle*, edited by Guy Picolet, 275–311. Paris: Editions du CNRS, 1987.

Loukissas, Yanni Alexander. "A Place for Big Data: Close and Distant Readings of Accessions Data from the Arnold Arboretum." *Big Data & Society* 3 (July–December 2016): 1–20.

Loukissas, Yanni Alexander. *All Data are Local: Thinking Critically in a Data-Driven Society*. Cambridge, MA: MIT Press, 2019.

Lourenço, Marta C., and Lydia Wilson. "Scientific Heritage: Reflections on Its Nature and New Approaches to Preservation, Study and Access." *Studies in History and Philosophy of Science* 44 (2013): 744–53.

Lourenço, Marta C., and José Pedro Sousa Dias. "'Time Capsules' of Science: Museums, Collections, and Scientific Heritage in Portugal." *Isis* 108 (2017): 390–98.

Love, Rosaleen. "Herman Boerhaave and the Element-Instrument Concept of Fire." *Annals of Science* 31 (1974): 547–59.

Loyseau, Charles. *A Treatise of Orders and Plain Dignities*, edited and translated by Howell A. Lloyd. Cambridge: Cambridge University Press, 1994 [1610].

MacDonald, Isobel. "Counting When, Who and How: Visualizing the British Museum's History of Acquisition Through Collection Data, 1753–2019." *Journal of the History of Collections*, 2022. https://doi.org/10.1093/jhc/fhac034.

Macdonald, Sharon, ed. *The Politics of Display: Museums, Science, Culture*. London: Routledge, 1998.

Macdonald, Sharon. *Behind the Scenes at the Science Museum*. Oxford: Berg, 2002.

Mahoney, Michael. "Reading a Machine," (1999). https://www.princeton.edu/~hos/h398/readmach/modeltfr.html.

Mairesse, François, and Renata F. Peters, eds. *What is the Essence of Conservation? Materials for a Discussion*. Paris: ICOFOM, 2019.

Mann, Peter Robert. "Working Evidence and the Destruction of Evidence in the Science Museum." *The International Journal of Museum Management and Curatorship* 8 (1989): 369–87.

Mann, Peter Robert. "The Restoration of Vehicles for Use in Research, Exhibition, and Demonstration." In *Restoration: Is It Acceptable?*, 131–8. London: British Museum, Department of Conservation, 1994.

Marcel-Dubois, Claudie. *Les Instruments de musique de l'Inde ancienne*. Paris: Presses Universitaires de France, 1941.

Marion, Jean-Luc. *Sur l'ontologie grise de Descartes. Science cartésienne et savoir aristotélicien dans les* Regulæ, 2nd edn. Paris: Vrin, 1981.

Marion, Jean-Luc. "Cartesian Metaphysics and the Role of the Simple Natures." In *The Cambridge Companion to Descartes*, edited by John Cottingham, 115–39. Cambridge: Cambridge University Press, 1992.

Marr, Alexander, ed. *The Worlds of Oronce Fine: Mathematics, Instruments and Print in Renaissance France*. Donington: Shaun Tyas, 2009.

Marr, Alexander. *Between Raphael and Galileo. Mutio Oddi and the Mathematical Culture of Late Renaissance Italy*. Chicago: The University of Chicago Press, 2011.

Martin, Luther H., Huck Gutman, and Patrick H. Hutton, eds. *Technologies of the Self. A Seminar with Michel Foucault*. Amherst: The University of Massachusetts Press, 1988.

Mary, André. *Le travail de l'habitus dans l'oeuvre de Pierre Bourdieu*. Paris: Karthala, 2022.

Massip, Catherine. "Facteurs d'instruments et maîtres à danser parisiens au XVIIe siècle." In *Instrumentistes et luthiers parisiens, XVIIe–XIXe siècles*, edited by Florence Gétreau, 17–34. Paris: Délégation à l'action artistique de la Ville de Paris, 1988.

Mauss, Marcel. "Techniques of the Body." *Economy and Society* 2 (1973): 70–88.

Mauss, Marcel. "Body Techniques." In *Sociology and Psychology: Essays*, edited by Ben Brewster, 97–135. London: Routledge, 1979.

Meadow, Mark A., and Bruce Robertson. *The First Museum Treatise on Museums: Samuel Quiccheberg's* Inscriptiones, *1565*. Los Angeles: The Getty Research Institute Publications Program, 2013.

Merleau-Ponty, Maurice. *Phénoménologie de la perception*. Paris: Gallimard, 1945.

Mersenne, Marin. *Harmonie universelle: The Books on Instruments*, translated by Roger E. Chapman. The Hague: M. Nijhoff, 1957.

Mersenne, Marin. *Les Prelvdes de l'Harmonie vniverselle, ov qvestions curievses Vtiles aux Predicateurs, aux Theologiens, aux Astrologues, aux Medecins & aux Philosophes*. Paris: Chez Henry Gvenon, 1634.

Mersenne, Marin. *Traité de l'Harmonie universelle*. Paris, 1627.

Mersenne, Marin. *Harmonie universelle, contenant la théorie et la pratique de la musique*, 3 vols. Paris: Centre national de la recherche scientifique, 1963 [1636–7].

Mersenne, Marin. *Harmonicorum libri XII*. Genève: Minkoff Reprint, 1972 [1636].

Mersenne, Marin. *Qvestions inovyes, ov recreation des sçavans*. Paris, 1634; facsimile Stuttgart, 1972.

Mersenne, Marin. *Correspondance du P. Marin Mersenne: religieux minime*, edited and annotated by Cornélis de Waard (with the collaboration of René Pintard), 17 vols. Paris: G. Beauchesne, 1933–1988.

Meyer, Morgan. "Researchers on Display: Moving the Laboratory into the Museum." *Museum Management and Curatorship* 26 (2011): 261–72.

Micraelius, Johannes. *Lexicon philosophicum*. Jena, 1653.

Millburn, John R. *Benjamin Martin: Author, Instrument-Maker, and 'Country Showman'*. Leiden: Noordhoff International Publishing, 1976.

Miller, Daniel. *The Comfort of Things*. Cambridge: Polity Press, 2008.

Miller, Daniel. *Stuff*. Cambridge: Polity Press, 2010.

Miller, Peter N. *Peiresc's Mediterranean World*. Cambridge, MA: Harvard University Press, 2015.

Mills, Allan. "Facsimiles: Benefits and Worries." *Bulletin of the Scientific Instrument Society* 25 (1990): 1.

Miniati, Mara. "Les *cistæ mathematicæ* et l'organisation des connaissances au XVIIe siècle." In *Studies in the History of Scientific Instruments*, edited by Christine Blondel, *et al.*, 43–51. London: Rogers Turner Books Ltd, 1989.

Mirambet-Paris, Agnès, and François Mirambet. "La conservation-restauration du patrimoine technique et industriel dans le cadre de la loi sur les Musées de France, une mission impossible?" *La Lettre de l'OCIM*, 135 (2011): 27–35.

Mitman, Gregg, Marco Armiero, and Robert S. Emmet. *Future Remains. A Cabinet of Curiosities for the Anthropocene*. Chicago: The University of Chicago Press, 2018.

de Monconys, Balthasar. *Journal des voyages de Monsieur Monconys, Conseiller du Roy en ses Conseils d'Estat & Privé, & Lieutenant Criminel au Siege Presidial de Lyon. Où les Sçavants trouveront un nombre infini de nouveautez, en Machines de Mathematique, Experiences Physiques, Raisonnemens de la belle Philosophie, curiositez de Chymie, & conversations des Illustres de ce Siecle* ... 3 vols. Lyon, 1665–6.

Monger, George. "Conservation or Restoration?" *The International Journal of Museum Management and Curatorship* 7 (1988): 375–80.

Morton, Alan Q., and Jane A. Wess. *Public & Private Science: The King George III Collection*. Oxford: Oxford University Press, 1993.

Moyer, Ann E. *Musica scientia: Musical Scholarship in the Italian Renaissance*. Ithaca: Cornell University Press, 1992.

Mukerji, Chandra. *Territorial Ambitions and the Gardens of Versailles*. Cambridge: Cambridge University Press, 1997.

Mukerji, Chandra. "The Cultural Power of Tacit Knowledge: Inarticulacy and Bourdieu's Habitus." *American Journal of Cultural Sociology* 2 (2014): 348–75.

de Munck, Bert, and Antonella Romano, eds. *Knowledge and the Early Modern City*. New York: Routledge, 2020.

Muñoz Viñas, Salvador. "'Who is Afraid of Cesare Brandi?' Personal Reflections on the *Teoria del restauro*." *CeROArt* [Online] (June 2015). https://journals.openedition.org/ceroart/4653.

Nagamachi, Yuji, S. J. *Selbstbezüglichkeit und Habitus. Die latente Idee der Geistmetaphysik bei Thomas von Aquin*. St. Ottilien: EOS Verlag Erzabtei, 1997.

Newman, William. *Newton the Alchemist: Science, Enigma, and the Quest for Nature's 'Secret Fire.'* Princeton: Princeton University Press, 2019.

Nicot, Jean. *Thresor de la langue française tant ancienne que moderne* ... Paris, 1606.

Nollet, Jean-Antoine. *Programme, ou Idée générale d'un cours de physique expérimentale, avec un Catalogue raisonné des instrumens qui servent aux expériences*. Paris: P. G. Le Mercier, 1738.

Nollet, Jean-Antoine. *L'Art des expériences, ou Avis aux amateurs de la physique sur le choix, la construction et l'usage des instrumens* ... 3 vols. Paris: Durand neveu, 1770.

Norton Wise, M. *Aesthetics, Industry & Science. Hermann von Helmholtz and the Berlin Physical Society*. Chicago: The University of Chicago Press, 2018.

Oddy, W. A. *Reversibility—Does It Exist?* London: The British Museum Press, 1999.

Olesko, Katherine. *Physics as a Calling: Discipline and Practice in the Königsberg Seminar for Physics*. Cornell: Cornell University Press, 1991.

Oosterhoff, Richard J. "*Idiotae*, Mathematics, and Artisans: The Untutored Mind and the Discovery of Nature in the Fabrist Circle." *Intellectual History Review* 24 (2014): 301–19.

Oosterhoff, Richard J. *Making Mathematical Culture. University and Print in the Circle of Lefèvre d'Étaples*. Oxford: Oxford University Press, 2018.

van Orden, Kate. *Music, Discipline and Arms in Early Modern France*. Chicago: The University of Chicago Press, 2005.

d'Orléans, Chérubin. *La Dioptrique oculaire, ou la théorique, la positive, et la méchanique, de l'oculaire dioptrique en toutes ses especes*. Paris, 1671.

O'Rourke Boyle, Marjorie. "Harvey in the Sluice: From Hydraulic Engineering to Human Physiology." *History and Technology* 24 (2008): 1–22.

Osler, Margaret J. "Whose Ends? Teleology in Early Modern Natural Philosophy." *Osiris*, 2nd series, 16 (2001): 151–68.

Palisca, Claude V. "Scientific Empiricism in Musical Thought." In *Seventeenth-Century Science and the Arts*, edited by H. H. Rhys, 91–137. Princeton: Princeton University Press, 1961.

Palisca, Claude V. *G. B. Doni's* Lyra Barberina. Bologna: A.M.I.S., 1981.

Palisca, Claude V. *The Florentine Camerata: Documentary Studies and Translations*. New Haven and London: Yale University Press, 1989.

Palisca, Claude V. "Mersenne *pro* Galilei *contra* Zarlino." *Nuova civiltà della macchine* 16 (1998): 74–80.

Palissy, Bernard. *Discours admirables de la nature des eaux et fontaines, etc.* In *Œuvres complètes de Bernard Palissy, édition conforme aux textes originaux imprimés du vivant de l'auteur*, edited by Paul-Antoine Cap. Paris: J. J. Dubochet et Cie, Éditeurs, 1844.

Pantalony, David. "What Remains: The Enduring Value of Museum Collections in the Digital Age." *Journal of the History of Science and Technology* 14 (2020): 160–82.

Pascal, Blaise. *Oeuvres complètes*, edited by Jean Mesnard, 4 vols. Paris: Desclée de Brouwer, 1964–1992.

Pascal, Blaise. *Pensées*, edited by Gérard Ferreyrolles. Paris: Le Livre de Poche, 2000.

Pastorino, Cesare. "The Mine and the Furnace: Francis Bacon, Thomas Russell and Early Stuart Mining Culture." *Early Science and Medicine* 14 (2009): 630–60.

Pautet, Sébastien. "Produire une élite savante et technicienne à l'École du génie de Mézières: dispositions techniques et scientifiques des élèves ingénieurs." *Artefact. Techniques, histoire et sciences humaines* 4 (2016): 119–33.

Payette, Serge, ed. *Flore nordique du Québec et du Labrador*, 3 vols. Québec: Presses de l'Université Laval, 2013–2018.

Peacock, Darren. "Making Ways for Change: Museums, Disruptive Technologies and Organisational Change." *Museum Management and Curatorship* 12 (2008): 333–51.

Pedretti, Erminia, and Ana Maria Navas Iannini. *Controversy in Science Museums. Re-Imagining Exhibition Spaces and Practice*. London: Routledge, 2020.

Perry, Curtis, ed. *Material Culture and Cultural Materialisms in the Middle Ages and Renaissance*. Turnhout, Belgium: Brepols, 2001.

Pesic, Peter. "Hearing the Irrational: Music and the Development of the Modern Concept of Number." *Isis* 101 (2010): 501–30.

Pesic, Peter. "Francis Bacon, Violence, and the Motion of Liberty: The Aristotelian Background." *Journal of the History of Ideas* 75 (2014): 69–90.

Pesic, Peter. *Music and the Making of Modern Science*. Cambridge, MA: MIT Press, 2014.

Pestre, Dominique. "La pratique de reconstitution des expériences historiques, une toute première réflexion." In *Restating Coulomb. Usages, controverses et réplications autour de la balance de torsion,* edited by Christine Blondel and Matthias Dörries, 17–30. Florence: Leo S. Olschki, 1994.

Phillipson, Tacye. "Collections Development in Hindsight: A Numerical Analysis of the Science and Technology Collections of National Museums Scotland since 1855." *Science Museum Group Journal* 12 (2019). http://dx.doi.org/10.15180/191205/001.

Pickering, Andrew. *The Mangle of Practice. Time, Agency & Science.* Chicago: The University of Chicago Press, 1995.

Pickering, Andrew. "Material Culture and the Dance of Agency." In *The Oxford Handbook of Material Culture Studies*, edited by Dan Hicks and Mary C. Beaudry, 191–208. Oxford: Oxford University Press, 2010.

Pickstone, John. *Ways of Knowing: A New History of Science, Technology and Medicine.* Chicago: The University of Chicago Press, 2000.

Plato. *The Dialogues of Plato*, translated and edited by Benjamin Jowett, 4th ed. Oxford: Oxford University Press, 1953.

Pollens, Stewart. *The Manual of Musical Instrument Conservation.* Cambridge: Cambridge University Press, 2015.

Pollnitz, Aysha. *Princely Education in Early Modern Britain.* Cambridge: Cambridge University Press, 2015.

Pomian, Krzysztof. *Collectionneurs, amateurs et curieux. Paris, Venise: XVIe–XVIIIe siècle.* Paris: Gallimard, 1987.

Pomian, Krzysztof. *Le musée, une histoire mondiale*, 3 vols. Paris: Gallimard, 2020–2022.

Poisson, Nicolas. *Traité de la mécanique composé par Monsieur Descartes ...* Paris, 1668.

Poulot, Dominique. *Surveiller et s'instruire: La Révolution française et l'intelligence de l'héritage historique*. Oxford: Voltaire Foundation [SVEC 344], 1996.

Poulot, Dominique. "Le musée et le patrimoine: une histoire de contextes et d'origines." In *La Muséologie, champ de théories et de pratiques*, edited by Anik Meunier and Jason Luckerhoff, 19–39. Montréal: Presses de l'Université du Québec, 2012.

Poulot, Dominique. *L'Art d'aimer les objets*. Québec: Presses de l'Université Laval, 2016.

Powell, Christopher, and Sophy Wills. "Conservation or Restoration? Treatment of an 18th-Century Clock." *V & A Conservation Journal* 39 (2001): 21–2.

Powers, John C. "Chemistry Without Principles: Herman Boerhaave on Instruments and Elements." In *New Narrative in Eighteenth-Century Chemistry*, edited by Lawrence M. Principe, 45–61. Dordrecht: Springer, 2007.

Powers, John C. *Inventing Chemistry: Herman Boerhaave and the Reform of the Chemical Arts*. Chicago: The University of Chicago Press, 2012.

Powers, John C. "Fire Analysis in the Eighteenth Century: Herman Boerhaave and Scepticism About the Elements." *Ambix* 61 (2014): 385–406.

Prætorius, Michael. *Syntagma musicum II: De organographia*, translated and edited by David Z. Crookes. Oxford: Clarendon Press, 1986.

Prætorius, Michael. *Syntagma musicum: Textes relatifs à l'orgue, comprenant aussi la basse générale ou continue*, translated and edited by Jacques Leguy. Chatenay Malabry: Éditions Ars Musicæ, 1999.

Price, Sally. *Paris Primitive. Jacques Chirac's Museum on the Quai Branly*. Chicago: The University of Chicago Press, 2007.

Prizer, William F. "Isabella d'Este and Lorenzo da Paria, 'Master Instrument-Maker'." *Early Music History* 2 (1982): 87–127.

Pye, Elizabeth. "Challenges of Conservation: Working Objects." *Science Museum Group Journal* [online journal] 6 (2006). http://journal.sciencemuseum.ac.uk/browse/issue-06/challenges-of-conservation/.

Quiccheberg, Samuel. *Inscriptions vel tituli theatri amplissimi*. Munich, 1565.

Rabinovitch, Oded. *The Perraults: A Family of Letters in Early Modern France*. Ithaca: Cornell University Press, 2018.

Radder, Hans, ed. *The Philosophy of Scientific Experimentation*. Pittsburgh: University of Pittsburgh Press, 2003.

Raleigh, Sir Walter. *The History of the World*. London, 1617.

Raphael, Renée. "Galileo's *Discorsi* and Mersenne's *Nouvelle pensées*: Mersenne as a Reader of Galilean 'Experience'." *Nuncius* 23 (2008): 7–36.

Ratcliff, Marc J. *The Quest for the Invisible: Microscopy in the Enlightenment*. Farnham, UK: Ashgate Publishing Ltd, 2009.

Ratte, Franz Josef. *Die Temperatur der Clavierinstrumente. Quellenstudien zu den theoretischen Grundlagen und praktischen Anwendungen von der Antike bis ins 17. Jahrhundert*. Kassel: Bärenreiter, 1991.

Reeves, Nicky. "'To Demonstrate the Exactness of the Instrument': Mountainside Trials of Precision in Scotland, 1774." *Science in Context* 22 (2009): 323–40.

Regel, Ben, *et al.* "Doping at the Science Museum: The Conservation Challenge of Doped Fabric Aircraft in the *Flight* gallery." *Science Museum Group Journal* [online journal], 6 (2006). http://journal.sciencemuseum.ac.uk/browse/issue-06/doping-at-the-science-museum/.

Reisinger, Yvette, and Carol J. Steiner. "Reconceptualizing Object Authenticity." *Annals of Tourism Research* 33 (2006): 65–86.

Restle, Conny. "Organology: The Study of Musical Instruments in the 17th Century." In *Instruments in Art and Science. On the Architectonics of Cultural Boundaries in the 17th Century*, edited by Helmar Schramm, Ludger Schwarte, and Jan Lazardzig, 257–68. Berlin: Walter de Gruyter, 2008.

Ribe, Neil M. "Cartesian Optics and the Mastery of Nature." *Isis* 88 (1997): 42–61.

Riede, Felix, *et al.* "The Role of Play Objects and Object Play in Human Cognitive Evolution and Innovation." *Evolutionary Anthropology* 27 (2018): 46–59.

Ripin, Edwin M. "The Early Clavichord." *The Musical Quarterly* 53 (1967): 518–38.

Robbins, Nina. "Why We Still Need Collections—Museums in the Business of Originality." *ICOFOM Study Series* 48–1 (2020): 183–92.

Roberts, Lissa, Simon Schaffer, and Peter Dear, eds. *The Mindful Hand. Inquiry and Invention from the Late Renaissance to Early Industrialisation.* Amsterdam: KNAW, 2007.

Roche, Daniel. *Histoire des choses banales. Naissance de la consommation dans les sociétés traditionnelles (XVIIe–XIXe siècle).* Paris: Fayard, 1997.

Roche, Daniel. *Humeurs vagabondes. De la circulation des hommes et de l'utilité des voyages.* Paris: Fayard, 2003.

Rodéhn, Cecilia. "Emotions in the Museum of Medical History. An Investigation of How Museum Educators Employ Emotions and What These Emotions Do." *International Journal of Heritage Studies* 26 (2020): 201–13.

Rodis-Lewis, Genevieve. "Descartes' Life and the Development of his Philosophy." In *The Cambridge Companion to Descartes*, edited by John Cottingham, 21–57. Cambridge: Cambridge University Press, 1992.

Rodrigues Barbosa, Catia. "L'esthétique dans les musées des sciences et des techniques." *La Lettre de l'OCIM* 95 (2004): 27–30.

Rolland-Villemot, Bénédicte. "Le traitement des collections industrielles et techniques, de la connaissance à la diffusion." *La Lettre de l'OCIM* 73 (2001): 13–18.

Rosen, Edward. *The Naming of the Telescope.* New York: Henry Schuman, 1947.

Rossi, Paolo. *Philosophy, Technology, and the Arts in the Early Modern Era*, translated by Salvator Attanasio. New York: Harper & Row, 1970.

Rowe, Jonathan P., *et al.* "Play in the Museum: Design and Development of a Game-Based Learning Exhibit for Informal Science Education." *International Journal of Gaming and Computer-Mediated Simulations* 9 (2017): 96–113.

Roy, Andréanne, Jacques Des Roches, and Yseult Riopelle, *Riopelle: à la rencontre des territoires nordique et des cultures autochtones.* Montréal: Musée des beaux-arts de Montréal, 2020.

Runia, Eelco. "Presence." *History and Theory* 45 (2006): 1–29.

Sainte-Marie, Maxime, Jean-François Gauvin, and Vincent Larivière. "Les collections muséales au service de la science. Une analyse de la Collection of Historical Scientific Instruments de l'université Harvard." *Document numérique* 2–3 (2017): 137–58.

Sakellariadis, Spyros. "Descartes's Use of Empirical Data to Test Hypotheses." *Isis* 78 (1982): 68–76.

Schaffer, Simon. "Natural Philosophy and Public Spectacle in the Eighteenth Century." *History of Science* 21 (1983): 1–43.

Schaffer, Simon. "Glass Works: Newton's Prisms and the Uses of Experiment." In *The Uses of Experiment. Studies in the Natural Sciences*, edited by D. Gooding, T. Pinch and S. Schaffer, 67–104. Cambridge: Cambridge University Press, 1989.

Schaffer, Simon. "Late Victorian Metrology and Its Instrumentation: A Manufactory of Ohms." In *Invisible Connections. Instruments, Institutions, and Science*, edited by Robert Bud and Susan E. Cozzens, 23–55. Bellingham, WA: SPIE Optical Engineering Press, 1992.

Schaffer, Simon. "Machine Philosophy: Demonstration Devices in Georgian Mechanics." *Osiris*, 2nd Series 9 (1994): 157–82.

Schaffer, Simon. "Experimenters' Techniques, Dyers's Hands, and the Electric Planetarium." *Isis* 88 (1997): 456–83.

Schaffer, Simon. "A Science Whose Business is Bursting: Soap Bubbles as Commodities in Classical Physics." In *Things that Talk: Object Lessons from Art and Science*, edited by Lorraine Daston, 147–92. New York: Zone Books, 2004.

Schaffer, Simon. "L'inventaire de l'astronome. Le commerce d'instruments scientifiques au XVIIIe siècle (Angleterre-Chine-Pacifique)." *Annales. Histoire, Sciences Sociales* 60 (2005): 791–815.

Schaffer, Simon. "Instruments as Cargo in the China Trade." *History of Science* 44 (2006): 217–46.

Schaffer, Simon. "Easily Cracked: Scientific Instruments in States of Disrepair." *Isis* 102 (2011): 706–17.

Schaffer, Simon. "The Laird of Physics." *Nature* 471 (March 2011): 289–91.

Schaffer, Simon. "Chronometers, Charts, Charisma: On Histories of Longitude." *Science Museum Group Journal* vol. 2 (2014). http://journal.sciencemuseum.ac.uk/browse/issue-02/chronometers-charts-charisma/.

Schama, Simon. *The Embarrassment of Riches: An Interpretation of Dutch Culture in the Golden Age*. New York: Vintage, 1997.

Schnapp, Jeffrey. "The Intimate Lives of Cultural Objects." In *The Routledge Companion to Media Studies and Digital Humanities*, edited by Jenteray Sayers, 423–32. New York: Routledge, 2018.

Schott, Gaspard. *Organum mathematicum libris IX explicatum*. Herbipoli [Würzburg], 1668.

Schouls, Peter A. *Descartes and the Possibility of Science*. Ithaca: Cornell University Press, 2000.

Schuster, John. *Descartes-Agonistes. Physico-mathematics, Methods & Corpuscular-Mechanism 1618–1633*. London: Springer, 2013.

Sennett, Richard. *The Craftsman*. New Haven: Yale University Press, 2008.

Serra, Laura, Anna Giatti, and Paolo Brenni. "The Pneumatic Pump of the Mariano College in Bergamo." In *Atti del XXXVI convegno annuale SISFA* (Naples, 2006), edited by Salvatore Esposito, 371–9. Pavia: Pavia University Press, 2017.

Shapin, Steven, and Simon Schaffer, *Leviathan and the Air-Pump. Hobbes, Boyle, and the Experimental Life*. Princeton: Princeton University Press, 1985.

Shapin, Steven. "Why the Public Ought to Understand Science-in-the-Making." *Public Understanding of Science* 1 (1992): 27–30.

Shapin, Steven. *A Social History of Truth: Civility and Science in Seventeenth-Century England*. Chicago: The University of Chicago Press, 1994.

Shapiro, Barbara. *A Culture of Fact: England, 1550–1720*. Cornell: Cornell University Press, 2000.

Sheets-Pyenson, Susan. *Cathedrals of Science: The Development of Colonial Natural History Museums During the Late Nineteenth Century*. Montréal: McGill-Queens University Press, 1988.

Shell, Hanna Rose. "Ceramic Nature." In *Materials and Expertise in Early Modern Europe*, edited by Ursula Klein and Emma C. Spary, 50–70. Chicago: The University of Chicago Press, 2010.

Shiner, Larry. *The Invention of Art: A Cultural History*. Chicago: The University of Chicago Press, 2001.

Sibum, Otto. "Reworking the Mechanical Value of Heat: Instruments of Precision and Gestures of Accuracy in Early Victorian England." *Studies in the History and Philosophy of Science* 26 (1995): 73–106.

Sibum, Otto. "Working Experiments: A History of Gestural Knowledge." *The Cambridge Review* 116 (1995): 25–37.

Simpson, James. *Under the Hammer: Iconoclasm in the Anglo-American Tradition*. Oxford: Oxford University Press, 2011.

Simpson, W. *Philosophical Dialogues Concerning the Principles of Natural Bodies Wherein the Principles of the Old and New Philosophy are Stated, and the New Demonstrated More Agreeable to Reason, from Mechanical Experiments and its Usefulness to the Benefit of Man-kind*. London, 1677.

Smith, Justin E. H. *Divine Machines: Leibniz and the Sciences of Life*. Princeton: Princeton University Press, 2011.

Smith, Pamela. *The Body of the Artisan. Art and Experience in the Scientific Revolution*. Chicago: The University of Chicago Press, 2004.

Smith, Pamela. *From Lived Experience to the Written Word: Reconstructing Practical Knowledge in the Early Modern World*. Chicago: The University of Chicago Press, 2022.

Smith, Pamela, and The Making and Knowing Project. "Historians in the Laboratory: Reconstruction of Renaissance Art and Technology in the Making and Knowing Project." *Art History* 39 (2016): 210–33.

Söderqvist, Thomas, Adam Bencard, and Camilla Mordhorst. "Between Meaning Culture and Presence Effects: Contemporary Biomedical Objects as a Challenge to Museums." *Studies in History and Philosophy of Science* 40 (2009): 431–38.

Sorbiere, Samuel. *Relations, lettres et discours de Mr de Sorbiere sur diverses matieres curieuses*. Paris, 1660.

Sorbiere, Samuel. *Lettres et discours de Mr de Sorbiere sur diverses matieres curieuses*. Paris, 1660.

Sprat, Thomas. *History of the Royal Society*. London, 1667.

Stanley Price, Nicholas, M. Kirby Talley Jr., and Alessandra Melucco Vaccaro, eds. *Historical and Philosophical Issues in the Conservation of Cultural Heritage*. Los Angeles: The Getty Conservation Institute, 1996.

Stansfield Eastwood, Bruce. "Descartes on Refraction: Scientific Versus Rhetorical Method." *Isis* 75 (1984): 481–502.

Starn, Randolph. "A Historian's Brief Guide to New Museum Studies." *American Historical Review* 110 (2005): 68–98.

Staubermann, Klaus, ed. *Reconstructions: Recreating science and technology of the past*. Edinburgh: National Museums of Scotland, 2011.

Sterne, Jonathan. *The Audible Past. Cultural Origins of Sound Reproduction*. Durham: Duke University Press, 2003.

Sterne, Jonathan. "Bourdieu, Technique and Technology." *Cultural Studies* 17 (2003): 367–89.

Stewart, Larry. "Science, Instruments, and Guilds in Early Modern Britain." *Early Science and Medicine* 10 (2005): 392–410.

Stewart, Larry. "Experimental Spaces and the Knowledge Economy." *History of Science* 45 (2007): 155–77.

Stoler, Laura, ed. *Imperial Debris. On Ruins and Ruination*. Durham: Duke University Press, 2013.

Strasser, Bruno. *Collecting Experiments. Making Big Data Biology*. Chicago: The University of Chicago Press, 2019.

Strasser, Bruno, *et al.* "'Citizen Science"? Rethinking Science and Public Participation." *Science and Technology Studies* 32 (2018): 52–76.

Stroup, Alice. "Wilhelm Homberg and the Search for the Constituents of Plants at the 17th-Century Académie royale des sciences." *Ambix* 26 (1979): 184–201.

Stroup, Alice. "Christiaan Huygens and the Development of the Air-Pump." *Janus* 68 (1981): 129–58.

Stroup, Alice. *A Company of Scientists. Botany, Patronage, and Community at the Seventeenth-Century Parisian Academy of Sciences*. Los Angeles: University of California Press, 1990.

Sturdy, David J. *Science and Social Status. The Members of the Académie des sciences, 1666–1750*. Woodbridge, UK: The Boydell Press, 1995.

Tabb, Kathryn. "Locke on Habituation and the Association of Ideas." In *The Lockean Mind*, edited by Jessica Gordon-Roth and Shelley Weinberg, 385–395. London: Routledge, 2022.

Tang Simpson, Mette, and Michael Huntley, eds. *La restauration des objets anciens.* Paris: Armand Colin, 1992.

Tari, Lujza. "Women, Musical Instruments and Instrumental Music." *Studia Musicologica Academiæ Scientiarum Hungaricæ* 40 (1999): 95–143.

Taub, Liba, ed. "On Scientific Instruments." *Studies in History and Philosophy of Science Part A* 40 (2009): 337–438.

Taub, Liba. "Introduction: Reengaging with Scientific Instruments." *Isis* 102 (2011): 689–96.

Taub, Liba. "What is a Scientific Instrument, Now?" *Journal of the History of Collections* 31 (2019): 453–67.

Tilley, Christopher, ed. *Reading Material Culture: Structuralism, Hermeneutics, and Post-Structuralism.* Cambridge, MA: B. Blackwell, 1990.

Tomsin, Philippe. "Les patrimoines mobiliers scientifique et technique: spécificités de leur restauration, de leur conservation et de leur valorisation." *CeROArt* [online], 1 (2007). http://journals.openedition.org/ceroart/370.

Tõugu, Pirko, *et al.* "Connecting Play Experiences and Engineering Learning in a Children's Museum." *Journal of Applied Developmental Psychology* 53 (2017): 10–19.

Trichet, Pierre. *Traité des instruments de musique (vers 1640)*, edited by François Lesure. Neuilly-sur-Seine: Société de musique d'autrefois, 1957.

Tresch, John, and Emily Dolan. "Toward a New Organology: Instruments of Music and Science." In *Music, Sound, and the Laboratory from 1750–1980*, edited by Alexandra Hui, Julia Kursell, and Myles Jackson, *Osiris* 28 (2013): 278–98.

Turkle, Sherry, ed. *Falling for Science: Objects in Mind.* Cambridge, MA: MIT Press, 2008.

Turkle, Sherry, ed. *The Inner History of Devices*. Cambridge, MA: MIT Press, 2008.

Turkle, Sherry, ed. *Evocative Objects: Things We Think With.* Cambridge, MA: MIT Press, 2011.

Turner, Anthony. "Paper, Print, and Mathematics. Philippe Danfrie and the Making of Mathematical Instruments in late 16th-Century Paris." In *Studies in the History of Scientific Instruments*, edited by Christine Blondel, *et al.*, 23–41. London: Rogers Turner Books Ltd, 1989.

Turner, Anthony. "From Mathematical Practice to the History of Science." *Journal of the History of Collections* 7 (1995): 135–50.

Turner, Anthony. "Mathematical Instrument-Making in Early Modern Paris." In *Luxury Trades and Consumerism in ancien régime Paris: Studies in the History of the Skilled Workforce*, edited by Robert Fox and Anthony Turner, 63–96. Aldershot: Ashgate, 1998.

Turner, Anthony. "John Dee, Louvain and the Origins of English Instrument-Making." In *Musa musei: Studies on Scientific Instruments and Collections in Honour of Mara Miniati*, edited by Marco Beretta, Paolo Galluzzi, and Carlo Triarico, 63–78. Firenze: L. S. Olschki, 2003.

Turner, Anthony. "Grollier de Servière, the Brothers Monconys: Curiosity and Collecting in Seventeenth-Century Lyon." *Journal of the History of Collections* 20 (2008): 205–15.

Turner, Gerard L'Estrange. *Essays on the History of the Microscope*. Oxford: Senecio Publication Co., c1980.

Turner, Gerard L'Estrange. "Presidential Address: Scientific Toys." *The British Journal for the History of Science* 20 (1987): 377–98.

Turner, Gerard L'Estrange. *Elizabethan Instrument Makers. The Origins of the London Trade in Precision Instrument Making*. Oxford: Oxford University Press, 2000.

Turner, Stephen P. *Understanding the Tacit*. New York: Routledge, 2014.

Uchacz, Tianna Helena. "Reconstructing Early Modern Artisanal Epistemologies and an 'Undisciplined' Mode of Inquiry." *Isis* 111 (2020): 606–13.

Ulrich, Laurel, Ivan Gaskell, Sara J. Schechner, Sarah Anne Carter, and Samantha van Gerbig. *Tangible Things. Making History Through Objects*. Oxford: Oxford University Press, 2015.

Vaccaro, Jean-Michel. *La Musique de luth en France au XVIe siècle*. Paris: Éditions du Centre National de la Recherche Scientifique, 1981.

Valleriani, Matteo. *Galileo Engineer*. Dordrecht: Springer [Boston Studies in the Philosophy of Science, vol. 269], 2010.

Valleriani, Matteo. "Galileo's Abandoned Project on Acoustic Instruments at the Medici Court." *History of Science* 50 (2012): 1–31.

Valleriani, Matteo, ed. *The Structures of Practical Knowledge*. Cham: Springer, 2017.

Valleriani, Matteo. "From the Quadrivium to Modern Science." *Journal of History of Science and Technology* 16, no. 1 (2022): 121–32.

Vendrix, Philippe. "La dichotomie vocal/instrumental dans la théorie musicale aux confins de la Renaissance et du Baroque." In *Le Concert des voix et des instruments à la Renaissance*, 71–81. Paris: C.N.R.S., 1995.

Vermeir, Koen, and Dániel Margócsy, eds. *States of Secrecy*, special issue of the *British Journal for the History of Science* 45, no.2 (2012): 153–280.

Viala, Alain. *Naissance de l'écrivain: sociologie de la littérature à l'âge classique*. Paris: Editions de Minuit, 1985.

Virdung, Sebastian. *Musica getutscht: A Treatise on Musical Instruments (1511) by Sebastian Virdung*, translated and edited by Beth Bullard. Cambridge: Cambridge University Press, 1993.

Wainwright, Jonathan, and Peter Holman, eds. *From Renaissance to Baroque: Change in Instruments and Instrumental Music in the Seventeenth Century*. Aldershot, UK: Ashgate, 2005.

Waquet, Françoise. *L'Ordre matériel du savoir: Comment les savants travaillent, XVIe–XXIe siècles*. Paris: CNRS Éditions, 2015.

Walker, Daniel P. "Joan Albert Ban and Mersenne's Musical Competition of 1640." *Music and Letters* 57 (1976): 233–55.

Walker Bynum, Caroline. *Christian Materiality. An Essay on Religion in Late Medieval Europe*. New York: Zone Books, 2011.

Wardhaugh, Benjamin. *Music, Experiments and Mathematics in England, 1653–1705*. Aldershot, UK: Ashgate, 2008.

Wardhaugh, Benjamin. *The* Compendium Musicae *of René Descartes: Early English Responses*. Brepols: Turnhout, 2013.

Warner, Deborah J. "What is a Scientific Instrument, When Did it Become One, and Why?" *British Journal for the History of Science* 23 (1990): 83–93.

Warner, Deborah J. "Replicating Historic Scientific Instruments." *Bulletin of the Scientific Instrument Society* 151 (2021): 10–13.

Warnier, Jean-Pierre. *Construire la culture matérielle. L'Homme qui pensait avec ses doigts*. Paris: Presses Universitaires de France, 1999.

Warnier, Jean-Pierre. "A Praxeological Approach to Subjectivation in a Material World." *Journal of Material Culture* 6 (2001): 5–24.

Watson, E. C. "The Early Days of the Académie des sciences as Portrayed in the Engravings of Sébastien Le Clerc." *Osiris* 7 (1939): 556–87.

Werrett, Simon. "Wonders Never Cease: Descartes's *Météores* and the Rainbow Fountain." *British Journal for the History of Science* 34 (2001): 129–47.

Werrett, Simon. *Thrifty Science. Making the Most of Materials in the History of Experiment*. Chicago: The University of Chicago Press, 2019.

Westfall, Richard. "Science and Technology during the Scientific Revolution: An Empirical Approach." In *Renaissance and Revolution: Humanists, Scholars, Craftsmen and Natural Philosophers in Early Modern Europe*, edited by J. V. Field and Frank A. J. L. James, 63–72. Cambridge: Cambridge University Press, 1993.

Wheatland, David P. *The Apparatus of Science at Harvard, 1765–1800*. Cambridge, MA: Harvard University Press, 1968.

Whitelaw, Mitchell. "Generous Interfaces for Digital Cultural Collections." *DHQ: Digital Humanities Quarterly* 9.1 (2015). http://www.digitalhumanities.org/dhq/vol/9/1/000205/000205.html.

Wilkins, John. *An Essay Towards a Real Character, and a Philosophical Language*. London, 1668.

Williams, Peter. "The Meaning of *Organum*: Some Case Studies." *Plainsong and Medieval Music* 10 (2001): 103–20.

Wilson, Catherine. *The Invisible World. Early Modern Philosophy and the Invention of the Microscope*. Princeton, N.J.: Princeton University Press, 1995.

Windhager, Florian, *et al.* "Visualization of Cultural Heritage Collection Data: State of the Art and Future Challenges." *IEEE Transactions on Visualization and Computer Graphics* 25 (2019): 2311–30.

Wolf, C. *Histoire de l'Observatoire de Paris, de sa fondation à 1793*. Paris: Gauthier-Villars, 1902.

Wolfe, Charles T., and Ofer Gal, eds. *The Body as Object and Instrument of Knowledge: Embodied Empiricism in Early Modern Science*. Dordrecht: Springer, 2010.

Wright, John P. "Ideas of Habit and Custom in Early Modern Philosophy." *Journal of the British Society for Phenomenology* 42 (2011): 18–32.

Wright, Michael. "The Antikythera Mechanism: Reconstruction as a Medium for Research and Publication." In *Reconstructions. Recreating Science and Technology of the Past*, edited by Klaus Staubermann, 1–20. Edinburgh: National Museums Scotland, 2011.

Wylie, Alison. *Thinking from Things. Essays in the Philosophy of Archaeology*. Berkeley: University of California Press, 2002.

Yates, Frances. *The French Academies of the Sixteenth Century*. London: Warburg Institute, 1947; reprinted in London: Routledge, 1988.

Young, Mark Thomas. "Manual Labor and 'Mean Mechaniks': Bacon's Mechanical History and the Deprecation of Craft Skills in Early Modern Science." *Perspectives on Science* 25 (2017): 521–50.

Zarlino, Gioseffo. *Sopplimenti musicali del rev. M. Gioseffo Zarlino da Chioggia* ... Venice, 1588.

Zarlino, Gioseffo. *Le Istitutioni harmoniche*. Venice, 1561; facsimile of the second edition, Arnaldo Forni Editore, 1999.

Ziegler, S. L. "Open Data in Cultural Heritage Institutions: Can We Be Better Than Data Brokers." *DHQ: Digital Humanities Quarterly* 14.2 (2020). http://digitalhumanities.org:8081/dhq/vol/14/2/000462/000462.html.

Index

www.ingramcontent.com/pod-product-compliance
Lightning Source LLC
Chambersburg PA
CBHW071325230926
51849CB00002B/317
* 9 7 8 9 0 0 4 5 0 4 6 0 8 *